Lecture Notes in Bioinformatics 9874

Subseries of Lecture Notes in Computer Science

More information about this series at http://www.springer.com/series/5381

Claudia Angelini
Paola M.V. Rancoita
Stefano Rovetta (Eds.)

Computational Intelligence Methods for Bioinformatics and Biostatistics

12th International Meeting, CIBB 2015
Naples, Italy, September 10–12, 2015
Revised Selected Papers

 Springer

Editors
Claudia Angelini
Istituto per le Applicazioni del Calcolo
Consiglio Nazionale delle Ricerche
Naples
Italy

Stefano Rovetta
DIBRIS
University of Genoa
Genova
Italy

Paola M.V. Rancoita
Center for Statistics in the Biomedical
  Sciences
Vita-Salute San Raffaele University
Milano
Italy

ISSN 0302-9743   ISSN 1611-3349   (electronic)
Lecture Notes in Bioinformatics
ISBN 978-3-319-44331-7   ISBN 978-3-319-44332-4   (eBook)
DOI 10.1007/978-3-319-44332-4

Library of Congress Control Number: 2016947208

LNCS Sublibrary: SL8 – Bioinformatics

Printed on acid-free paper

This Springer imprint is published by Springer Nature
The registered company is Springer International Publishing AG Switzerland

Preface

This volume contains a revised and selected version of the proceedings of the International Meeting on Computational Intelligence Methods for Bioinformatics and Biostatistics (CIBB 2015), which was in its 12th edition this year.

CIBB is a meeting with more than 10 years of history. Its main goal is to provide a forum open to researchers from different disciplines to present problems concerning computational techniques in bioinformatics, systems biology and medical informatics, to discuss cutting-edge methodologies and accelerate life science discoveries. Following this tradition and roots, this year's meeting brought together more than 80 researchers from the international scientific community interested in this field to discuss the advancements and the future perspectives in bioinformatics and biostatistics. Moreover, applied biologists participated in the conference in order to propose novel challenges aimed at having high impact on molecular biology and translational medicine. CIBB maintains a large Italian participation in terms of authors and conference venues but it has progressively become more international and more important in the current landscape of bioinformatics and biostatistics conferences.

This year the conference was organized in Naples (Italy) in the CNR research area during September 10–12, 2015. The topics of the conferences have kept pace with the appearance of new types of challenges in biomedical computer science, particularly with respect to a variety of molecular data and the need to integrate different sources of information. About 40 contributed papers were selected for presentation at the conference in the form of extended or short talks, either in the two main topic areas (i.e., bioinformatics and biostatistics) or in the five special sessions (The EDGE, enhanced definition of genomic entities for systems biomedicine in oncology; Multi-Omic metabolic models and statistical Bioinformatics of adaptations and biological associations; Large-Scale and HPC data analysis in bioinformatics: intelligent methods for computational, systems and synthetic biology; New knowledge from old data: power of data analysis and integration methods; Regularization methods for genomic data analysis). Each contributed paper received two reviews or more. Moreover, seven invited papers were presented in form of keynote talks. We deeply thank our invited speakers Michele Ceccarelli, Dario Greco, Dirk Husmeier, Wessel Van Wieringen, Cinzia Viroli, and Daniel Yekutieli. We are also indebted to the chairs of the very interesting and successful special sessions, which attracted very interesting contributions and attention.

All authors of contributed and invited papers were asked to submit an extended and revised paper for this volume. Afterward a further reviewing process took place, which led to the 21 papers that were selected to appear in this volume. The authors are spread over more than ten countries.

The editors would like to thank all the Program Committee members and the external reviewers of both the conference and post-conference versions of the papers for their valuable work.

A big thanks also to the sponsors, Gruppo Nazionale per il Calcolo Scientifico—GNCS INdAM, Bioinformatics Italian Society, Genomix4Life S.r.l., BMR Genomics S.R.L., M&M Biotech S.C.A.R.L., and in particular to the Istituto per le Applicazioni del Calcolo M. Picone and Institute of Genetics and Biophysics A. Buzzati Traverso that made this event possible. Finally, the editors would also like to thank all the authors for the high quality of the papers they contributed and for the interesting and stimulating discussion we had in Naples.

June 2016

Claudia Angelini
Paola Maria Vittoria Rancoita
Stefano Rovetta

Organization

CIBB2015 was jointly organized by: Istituto per le Applicazioni del Calcolo Mauro Picone, Istituto di Genetica e Biofisica Adriano Buzzati Traverso, and INNS International Neural Network Society SIG Bioinformatics, INNS International Neural Network Society SIG Bio-pattern, and the IEEE-CIS-BBCT Task Forces on Neural Networks and Evolutionary Computation.

General Chairs

Claudia Angelini	Istituto per le Applicazioni del Calcolo, Italy
Adriano Decarli	University of Milan, Italy
Erik Bongcam-Rudloff	Swedish University of Agricultural Sciences, Sweden

Biostatistics Technical Chair

Paola M.V. Rancoita	Vita-Salute San Raffaele University, Italy

Bioinformatics Technical Chair

Stefano Rovetta	University of Genoa, Italy

Special Session Organizers

Elia Biganzoli	University of Milan, Italy
Clelia Di Serio	Vita-Salute San Raffaele University, Italy
Anagha Joshi	Roslin Institute, University of Edinburgh, UK
Tom Michoel	Roslin Institute, University of Edinburgh, UK
Franck Picard	CNRS LBBE, Lyon 1, France
Andrea Bracciali	University of Stirling, UK
Mario Rosario Guarracino	High Performance Computing and Networking Institute, CNR, Italy
Ivan Merelli	Institute for Biomedical Technologies, CNR, Italy
Claudio Angione	University of Cambridge, UK
Pietro Liò	University of Cambridge, UK
Sandra Pucciarelli	University of Camerino, Italy
Barbara Simionati	BMR Genomics, Italy

Program Committee

Fentaw Abegaz	University of Groningen, The Netherlands
Federico Ambrogi	University of Milan, Italy
Sansanee Auephanwiriyakul	Chiang Mai University, Thailand
Mario Cannataro	University Magna Grecia of Catanzaro, Italy

Hailin Chen	Qiagen, Inc., USA
Davide Chicco	University of Toronto, Canada
Federica Cugnata	University of Vita-Salute San Raffaele, Italy
Antonio Eleuteri	The Royal Liverpool and Broadgreen University Hospitals, UK
Enrico Formenti	University of Nice-Sophia Antipolis, France
Arief Gusnanto	University of Leeds, UK
Raffaele Giancarlo	University of Palermo, Italy
Javier Gonzalez	University of Sheffield, UK
Yin Hu	Sage Bionetworks, USA
Pawel Labaj	BOKU Vienna, Austria
Paulo Lisboa	Liverpool John Moores University, UK
Giosué Lo Bosco	University of Palermo, Italy
Anna Marabotti	Uviversity of Salerno, Italy
Giancarlo Mauri	University Bicocca of Milan, Italy
Luciano Milanesi	ITB-CNR, Italy
Marta Milo	University of Sheffield, UK
Paola Paci	IASI-CNR, Italy
Marianna Pensky	University of Central Florida, USA
Davide Risso	University of California, Berkeley, USA
Riccardo Rizzo	ICAR-CNR, Italy
Paolo Tieri	IAC-CNR, Italy
Maurizio Urso	ICAR-CNR, Palermo, Italy
Veronica Vinciotti	Brunel University, UK
Blaz Zupan	University of Ljubljana, Slovenia

Steering Committee

Pierre Baldi	University of California, USA
Elia Biganzoli	University of Milan, Italy
Clelia Di Serio	Vita-Salute San Raffaele University, Italy
Alexandru Floares	Oncological Institute Cluj-Napoca, Romania
Jon Garibaldi	University of Nottingham, UK
Nikola Kasabov	Auckland University of Technology, New Zealand
Francesco Masulli	University of Genoa, Italy and Temple University, USA
Leif Peterson	TMHRI, USA
Roberto Tagliaferri	University of Salerno, Italy

Local Organizing Committee

Claudia Angelini	Istituto per le Applicazioni del Calcolo, Italy
Valerio Costa	Institute of Genetics and Biophysics, Italy
Italia De Feis	Istituto per le Applicazioni del Calcolo, Italy
Angelo Facchiano	Istituto di Scienze dell'Alimentazione, Italy

Secretary and Administrative Chair

Patrizia Montanaro   Istituto per le Applicazioni del Calcolo, Italy

Additional Reviewers

Abbruzzo, A.	Facchiano, A.	Pensky, M.
Alfieri, R	Fondi, M.	Rancoita, P.M.V.
Angelini, C.	Formenti, E.	Risso, D.
Auephanwiriyaku, S.	Hu, Y.	Romano, P.
Brilli, M.	Labaj, P.	Rovetta, S.
Cannataro, M.	Lisboa, P.	Stingo, F.
Chen, H.	Magillo, P.	Tagliaferri, R.
Chicco, D.	Mahmoud, H.	Tarazona, S.
Cugnata, F.	Marabotti, A.	Tieri, P.
Cutillo, L.	Masulli, F.	Zupan, B.
De Canditiis, D.	Mauri, G.	
De Feis, I.	Miglio, R.	

Sponsoring Institutions

Istituto per le Applicazioni del Calcolo M. Picone, Italy
Institute of Genetics and Biophysics A. Buzzati Traverso, Italy
Department of Clinical Sciences and Community Health, University of Milan, Italy
Gruppo Nazionale per il Calcolo Scientifico - GNCS INdAM
Bioinformatics Italian Society
Genomix4Life S.r.l.
BMR Genomics S.R.L.
M&M Biotech S.C.A.R.L.

Contents

A Commentary on a Censored Regression Estimator

Antonio Eleuteri[(✉)]

Department of Medical Physics and Clinical Engineering,
Royal Liverpool and Broadgreen University Hospital Trusts,
Daulby Street L7 8XP, Liverpool, UK
antonio.eleuteri@liv.ac.uk

Abstract. In this note we evaluate the properties and performance of a censored median regression estimator, as presented in literature by different authors in the context of support vector regression. This estimator is based on minimisation of an inequality constrained loss in a linear program formulation. Using a theoretical argument, we conjecture that the estimator is not consistent, and we compare its performance on simulated and real data in the one-sample case, with the Kaplan-Meier estimator and an inverse probability weighted estimator. We also compare the performance of the estimator on simulated and real data in the censored median regression setting, with the Portnoy estimator and the inverse probability weighted estimator.

Keywords: Censoring · Quantile regression · Survival analysis · Support vector machines

1 Scientific Background

Let us consider a sample of pairs $\{(T_i, C_i) : i = 1, \cdots, n\}, T_i \sim F, T_i$ and C_i conditionally independent. Let us also consider the case of right censoring, so what we observe are actually the variables $Y_i = \min\{T_i, C_i\}$ and $\delta_i = I(T_i < C_i)$, where $I(.)$ is the set indicator function. We consider the case of median estimation, and for simplicity we focus our attention on the one-sample problem; we'll address the regression case later.

The relevance of modeling the censoring phenomenon stems from important applications in many fields: from medical statistics to industrial life testing, it is often the case that not all statistical units are observed until the realization of an event of interest; for example some units may be lost to follow-up. (a very common case in survival time modeling.)

We will denote by θ the median to be estimated. The basic idea behind median (and generally, quantile) estimation derives from observing that minimisation of the ℓ_1 loss for location estimates results in the median [3]. We denote the residual for the i-th observation in the uncensored case as $r_i = T_i - \theta$. The median loss function ℓ_1 (see Fig. 1) can then be written:

© Springer International Publishing Switzerland 2016
C. Angelini et al. (Eds.): CIBB 2015, LNBI 9874, pp. 1–13, 2016.
DOI: 10.1007/978-3-319-44332-4_1

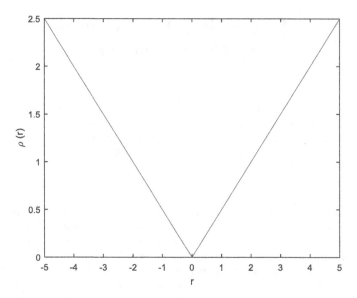

Fig. 1. Median loss

$$\rho(r) = r\{1/2 - I(r<0)\}. \tag{1}$$

Estimation of the median given a sample of observed points leads to minimisation of a piecewise linear empirical risk function:

$$\min_{\theta \in \mathbb{R}} \sum_i \rho(r_i). \tag{2}$$

Due to the discontinuous nature of the median loss, a linear programming problem formulation is used in practice, by introducing $2n$ slack variables [3]:

$$\min_{(\theta,u,v) \in \mathbb{R} \times \mathbb{R}_+^{2n}} \frac{1}{2} \sum_i u_i + \frac{1}{2} \sum_i v_i \tag{3}$$

$$\text{s.t.} \quad \theta + u_i - v_i = Y_i, \ \forall i = 1 \dots n$$

In the censored case, we denote the residual for the i-th observation as $r_i = Y_i - \theta$. The loss function of the censored median estimator proposed in literature (see e.g. [1, 2]) in this case is defined as (see Fig. 2):

$$\rho_I(r) = \delta\rho(r) + 1/2(1 - \delta)rI(r > 0). \tag{4}$$

Note that this loss, when evaluated on censored observations (i.e. when $\delta = 0$ in Eq. 4) is one-sided, and it reaches its minimum zero when $\theta > Y_i = C_i$ (i.e. the residual is negative, resulting in a zero loss.) In this way, estimates larger than the censored observations are "encouraged".

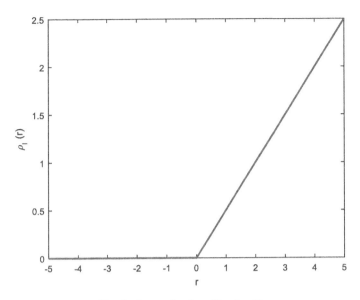

Fig. 2. Inequality loss (for $\delta = 0$)

Similarly to the uncensored case, estimation of the median in this case requires minimisation of a piecewise linear empirical risk function:

$$\min_{\theta \in \mathbb{R}} \sum_i \rho_I(r_i), \tag{5}$$

that translates into the following linear programming problem:

$$\min_{(\theta,u,v) \in \mathbb{R} \times \mathbb{R}^{2n}_+} \frac{1}{2}\sum_i u_i + \frac{1}{2}\sum_i v_i$$
$$\text{s.t.} \quad \begin{aligned} \theta + u_i - v_i &= Y_i \,, \; \forall i : \delta_i = 1 \\ \theta + u_i - v_i &\geq Y_i \,, \; \forall i : \delta_i = 0 \end{aligned} \tag{6}$$

Note the set of inequality constraints in correspondence to censored observations. The intuition behind this approach is simple: try to estimate the median by taking into account that censored observations provide a lower bound for the "true" unobserved points.

1.1   Median Regression

In analogy to median estimation, the aim of median regression is identification of a group of p observations (where p is the number of independent variables) that define a hyperplane that optimally represents the conditional median function [3].

In general there are several complications which arise in the regression setting, that are not present in the simpler one-sample case, for example [5]: the weaker nature of monotonicity in a p-dimensional setting, which can lead to "crossing" of conditional quantile functions; or the impossibility to estimate the parameters without further assumptions. (e.g. Y_i and C_i to be unconditionally independent.)

Formally, the more general case of median regression can be handled by noting that assuming a linear model, the residual can be written $r_i = T_i - x_i^T \theta, \theta \in \mathbb{R}^{p+1}$. The intercept is assumed to be part of the parameter vector, corresponding to a fixed regressor of constant value 1. With this definition in mind, we can write the problem in Eq. 6 as:

$$\min_{(\theta,u,v)\in\mathbb{R}^{p+1}\times\mathbb{R}_+^{2n}} \frac{1}{2}\sum_i u_i + \frac{1}{2}\sum_i v_i$$
$$\text{s.t.} \quad \begin{array}{l} x_i^T\theta + u_i - v_i = Y_i \;, \; \forall i : \delta_i = 1 \\ x_i^T\theta + u_i - v_i \geq Y_i \;, \; \forall i : \delta_i = 0 \end{array} \tag{7}$$

2 Materials and Methods

We will show now how the intuition behind the inequality constraint approach proposed in [1, 2] ignores some intrinsic and not readily evident aspects of the censoring process.

First note that in ordinary median estimation the contribution of each point to the subgradient[1] condition only depends on the sign of the residuals $r_i = T_i - \theta$ [3]. So a correct evaluation of the sign of the residuals is fundamental for any estimation procedure to work. Let us consider the two cases of uncensored and censored points.

For uncensored data we can observe both $Y_i = T_i < C_i$ and $I(r_i < 0)$; note that in this case the residuals can be either negative or positive.

For censored data, in the case $\theta < Y_i = C_i$, by the definition of right censoring we have $T_i > C_i$, hence we can observe $I(r_i < 0) = 0$.

However, if $\theta > Y_i = C_i$ there is an ambiguity: *we cannot observe the sign of the residual at all*, since we can have either $\theta > T_i$ or $\theta \leq T_i$, i.e. the residual can be negative *or* positive. In contrast, in the inequality loss formulation in Eq. 4, this case *always* results in a negative residual. As we will see with simulations, this fact impacts the performance of the estimator, resulting in an "excess" of negative residuals which "pull" the estimation process towards biased-low results.

What can we say about a residual when we cannot observe it? We can evaluate the following conditional expectation (with respect to the measure F):

$$\mathbf{E}[I(r_i < 0)|T_i > C_i] = \frac{\Pr\{C_i < T_i < \theta\}}{\Pr\{C_i < T_i\}} = \frac{F(\theta) - F(C_i)}{1 - F(C_i)} = \frac{1/2 - F(C_i)}{1 - F(C_i)}. \tag{8}$$

---

[1] The partial derivatives in the positive and negative directions.

The above quantity (calculated for $F(C_i) < 1/2$ since we are interested in the median) gives a measure of the "weight" attached to ambiguous censored observations.

This suggests a weighting scheme originally proposed by Efron [4] and adapted by Portnoy [5] to quantile regression. The key observation is that the weight we assign is split into two pieces: a part of the probabilistic mass is left in its position at the censored observation, but the remainder is shifted to the right to an unspecified, indefinitely large observation Y_∞ (which in practice can be set to, say, ten times the largest observation in the data). If we denote by w_i the expectation defined in Eq. 8, we can write the minimization problem as:

$$\min_{\theta \in \mathbb{R}} \sum_i \delta_i \rho(r_i) + (1 - \delta_i)[w_i \rho(r_i) + (1 - w_i)\rho(Y_\infty - \theta)]. \qquad (9)$$

Note that the weights depend on knowledge of the true distribution of the observations, which is usually not available; however, as shown in [5] these can be estimated non-parametrically using the Kaplan-Meier estimator of F. In general, Kaplan-Meier quantiles can be framed as solutions of the above problem (with w_i depending on the quantile of interest). Comparison of Eq. 9 with Eq. 4 suggests that the inequality loss estimator may not be consistent.

Portnoy's estimator [5] extends Eq. 9 to the more general quantile regression framework, and it produces consistent estimates of the parameters.

An alternative estimator, first proposed in [6, 7] in the context of mean regression, and recently adapted in [8] to support vector quantile regression, is based on the idea of inverse probability reweighting. The weighted estimator takes the form:

$$\min_{\theta \in \mathbb{R}} \sum_i \frac{\delta_i}{\hat{G}(Y_i)} \rho(Y_i - \theta), \qquad (10)$$

where $\hat{G}(Y_i)$ is the Kaplan-Meier estimate of the survival distribution of the censoring process. Note that this estimation method is not applicable with fixed and constant censoring, except in the special case of no censoring (where it reduces to the canonical median estimator.) It is known [6, 7] that this estimator produces consistent estimates.

3   Results

In the following sections we report results on simulated and real data, both for the one-sample and regression cases. In most cases we will provide comparisons with the weighted and Portnoy estimators as they naturally arise from the above discussion.

Both the inequality and weighted estimators have been implemented using MATLAB® and the Optimization Toolbox (function *linprog*). The Portnoy estimator is available in the R *quantreg* package (function *crq*.) All experiments were run using MATLAB® R2015a and R 3.2.2.

3.1 One-Sample Experiments

3.1.1 Simulations

We performed a series of simulation experiments to compare the finite sample performance of some estimates of the median (denoted by $\hat{\theta}$) in a censored one-sample setting.

We assume the distribution of events as standard lognormal with median $\theta = 1$, and the censoring distribution as exponential with mean 4. This results in approximately 30 % censored observations. We follow the experimental setup in [9]. For each problem instance the estimate was calculated 1000 times and the results averaged. We also report the performance of the (infeasible) sample median (estimated on uncensored data) and the naïve estimator (i.e. the sample median ignoring the censored observations.)

In Table 1 we report the bias $\hat{\theta} - \theta$ of the estimates, and in Table 2 the mean squared error (MSE) of the estimates (scaled to the sample size, to conform to the asymptotic variance calculations [9] and denoted by "n = ∞" in the last row.)

Table 1. Bias

	Sample Median (infeasible)	Kaplan-Meier	Inequality loss	Weighted	Naïve
n = 50	0.0138	−0.0503	−0.0503	−0.0142	−0.214
n = 200	0.00641	−0.00892	−0.0603	−0.0094	−0.221
n = 500	−0.00159	−0.00750	−0.0674	−0.0123	−0.223
n = 1000	−0.000160	−0.00206	−0.0659	−0.0073	−0.224

Table 2. Scaled mean squared error

	Sample Median (unfeasible)	Kaplan-Meier	Inequality	Weighted	Naïve
n = 50	1.674	1.756	1.555	2.020	3.424
n = 200	1.780	2.023	2.268	2.153	10.860
n = 500	1.565	1.902	3.693	2.092	25.955
n = 1000	1.445	1.716	5.612	1.818	50.421
n = ∞	1.571	1.839	–	–	–

From the tables we can see that the inequality estimator behaves qualitatively in a similar way to the naïve estimator, in that the bias is roughly constant independently of the sample size; and the scaled MSE increases with sample size (although at a different rate.) We extended the experiment for the inequality and weighted estimators up to a sample size of 10000. The results support our conjecture that the inequality estimator is not consistent (Table 3).

Table 3. Performance for large sample sizes

	Bias	SMSE
Inequality		
n = 2000	−0.0653	9.904
n = 5000	−0.0668	23.715
n = 10000	−0.0662	45.295
Weighted		
n = 2000	−0.00411	1.914
n = 5000	−0.00384	1.999
n = 10000	−0.00204	2.099

In Fig. 3 we compare the naïve empirical risk, inequality empirical risk and weighted empirical risk with the true (unfeasible) empirical risk. The weighted empirical risk is the closest to the true empirical risk.

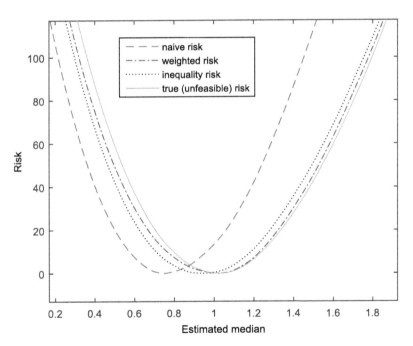

Fig. 3. Empirical risk functions for lognormal data (median 1) with exponential censoring (mean 4). n = 2000

3.2 Regression Experiments

In these experiments we will deal with conditional censored median estimation under different censoring mechanism. In the case of simulated data we will report the bias of the estimate, the median absolute error (MAE) and the root mean squared error (RMSE); note we won't report the scaled mean square error as it was done in the one-sample case, since asymptotic variance results aren't generally available for conditional estimation. For real world data we will report parameter estimates and standard errors in accordance to common practice (we cannot calculate the true bias and MAE as the data generating process is not known.)

3.2.1   Simulation: IID Error, Input-Dependent Censoring

In this experiment event times were generated according to the IID linear model:

$$T_i = 5 + x_i + u_i \quad x_i \sim U[0,2], \ u_i \sim N(0, 0.39^2).$$

Censoring times were generated according to the linear model:

$$C_i = 5.5 + .75x_i + v_i \quad v_i \sim N(0, 0.3^2).$$

The above model results in a proportion of censored observations of roughly 30 %.

Note that censoring depends on the covariate x; this means that in the case of the weighted estimator, use of Eq. 10 corresponds to estimation under misspecification of the model.

In Table 4 we report the performance of the Portnoy, weighted and inequality estimators for different sample sizes.

Table 4. IID error and input-dependent censoring

	Intercept (= 5)			Slope (= 1)		
	Bias	MAE	RMSE	Bias	MAE	RMSE
Portnoy						
n = 100	−0.0042	0.065	0.094	0.0024	0.059	0.087
n = 400	−0.0025	0.037	0.054	−0.0009	0.032	0.047
n = 1000	−0.0025	0.021	0.031	0.0006	0.019	0.028
Inequality						
n = 100	−0.0019	0.065	0.095	−0.032	0.057	0.085
n = 400	−0.0034	0.031	0.047	−0.035	0.038	0.053
n = 1000	−0.00073	0.019	0.029	−0.035	0.035	0.043
n = 5000	−0.00065	0.0087	0.013	−0.036	0.036	0.037
n = 10000	−0.0016	0.0062	0.0094	−0.035	0.035	0.036
Weighted						
n = 100	−0.11	0.11	0.15	0.028	0.067	0.11
n = 400	−0.12	0.12	0.13	0.042	0.050	0.070
n = 1000	−0.12	0.12	0.13	0.049	0.049	0.060
n = 5000	−0.13	0.13	0.13	0.056	0.056	0.059
n = 10000	−0.13	0.13	0.13	0.059	0.059	0.060

To emphasize the properties of the inequality estimator and its closer analogue the weighted estimator, we ran simulations to larger sample sizes than the Portnoy estimator. The results in Table 4 show that the bias of the slope remains practically constant and negative, thus corroborating the results of the theoretical analysis. Note also how the performance of the weighted estimator suffer from the misspecification.

3.2.2   Simulation: IID Error, IID Censoring

In this experiment event times were generated according to the same IID linear model as in the previous paragraph, but censoring times were generated independently from the input, according to the model:

$$C_i = 6.25 + v_i \quad v_i \sim N(0, 0.3^2).$$

Note that this has been obtained by replacing the dependence from the input, with its expected value 1. In Table 5 we report the performance of the three estimators. We again see how the bias of the slope in the case of the inequality estimator remains essentially constant negative with increasing sample sizes, whereas it decreases for the other two estimators.

3.2.3   Simulation: IID Error, Constant Censoring

In this experiment event times were generated according to the previous experiment, but censoring times were constant for all observations:

$$C_i = 6.5.$$

Table 5. IID error and IID censoring

	Intercept (= 5)			Slope (= 1)		
	Bias	MAE	RMSE	Bias	MAE	RMSE
Portnoy						
n = 100	−0.0042	0.065	0.094	0.0024	0.059	0.087
n = 400	−0.0025	0.037	0.054	−0.0009	0.032	0.047
n = 1000	−0.0025	0.021	0.031	0.0006	0.019	0.028
Inequality						
n = 100	0.051	0.088	0.12	−0.11	0.12	0.15
n = 400	0.055	0.058	0.077	−0.13	0.13	0.13
n = 1000	0.053	0.054	0.062	−0.12	0.12	0.12
n = 5000	0.054	0.053	0.056	−0.12	0.12	0.12
n = 10000	0.053	0.052	0.054	−0.12	0.12	0.12
Weighted						
n = 100	0.066	0.11	0.15	−0.13	0.17	0.21
n = 400	0.048	0.073	0.098	−0.091	0.12	0.15
n = 1000	0.043	0.055	0.070	−0.079	0.096	0.11
n = 5000	0.031	0.038	0.047	−0.055	0.063	0.076
n = 10000	0.027	0.032	0.038	−0.048	0.055	0.063

Table 6. IID error and constant censoring

	Intercept (= 5)			Slope (= 1)		
	Bias	MAE	RMSE	Bias	MAE	RMSE
Portnoy						
n = 100	−0.0032	0.064	0.094	0.0024	0.070	0.11
n = 400	−0.0066	0.041	0.054	-0.0036	0.039	0.059
n = 1000	−0.0022	0.022	0.031	0.0006	0.023	0.034
Inequality						
n = 100	0.067	0.089	0.13	−0.12	0.12	0.15
n = 400	0.069	0.070	0.087	−0.12	0.12	0.13
n = 1000	0.061	0.060	0.068	−0.11	0.11	0.11
n = 5000	0.066	0.066	0.067	−0.11	0.11	0.11
n = 10000	0.066	0.066	0.066	−0.11	0.11	0.11

The results in Table 6 show that the bias of the slope remains practically constant and negative, again corroborating the results of the theoretical analysis. Note that the weighted estimator cannot be applied in this instance since censoring is constant.

3.2.4   Stanford Heart Transplant Data

We assessed the performance of the median estimators on the Stanford Heart Transplant data (available in the R *survival* package as *stanford2*.) The sample size is 184, with 71 censored observations (38.6 % censoring). The model is a regression of survival days (expressed as logarithm base 10) vs. a 2nd degree polynomial in the variable age (following similar uses of this data set in literature.)

In Table 7 we report the parameter estimates of Cox, inequality, weighted and Portnoy estimators, with standard errors in brackets. The standard errors for the inequality and weighted estimators were calculated as the median absolute deviation of the bootstrap c.d.f. (based upon 1000 replications) divided by 0.67 (so that the estimate is consistent at the standard normal distribution and approximately equal to one.)

Table 7. Stanford Heart Transplant data

	Cox	Inequality	Weighted	Portnoy
Constant	–	1.822	0.860	2.067
		(1.292)	(1.395)	(0.687)
age	0.121	0.0837	0.108	0.0836
	(0.0527)	(0.0612)	(0.074)	(0.0514)
age^2	−0.0020	−0.0014	−0.0015	−0.0015
	(0.0007)	(0.0007)	(0.0009)	(0.0008)
Median survival (days) at mean age	729	733	494	1005

3.2.5 German Breast Cancer Study Group 2 Data

As a further test, we analysed the data from the German Breast Cancer Study Group 2 [10]. The sample size is 686, with 299 censored observations (43.6 % censoring). The model is a regression of survival days (expressed as logarithm base 10) vs. eight prognostic factors.

In Table 8 we report the coefficient estimates of Cox, inequality, weighted and Portnoy estimators, with standard errors in brackets. The standard errors for the inequality and weighted estimators were calculated by bootstrap. (same setup as in the previous analysis.)

Table 8. German Breast Cancer Study Group 2 data

	Cox	Inequality	Weighted	Portnoy
Constant	–	3.0878 (0.134)	2.757 (0.248)	3.126 (0.332)
horTh	0.337 (0.129)	0.107 (0.0344)	0.0716 (0.110)	0.150 (0.0865)
age	0.00939 (0.00927)	0.00498 (0.00260)	−0.0111 (0.00485)	0.00657 (0.00489)
menostat	−0.267 (0.183)	−0.0885 (0.0467)	−0.0326 (0.127)	−0.145 (0.112)
tsize	−0.00772 (0.00395)	−0.00282 (0.00141)	0.0000547 (0.00452)	−0.00278 (0.00302)
tgrade	−0.280 (0.106)	−0.0517 (0.0302)	−0.146 (0.0525)	−0.0424 (0.0573)
pnodes	−0.0499 (0.00741)	−0.0178 (0.00455)	−0.0145 (0.00868)	−0.0244 (0.00832)
progrec	0.00224 (0.000576)	0.000559 (0.000130)	0.000409 (0.000363)	0.00059 (0.00024)
estrec	−0.000168 (0.000448)	−0.00000166 (0.000165)	−0.0000779 (0.000290)	−0.00002 (0.00033)
Median survival (days) at mean age	1814	1316	1018	1639

3.2.6 Drug Relapse Data from Hosmer and Lemeshow

We estimated the conditional median of the data reported in [11], and also available in R under the name *uis*. The sample size is 575 with 111 censored observations ($\sim$ 19 % censoring.)

The model is a regression of the natural logarithm of the days to relapse to drug abuse of subjects in a drug treatment program vs. eight prognostic factors.

In Table 9 we report the coefficient estimates of Cox, inequality, weighted and Portnoy estimators, with standard errors in brackets. The standard errors for the inequality and weighted estimators were calculated by bootstrap (same setup as in the previous analysis.)

Table 9. Hosmer and Lemeshow data

	Cox	Inequality	Weighted	Portnoy
Intercept	–	2.367	3.0652	2.375
		(0.323)	(0.262)	(0.373)
ND1	0.511	0.343	0.0889	0.328
	(0.129)	(0.111)	(0.105)	(0.145)
ND2	0.191	0.129	0.0269	0.122
	(0.0499)	(0.0418)	(0.0451)	(0.0541)
IV3	−0.371	−0.145	−0.109	−0.168
	(0.108)	(0.0965)	(0.0703)	(0.123)
TREAT	0.490	0.662	0.640	0.665
	(0.0963)	(0.0895)	(0.0849)	(0.0888)
FRAC	1.253	1.346	1.214	1.382
	(0.103)	(0.108)	(0.0815)	(0.133)
RACE	0.349	0.442	0.311	0.446
	(0.115)	(0.143)	(0.149)	(0.138)
AGE	0.0380	0.0238	0.0126	0.0242
	(0.0100)	(0.00922)	(0.00725)	(0.0132)
SITE	0.870	0.953	0.447	0.894
	(0.519)	(0.422)	(0.430)	(0.579)
AGE:SITE interaction	−0.0414	−0.0430	−0.0274	−0.0410
	(0.157)	(0.0132)	(0.0137)	(0.0181)
Median survival (days) at mean age	167	141	123	144

4 Conclusions

In this paper we have shown that a censored median regression estimator independently proposed in literature by different authors doesn't appropriately take into account all the aspects of the censoring phenomenon.

Comparison results in the one-sample case with the Kaplan-Meier estimator (which is known to be consistent) through simulation seems to suggest the estimator is not consistent.

We compared the performance of the estimator in the more general regression setting with Portnoy's censored quantile regression estimator on simulated data, and we observed further evidence of inconsistency.

At the same time, the weighted estimator, although theoretically sound, suffers from possible misspecification of the censoring model.

Although the inequality estimator has an intuitive description and is simple to implement, we suggest that care should be taken when using it to analyse real world data.

References

1. Pelckmans, K., De Brabanter, J., Suykens, J.A.K., De Moor, B.: Risk Scores, Empirical Z-estimators and its application to Censored Regression. Technical Report kp06–105 (2006)
2. Shivaswamy, P., Chu, W., Jansche, M.: A support vector approach to censored targets. In: Proceedings of the 2007 Seventh IEEE International Conference on Data Mining (2007)
3. Koenker, R.: Quantile Regression. Cambridge University Press, Cambridge (2005)
4. Efron, B.: The two sample problem with censored data. In: Proceedings of the 5th Berkeley Symposium on Mathematical Statistics and Probability, Prentice-Hall, New York (1967)
5. Portnoy, S.: Censored quantile regression. J. Am. Statist. Assoc. **98**, 1001–1012 (2003)
6. Koul, H., Susarla, V., Van Ryzin, J.: Regression analysis with randomly right censored data. Ann. Stat. **9**, 1276–1288 (1981)
7. Robins, J.M., Rotnitzky, A., Zhao, L.P.: Estimation of regression coefficients when some regressors are not always observed. J. Am. Statist. Assoc. **89**(427), 846–866 (1994)
8. Eleuteri, A., Taktak, A.F.: Support vector machines for survival regression. In: Biganzoli, E., Vellido, A., Ambrogi, F., Tagliaferri, R. (eds.) CIBB 2011. LNCS, vol. 7548, pp. 176–189. Springer, Heidelberg (2012)
9. Koenker, R.: Censored quantile regression redux. J. Stat. Softw. **27**(6), 1–25 (2008)
10. Schumacher, M., Basert, G., Bojar, H., Huebner, K., Olschewski, M., Sauerbrei, W., Schmoor, C., Beyerle, C., Neumann, R.L.A., Rauschecker, H.F., German Breast Cancer Study Group: Randomized 2x2 trial evaluating hormonal treatment and the duration of chemotherapy in node-positive breast cancer patients. J. Clin. Oncol. **12**, 2086–2093 (1994)
11. Hosmer, D., Lemeshow, S.: Applied Survival Analysis: Regression Modeling of Time to Event Data. John Wiley & Sons, New York (1999)

Selecting Random Effect Components in a Sparse Hierarchical Bayesian Model for Identifying Antigenic Variability

Vinny Davies[✉], Richard Reeve, William T. Harvey, and Dirk Husmeier

University of Glasgow, Glasgow, Scotland, UK
v.davies.1@research.gla.ac.uk

Abstract. In Foot-and-Mouth Disease Virus (FMDV), understanding how viruses offer protection against related emerging strains is vital for creating effective vaccines. With testing large numbers of vaccines being infeasible, the development of an *in silico* predictor of cross-protection between virus strains has been a vital area of recent research. The current paper reviews a recent contribution to this area, the SABRE method, a sparse hierarchical Bayesian model which uses spike and slab priors to identify key antigenic sites within FMDV serotypes. WAIC is then combined with the SABRE method and its ability to approximate Bayesian Cross Validation performance in terms of correctly selecting random effect components analysed. WAIC and the SABRE method have then been applied to two FMDV datasets and the results analysed.

Keywords: Model selection · Spike and slab prior · Foot-and-Mouth Disease Virus · Bayesian hierarchical models · WAIC · Cross Validation

1 Introduction

In Foot-and-Mouth Disease Virus (FMDV) where new virus strains continuously emerge, choosing effective vaccines is vital. However FMDV has high genetic variability due to changes in the virus proteins which affect recognition by the host immune system. With the high antigenic variability, FMDV vaccines are only effective against strains that are closely related genetically and antigenically similar to the vaccine strain. As a result it is important to estimate antigenic similarity between different strains and understand how one strain can confer protection against another. The South African Territories types 1 and 2 (SAT1 and SAT2) serotypes both show significant levels of antigenic variability and can be used to explore the relationship between antigenic variation and changes in the protein structure.

In order to understand the relationship between antigenic variation and changes in the protein structure, we need a measure of the antigenic similarity between any two virus strains. Virus Neutralisation (VN) titre is *in vitro* measure which approximates the extent one strain confers protection on another by examining how well one strain (the challenge strain) is able to neutralise a

© Springer International Publishing Switzerland 2016
C. Angelini et al. (Eds.): CIBB 2015, LNBI 9874, pp. 14–27, 2016.
DOI: 10.1007/978-3-319-44332-4_2

second strain (the protective strain). Higher values of VN titre indicate that the protective strain offers a higher level of protection against the challenge strain and that the strains are more antigenically similar.

The antigenic differences between virus strains can be explained by changes in the protein structure on the surface of the virus shell. While many changes can occur, only some of these affect recognition by the host immune systems and result in a reduction in the observed VN titre. Identifying the individual areas of the surface exposed proteins, residues, that are considered to be key antigenic regions is critical to understanding the antigenic similarities between viruses. Similarly, understanding how antigenicity is affected by the evolutionary history of the virus strains is important and must be accounted for.

Predicting VN titre, the *in vitro* measure of antigenic similarity, based on the changes in the virus proteins and the shared evolutionary history of the virus strains is complicated by the presence of variation in the VN test, the test to determine VN titre. It is possible that certain virus strains will produce higher or lower VN titre measurements against all other virus strains due a reactivity or immunogenic effect caused by non-antigenic properties of the challenge or protective strains. Additionally the serum used as part of the VN test and the date of the experiment, a proxy for lab conditions, can affect the measured VN titre. Where available, see Sect. 5, the challenge strain, protective strain, serum and date are included as potential random effects and choosing which of these should be used in the analysis is an important problem as including irrelevant components will introduce unnecessary variation into the models.

To account for both the random and fixed effects, Reeve et al. (2010) used classical mixed-effects models, e.g. Pinheiro and Bates (2000), to predict the antigenic similarity between any two virus strains. The authors firstly selected the random effect components and then added terms to account for the evolutionary history of the virus using a forward inclusion algorithm. A univariate test for significance was then carried out on the residue variables with a p-value of less than 0.05 corresponding to an antigenically significant residue. Davies et al. (2014) then introduced a sparse hierarchical Bayesian model for detecting relevant antigenic sites in virus evolution (SABRE) method which was shown to outperform the method of Reeve et al. (2010) in terms variable selection. The first aim of the current work is to review the SABRE method of Davies et al. (2014), propose a slight methodological improvement and show how the SABRE method outperforms both classical mixed-effects models and the mixed-effects Least Absolute Shrinkage and Selection Operator (LASSO) (Tibshirani 1996; Schelldorfe et al. 2011) in terms of variable selection.

The SABRE method of Davies et al. (2014) combines a Bayesian hierarchical mixed-effects model with spike and slab priors. Hierarchical models allow for consistent inference of all parameters and hyper-parameters with the inferences borrowing strength from the sharing and combination of information; see Gelman et al. (2013). The introduction of spike and slab priors into the model allows for simultaneous model selection not offered by the classical mixed-effects models of Reeve et al. (2010) and improved variable selection over the ℓ_1 regularisation offered by the mixed-effects LASSO (Mohamed et al. 2012).

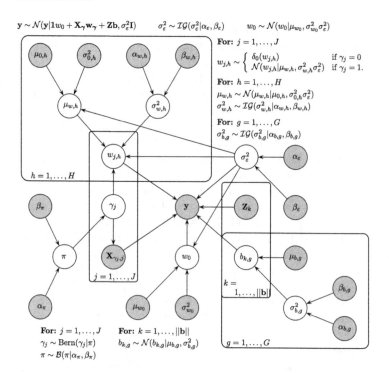

Fig. 1. Compact representation of the SABRE method as a PGM. The *grey* circles refer to the data and fixed (higher-order) hyperparameters, while the *white* circles refer to parameters and hyperparameters that are inferred.

The second and new contribution of the current work is to investigate how best to choose the random effect components that should be included in the SABRE method for each dataset. In larger datasets, where the SABRE method is computationally expensive, using Bayesian Cross Validation (CV) methods is computationally infeasible. The current work investigates whether the Widely Applicable Information Criterion (WAIC) (Watanabe 2010) can be used as a less computationally intensive alternative to Bayesian CV. While WAIC is asymptotically justified, it is unlikely to provide as accurate performance as Bayesian CV in terms of correctly including or excluding random effect components. The purpose of the current study is to understand the size of this reduction in accuracy and assess the suitability of WAIC to be used in larger more computationally demanding datasets, e.g. Harvey et al. (2015).

The final contribution of the current work is to give examples of the SABRE method and WAIC applied to two FMDV datasets. We apply the SABRE method with each possible combination of random effect components and then apply WAIC to find the best choice of model. The results are then analysed in terms of selecting relevant antigenic sites (residues).

2   SABRE Method

In this section we mathematically describe the SABRE method proposed in Davies et al. (2014) with the addition of a separate intercept parameter and increased conjugacy, where the Probabilistic Graphical Model (PGM) is shown in Fig. 1. The model parameters are sampled from the posterior distribution using Markov chain Monte Carlo (MCMC), using the distributions in Sect. 2.5.

2.1   Likelihood

The likelihood of the SABRE method is similar to that of classical mixed-effects models, e.g. Pinheiro and Bates (2000), where the response $\mathbf{y} = (y_1, \ldots, y_N)^\top$ is taken as the log VN titre. In classical mixed-effects models, the response, $\mathbf{y}$, is modelled by a combination of the intercept, w_0, the explanatory variables, $\mathbf{X}$, and corresponding regression coefficients, $\mathbf{w}$, as well as random effects, $\mathbf{b}$, and the design matrix, $\mathbf{Z}$. The SABRE method uses a similar structure as can be seen in Fig. 1, however it only includes the relevant explanatory variables, $\mathbf{X}_\gamma$, and regression coefficients, $\mathbf{w}_\gamma$:

$$p(\mathbf{y}|w_0, \mathbf{w}_\gamma, \mathbf{b}, \sigma_\varepsilon^2, \mathbf{X}_\gamma, \mathbf{Z}) = \mathcal{N}(\mathbf{y}|\mathbf{1}w_0 + \mathbf{X}_\gamma \mathbf{w}_\gamma + \mathbf{Z}\mathbf{b}, \sigma_\varepsilon^2 \mathbf{I}). \tag{1}$$

The relevance of the jth column of $\mathbf{X}$ is determined by $\gamma_j \in \{0, 1\}$, where feature j is said to be relevant if $\gamma_j = 1$, giving $\boldsymbol{\gamma} = (\gamma_1, \ldots, \gamma_J)^\top \in \{0, 1\}^J$. We then define $\mathbf{X}_\gamma$ to be the matrix of relevant explanatory variables with $||\boldsymbol{\gamma}||$ columns and N rows, where $||\boldsymbol{\gamma}|| = \sum_{j=1}^{J} \gamma_j$ is the number of non-zero elements of $\boldsymbol{\gamma}$. Similarly $\mathbf{w}_\gamma$ is given as the column vector of regressors, where the inclusion of each parameter is dependent on $\boldsymbol{\gamma}$.

2.2   Noise and Intercept Priors

As with classical mixed-effects models, we assume iid Gaussian noise, σ_ε^2, for the log VN titre, $\mathbf{y}$. σ_ε^2 is then given a conjugate Inverse-Gamma prior:

$$\sigma_\varepsilon^2 \sim \mathcal{IG}(\sigma_\varepsilon^2|\alpha_\varepsilon, \beta_\varepsilon) \tag{2}$$

where α_ε and β_ε are fixed, as indicated by the grey nodes in Fig. 1.

In addition to being used in the likelihood, (1), σ_ε^2 is also included in the distributions for w_0, $\mathbf{w}_\gamma$, $\boldsymbol{\mu}_\mathbf{w} = (\mu_{w,1}, \ldots, \mu_{w,H})^\top$ in Sect. 2.3. These additional relationships, indicated by the edges in Fig. 1, increase information sharing and mean that the error variance in terms of model fit is reflected in the distribution of the regression coefficients. Including these relationships also makes the model conjugate rather than semi-conjugate, see Chap. 3 of Gelman et al. (2013), and allows the creation of an improved sampling strategy based on using collapsed Gibbs sampling, e.g. Andrieu and Doucet (1999).

We also require a distribution for the intercept, w_0:

$$w_0 \sim \mathcal{N}(w_0|\mu_{w_0}, \sigma_{w_0}^2 \sigma_\varepsilon^2). \tag{3}$$

We treat the intercept differently from the remaining regressors, wishing to use vague prior settings so as not to penalise this term and effectively make the model scale invariant (Hastie et al. 2009).

2.3   Spike and Slab Priors

Spike and slab priors are known to outperform ℓ_1 methods such as the LASSO both in terms of variable selection and out-of-sample performance (Mohamed et al. 2012). They have been used in a number of forms, but were originally proposed by Mitchell and Beauchamp (1988) as a mixture of a Gaussian distribution and a Dirac spike, as used for the SABRE method in (4). Alternatives to the specification of Mitchell and Beauchamp (1988) include the mixture of two Gaussian distributions proposed by George and McCulloch (1993) and the Binary mask model, e.g. Jow et al. (2014).

The spike and slab prior reflects the relevance of each variable $w_{j,h}$ based on the value of the corresponding latent indicator variable, γ_j. If $\gamma_j = 0$, i.e. the jth variable, $\mathbf{X}_j$, is irrelevant, then we expect that $w_{j,h} = 0$. Conversely if $\gamma_j = 1$, we think the jth variable is relevant and the corresponding regression coefficient should be non-zero, $w_{j,h} \neq 0$, and we specify a conjugate Gaussian prior. To increase generality we allow the models to have multiple groups of variables $h \in \{1, \ldots, H\}$ which are defined by j, i.e. $w_{j,h}$ is shorthand for w_{j,h_j}, but only a single group is used for the results in Sects. 7 and 8.

$$p(w_{j,h}|\gamma_j, \mu_{w,h}, \sigma_{w,h}^2, \sigma_\varepsilon^2) = \begin{cases} \delta_0(w_{j,h}) & \text{if } \gamma_j = 0 \\ \mathcal{N}(w_{j,h}|\mu_{w,h}, \sigma_{w,h}^2 \sigma_\varepsilon^2) & \text{if } \gamma_j = 1 \end{cases} \qquad (4)$$

for $j \in 1, \ldots, J$ and where δ_0 is the delta function. Here we have a spike at 0 and as $\sigma_{w,h}^2 \sigma_\varepsilon^2 \to \infty$ the distribution, $p(w_{j,h}|\gamma_j = 1)$, approaches a uniform distribution, a slab of constant height.

We give the hyper-parameters of (4) conjugate priors, specifying $\sigma_{w,h}^2$ to have an Inverse-Gamma prior with fixed hyper-parameters $\alpha_{w,h}$ and $\beta_{w,h}$, and $\mu_{w,h}$ a Gaussian prior with fixed hyper-parameters $\mu_{0,h}$ and $\sigma_{0,h}^2$:

$$\sigma_{w,h}^2 \sim \mathcal{IG}(\sigma_{w,h}^2|\alpha_{w,h}, \beta_{w,h} \qquad \mu_{w,h} \sim \mathcal{N}(\mu_{w,h}|\mu_{0,h}, \sigma_{0,h}^2 \sigma_\varepsilon^2) \qquad (5)$$

where σ_ε^2 is again included in the variance of $\mu_{w,h}$ for further conjugacy. We allow $\mu_{w,h}$ to vary in order to reflect our biological understanding of the problem. In the FMDV data we are likely to observe a comparatively large intercept, with negative regression coefficients, $w_{j,h}$, reflecting the fact that any mutational or evolutionary changes are likely to reduce the similarity between virus strains, therefore reducing the measured VN titre.

For convenience we define $\mathbf{w}_\gamma^* = (w_0, \mathbf{w}_\gamma^\top)^\top$ with the following distribution:

$$\mathbf{w}_\gamma^* \sim \mathcal{N}(\mathbf{w}_\gamma^*|\mathbf{m}_\gamma, \sigma_\varepsilon^2 \boldsymbol{\Sigma}_{\mathbf{w}_\gamma^*}) \qquad (6)$$

where $\mathbf{m}_\gamma = (\mu_{w_0}, \mu_{w,1}, \ldots, \mu_{w,1}, \mu_{w,2}, \ldots, \mu_{w,H})^\top$ and $\boldsymbol{\Sigma}_{\mathbf{w}_\gamma^*} = diag(\boldsymbol{\sigma}_{\mathbf{w}^*}^2)$ with $\boldsymbol{\sigma}_{\mathbf{w}^*}^2 = (\sigma_{w_0}^2, \sigma_{w,1}^2, \ldots, \sigma_{w,1}^2, \sigma_{w,2}^2, \ldots, \sigma_{w,H}^2)^\top$. Each $\mu_{w,h}$ and $\sigma_{w,h}^2$ is repeated with length $\|\mathbf{w}_{\gamma,h}\|$ dependent on $\boldsymbol{\gamma}$.

The priors related to the latent inclusion parameters, $\boldsymbol{\gamma}$, are given by:

$$p(\boldsymbol{\gamma}|\pi) = \prod_{j=1}^{J} \text{Bern}(\gamma_j|\pi) \qquad \pi \sim \mathcal{B}(\pi|\alpha_\pi, \beta_\pi) \qquad (7)$$

where we define π to be the probability of an individual variable being relevant. Given we do not a-priori know the value of π, it is given a conjugate Beta prior where α_π and β_π are fixed to represent our vague knowledge that only a small proportion of variables should be included in the model; see Sect. 6.

2.4   Random-Effects Priors

The random-effect coefficients are given as $\mathbf{b} = (\mathbf{b}_1^\top, \ldots, \mathbf{b}_G^\top)^\top$, where each $\mathbf{b}_g$ relates to a vector of coefficients related to different levels with in a particular random effect component or group, $g \in \{1, \ldots, G\}$, e.g. challenge strain. Each $\mathbf{b}_g$ has $\|\mathbf{b}_g\|$ coefficients and follows a zero mean Gaussian distribution with a group dependent variance, $\mathbf{b}_g \sim \mathcal{N}(\mathbf{b}_g|\mathbf{0}, \sigma_{b,g}^2\mathbf{I})$, where $\mathbf{I}$ is the identity matrix. From this we then define all the random-effect coefficients to have a joint distribution $\mathbf{b} \sim \mathcal{N}(\mathbf{b}|\mathbf{0}, \boldsymbol{\Sigma}_\mathbf{b})$, where we define $\boldsymbol{\Sigma}_\mathbf{b}$ to be a diagonal matrix with $(\sigma_{b,1}^2, \ldots, \sigma_{b,1}^2, \sigma_{b,2}^2, \ldots, \sigma_{b,G}^2)^\top$ on the diagonal with each $\sigma_{b,g}^2$ being repeated $\|\mathbf{b}_g\|$ times.

We give $b_{k,g}$, the kth coefficient of $\mathbf{b}$, a Gaussian distribution with a fixed zero mean, $\mu_{b,g} = 0$, and a group dependant variance parameter, $\sigma_{b,g}^2$, which is in turn given an Inverse-Gamma prior:

$$b_{k,g} \sim \mathcal{N}(b_{k,g}|\mu_{b,g}, \sigma_{b,g}^2) \qquad \sigma_{b,g}^2 \sim \mathcal{IG}(\sigma_{b,g}^2|\alpha_{b,g}, \beta_{b,g}). \qquad (8)$$

The group g is defined by k, i.e. $b_{k,g}$ is shorthand for b_{k,g_k} and the hyper-parameters $\alpha_{b,g}$ and $\beta_{b,g}$ are fixed for each g.

2.5   Posterior Inference

To sample from the posterior distribution we have used an MCMC algorithm. As we have chosen mainly conjugate priors (see Sect. 2), we can use a Gibbs sampling scheme. The conditional dependence relations are shown in the graphical model of Fig. 1, and the detailed forms of the conditional distributions are available from Sect. 10.

Sampling $\boldsymbol{\gamma}$ is more difficult, as it does not naturally take a distribution of standard form. However we can still get a valid conditional distribution and use a variety of techniques to sample from it. Here we have used collapsing methods to achieve faster mixing and convergence:

$$p(\boldsymbol{\gamma}|\boldsymbol{\theta}', \mathbf{X}_{\boldsymbol{\gamma}}, \mathbf{Z}, \mathbf{y}) \propto \int p(\boldsymbol{\gamma}, \pi, \sigma_\varepsilon^2, \mathbf{w}_{\boldsymbol{\gamma}}^*, \boldsymbol{\mu}_\mathbf{w}|\boldsymbol{\theta}', \mathbf{X}_{\boldsymbol{\gamma}}, \mathbf{Z}, \mathbf{y})d\boldsymbol{\mu}_\mathbf{w}d\mathbf{w}_{\boldsymbol{\gamma}}^*d\pi d\sigma_\varepsilon^2 \qquad (9)$$

where the fixed hyper-parameters (given as grey circles in Fig. 1) have been dropped to improve notational clarity. In the current work we update multiple γ_j

simultaneously via a Metropolis-Hastings step (Metropolis et al. 1953; Hastings 1970), which Davies et al. (2014) found to be more computationally efficient than the more established component-wise Gibbs sampler. In addition to being used within the conditional distribution of $\boldsymbol{\gamma}$, collapsing steps are also used for $\mathbf{w}_{\boldsymbol{\gamma}}^*$, $\boldsymbol{\mu}_{\mathbf{w}}$, σ_ε^2 and π. These steps are not detailed here, but using them leads to improved mixing and convergence, e.g. Andrieu and Doucet (1999).

3 Random Effect Selection Methods

3.1 Cross Validation

Bayesian CV methods are reliable, if computationally expensive, techniques for measuring the out-of-sample performance of different models. CV methods work by partitioning the data into K groups and then analysing the predictive performance of a given model on each of the K different groups using the remainder of the data for training. In this sense CV methods estimates out-of-sample predictive performance while still making use of all of the available data.

Various CV methods can be used to analyse the performance of different models. Leave-One-Out CV (LOO-CV) uses each observation as an individual group, i.e. $K = N$, with the advantage of making maximum use of the available data at every step. However LOO-CV is computational infeasible for many models, as it requires fitting the model N times. As a compromise 10-fold CV is often used, where $K = 10$, as it only involves fitting 10 models and this method has been used here.

To calculate the 10-fold Bayesian CV performance of a model, we apply the SABRE method to partial data, $\mathbf{y}_{-k}$, $\mathbf{X}_{\boldsymbol{\gamma},-k}$ and $\mathbf{Z}_{-k}$, and use thinned samples of the model parameters, $\boldsymbol{\theta}^\iota$, for $\iota \in \{1, \ldots, I\}$, from $p(\boldsymbol{\theta}|\mathbf{y}_{-k}, \mathbf{X}_{\boldsymbol{\gamma},-k}, \mathbf{Z}_{-k})$, to estimate the performance on the remaining data, $\mathbf{y}_k$, $\mathbf{X}_{\boldsymbol{\gamma},k}$ and $\mathbf{Z}_k$, using (1). Doing this for each of the K groups gives the 10-fold Bayesian CV performance:

$$p_{CV} = \frac{1}{K} \sum_{k=1}^{K} \log \frac{1}{I} \sum_{\iota=1}^{I} p(\mathbf{y}_k|\boldsymbol{\theta}^\iota, \mathbf{X}_{\boldsymbol{\gamma},k}, \mathbf{Z}_k). \tag{10}$$

3.2 WAIC

WAIC, as proposed in Watanabe (2010) is a useful criterion for selecting the correct model when the underlying model is singular, e.g. the SABRE method. Additionally WAIC has the desirable property of averaging over the posterior distribution, as opposed to the Deviance Information Criterion (DIC) which uses a point estimate. Watanabe (2010) showed how WAIC is asymptotically equivalent to Bayesian LOO-CV and can be computed using the thinned parameter samples, $\boldsymbol{\theta}^\iota$, from the posterior distribution of the full dataset, $p(\boldsymbol{\theta}|\mathbf{y}, \mathbf{X}_{\boldsymbol{\gamma}}, \mathbf{Z})$, meaning we only have to sample the model parameters once:

$$p_{WAIC} = -2\sum_{i=1}^{N} \left(\log \left(\frac{1}{I} \sum_{\iota=1}^{I} p(y_i|\boldsymbol{\theta}^\iota, \mathbf{X}_{\boldsymbol{\gamma},i}, \mathbf{Z}_i) \right) - \mathrm{Var}\left(\log(p(y_i|\boldsymbol{\theta}^\iota, \mathbf{X}_{\boldsymbol{\gamma},i}, \mathbf{Z}_i))\right) \right). \tag{11}$$

3.3 Multiple Parameter Spike and Slab Prior

Using a spike and slab prior to include or exclude all random effect coefficients, $\mathbf{b}_g$, from a particular random effect component, g, is an alternative to both WAIC and 10-fold Bayesian CV. While WAIC and 10-fold Bayesian CV would be applied to each combination of random effect components separately, spike and slab priors would only require one model to be fitted. However, using spike and slab priors for selecting the random effects will come at a large computational cost. Some of the random effect components from the FMDV datasets contain between 30 to 50 different levels and this would mean including or excluding 30 to 50 parameters simultaneously at each proposal step of the MCMC sampling scheme. This is likely to lead to poor mixing as the difference in log-likelihood for the inclusion and exclusion of a random effect component is likely to be large. Poor mixing leads to the possibility of not sampling the optimal combination of fixed and random effects, as the proposals will struggle to move between different combinations of random effect components. Therefore in order to ensure the optimal selection of fixed and random effects is found it would be necessary to sample the model for a large number of iterations. Due to the computational inefficiency of this inter-model approach, we have used an intra-model approach and run MCMC simulations for a relatively small number of models in parallel to compute WAIC and 10-fold Bayesian CV scores for each plausible candidate model separately.

4 Simulated Data

To show that the SABRE method proposed in Davies et al. (2014), with the addition of a separate intercept parameter and increased conjugacy, still outperforms classical mixed-effects models and the mixed-effects LASSO in terms of variables selection, we generated 100 simulated datasets. Each of these datasets were given 40 possible variables, where the corresponding coefficients were set to be non zero with probability $\pi \sim \mathcal{U}(0.2, 0.4)$. Additionally 2 random effect components were added, each with 8 levels.

Additionally, to compare WAIC and 10-fold Bayesian CV, we generated 20 datasets each with 500 observations and 50 possible variables. The data was generated with 10 viruses, with every virus used as both the challenge and protective strains and for any given pair of viruses the variables remain identical as in the real FMDV datasets. Possible random effects were the protective and challenge strains and 2 generic random effects with 8 levels. The random effects were given a variance of zero, i.e. set to be irrelevant, with probability 0.5.

5 FMDV Data

Davies et al. (2014) analysed a dataset from the SAT1 serotype of FMDV, which was originally used in Reeve et al. (2010). Since that analysis, additional data has been collected and been analysed using mixed-effects models in

(Maree et al. 2015). We call this the extended SAT1 dataset and it contains 2125 VN titre measurements with 5 protective and 42 challenge strains, and 221 variables related to the residues and phylogenetic structure. Possible random effects included the serum used to get the VN titre measurement, the challenge strain, the protective strain and the date of the experiment (a proxy to lab conditions).

Reeve et al. (2010) also used a dataset on the SAT2 serotype, although it was not analysed in Davies et al. (2014). The SAT2 dataset contains 320 VN titre measurements from 4 protective and 22 challenge strains. In total there are 148 variables when the residues and phylogenetic data are combined. Possible random effects include the serum used to get the VN titre measurement, the challenge strain and the protective strain.

6   Computational Inference

Our code has been implemented in R, using the packages *lme4* (Bates et al. 2013) and *lmmlasso* (Schelldorfer et al. 2011) for the comparison with classical mixed-effects models and mixed-effects LASSO. For the mixed-effects models, as in Reeve et al. (2010), forward variable inclusion was used adjusting for multiple testing using the Holm-Bonferroni correction.

We ran MCMC simulations for 10,000 and 15,000 samples respectively for the simulated and real datasets removing an appropriate proportion for burn-in based on convergence diagnostics. Convergence was determined by computing the potential scale reduction factor (PSRF) (Gelman and Rubin 1992), where a PSRF ≤ 1.05 for 95 % of the variables was taken as the threshold for convergence. The latent inclusion parameters were sampled using a block Metropolis-Hastings algorithm following Davies et al. (2014).

In general, the fixed hyper-parameters, shown as grey nodes in Fig. 1, were set to give a vague distribution for the flexible (hyper-)parameters, shown as white nodes. The only exception was the prior on π, defined in (7), which was set to be weakly informative such that $\alpha_\pi = 1$ and $\beta_\pi = 4$. This corresponds to prior knowledge that only a small number of residues or branches have a significant antigenic effect. The remaining hyper-parameters, shown as grey nodes in Fig. 1, are fixed to give vague distributions: $\alpha_{b,g} = \beta_{b,g} = \alpha_{\eta,g} = \beta_{\eta,g} = 0.001$ and $\mu_{b,g} = \mu_{\eta,g} = 0$ for all g, $\alpha_{w,h} = \beta_{w,h} = 0.001$, $\mu_{0,h} = 0$ and $\sigma^2_{0,h} = 100$ for all h, $\mu_\xi = 0$, $\sigma^2_\xi = 100$, $\mu_{w_0} = max(\mathbf{y})$, $\sigma^2_{w_0} = 100$ and $\alpha_\varepsilon = \beta_\varepsilon = 0.001$. The only informative choice is $\mu_{w_0} = max(\mathbf{y})$ which follows from us expecting a high intercept with the regression coefficients then having a negative effect on the response. This is a result of strains having high reactivity with themselves, and any changes making the strains less similar, reducing their reactivity.

7   Simulation Study Results

To compare the variable selection accuracy of the SABRE method compared to the mixed-effects LASSO and classical mixed-effects models we produced receiver

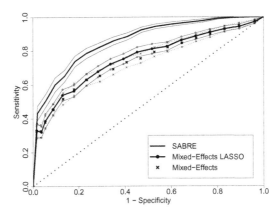

Fig. 2. ROC curves and 95 % confidence intervals for classical mixed-effects models (crosses), the mixed-effects LASSO (solid black and points), and the SABRE method (solid black) when applied to the simulated data in Sect. 4.

operating characteristic (ROC) curves for each of the methods by ordering the inclusion of variables. For the SABRE method we ordered the marginal posterior inclusion probabilities of each variable. For the mixed-effects LASSO, the model was run for different values of the penalty parameter λ and then the so-called LASSO path created (Hastie et al. 2009). Finally for the classical mixed-effects models we ran a forward inclusion algorithm with no stopping point, ranking the variables based on when they were included in the model. For each method the moving average (mean) and standard deviation of the ROC curves for each of the 100 simulated datasets was taken. The mean ROC curve for each method was then plotted with the corresponding 95 % confidence interval in Fig. 2. Using ROC curves to compare the methods gives a more general indication of performance than simply looking at the performance of a specific cut-off point.

Figure 2 shows that the SABRE method outperforms both the mixed-effects LASSO and classical mixed-effects models across all cut-offs. The improved performance is shown by the Area Under the ROC (AUROC) value, where the SABRE method achieves an AUROC value and 95 % confidence interval of 0.87 (0.86,0.88) compared to 0.76 (0.74,0.78) and 0.75 (0.73,0.77) for the mixed-effects LASSO and classical mixed-effects models. One-sided paired t-tests showed that the SABRE method performed better in terms of AUROC value than both the mixed-effects LASSO (p-value < 0.001) and standard mixed-effects models (p-value < 0.001). Similarly the mixed-effects LASSO performed better than standard mixed effects models (p-value $= 0.035$).

To analyse the performance of WAIC in comparison to 10-fold Bayesian CV, we looked at how accurate each method was at correctly selecting the random effect components used to generate the datasets simulated in Sect. 4. We applied both methods to each of the 16 possible models for each dataset and selected the best model in each case. We then analysed the ability of the best models to correctly include or exclude the random effect components that were used or

Table 1. Results comparing the model selection performance of WAIC compared to 10-fold Bayesian CV. The mean and 95 % confidence intervals are given in terms of correctly including or excluding random effect components in the simulated datasets described in Sect. 4.

	10-fold Bayesian CV	WAIC
Sensitivity	0.91 (0.85,0.97)	0.78 (0.69,0.87)
Specificity	0.63 (0.52,0.73)	0.77 (0.68,0.86)
Predictive accuracy	0.79 (0.70,0.88)	0.78 (0.68,0.87)
F1-score	0.83 (0.75,0.91)	0.80 (0.71,0.88)

not used to generate each of the datasets. Table 1 gives the results in terms of sensitivity, specificity, predictive accuracies and F-scores.

The results of Table 1 show that WAIC performs similarly to 10-fold Bayesian CV in terms of correctly selecting random effect components. While 10-fold Bayesian CV gets an increased sensitivity, WAIC has a better specificity and both perform similarly in their predictive accuracy and F1-score. However WAIC is much more computationally effective and to run the MCMC simulations for the WAIC took on average 87 min, as opposed to 761 min for 10-fold Bayesian CV.

8   FMDV Results

Having tested the use of WAIC on simulated datasets in Sect. 7, we have then used WAIC to find the best choice of random effect components for two FMDV datasets. After applying WAIC to the extended SAT1 dataset, the best model was found to contain only the protective strain and the serum as random effect components. Choosing these random effect components is an interesting result as the work of Davies et al. (2014), on the original SAT1 dataset of Reeve et al. (2010), was based on using the challenge strain and the serum. The results suggest that it is important to effectively chose the random effect components rather than simply choosing them based on biological prior knowledge.

Choosing the most appropriate random effect components will have an affect on which of the fixed effects are selected by the SABRE method. Based on the model selected by WAIC and using $\hat{\pi} \times J$ as the cut-off, the SABRE method found a total of 9 proven and 24 plausible residues or branches. Classification is based on the residue being experimentally validated in the SAT1 serotype or validated in 4 or more FMDV serotypes (proven), being experimentally validated in 3 or less serotypes (plausible), or from a region not known to be antigenic in any of the FMDV serotypes (implausible). The proven residues come from the 4 known antigenic regions (Grazioli et al. 2006); VP1 C-terminus, VP1 G-H, VP2 B-C and VP3 B-C. Additionally, other residues which are not known to be antigenic were also found in these regions and should be experimentally investigated.

Applying WAIC to the SAT2 dataset resulted in all of the random effect components being included in the model; challenge strain, protective strain and serum. As less is known about the SAT2 serotype we do not classify the branches and residues into different categories, and instead treat the best model as a tool for hypothesis generation. While we do not discuss the results in detail here due to space restrictions and the lack of biological prior knowledge that could be used for assessment, it is worth noting that residues were selected from 3 out of the 4 regions identified in the SAT1 serotype above; VP1 C-terminus, VP1 G-H and VP3 B-C. These residues included a large number in close proximity to each other on the VP1 G-H loop and this area could be of experimental interest.

9   Discussion

In the current work we described and improved the SABRE method and shown how it outperforms established alternatives in terms of variable selection in Fig. 2. In addition we have compared the performance of WAIC and 10-fold Bayesian CV in the context of correctly selecting random effect components. The results, given in Table 1, show that in terms of model selection (concerning the random effects to be included) WAIC achieves a similar performance at a lower computational cost to 10-fold Bayesian CV. We have quantified both the difference in performance and the reduction in computational cost. Finally we have applied the SABRE method with WAIC to two FMDV datasets, identifying a number of antigenically important locations on the surface of the virus shell and a number of residues worthy of investigation.

Further work will develop the SABRE method to better take into account the structure of the data. For any given pair of virus strains tested, the fixed effects will remain the same. By introducing a latent structure into the model we can more precisely account for the data generation process. An additional computational advantage can also be gained for larger datasets, e.g. Harvey et al. (2015), as often the datasets will have far more VN titre measurements than tested virus pairs. A latent variable model could take advantage of the structure and reduce the computational complexity of the conditional distribution for γ.

10   Appendix

For the Gibbs sampling we sample the intercept and regression coefficients together and define $\mathbf{w}_{\gamma}^{*} = (w_0, \mathbf{w}_{\gamma}^{\top})^{\top}$, $\mathbf{X}_{\gamma}^{*} = (\mathbf{1}, \mathbf{X}_{\gamma})$, $\mathbf{m}_{\gamma} = (\mu_{w_0}, \mu_{w,1}, \dots, \mu_{w,1},$ $\mu_{w,2}, \dots, \mu_{w,H})^{\top}$ and $\boldsymbol{\Sigma}_{\mathbf{w}_{\gamma}^{*}} = diag(\boldsymbol{\sigma}_{\mathbf{w}^{*}}^{2})$ with $\boldsymbol{\sigma}_{\mathbf{w}^{*}}^{2} = (\sigma_{w_0}^{2}, \sigma_{w,1}^{2}, \dots, \sigma_{w,1}^{2}, \sigma_{w,2}^{2},$ $\dots, \sigma_{w,H}^{2})^{\top}$. Each $\mu_{w,h}$ and $\sigma_{w,h}^{2}$ is repeated with length $||\mathbf{w}_{\gamma,h}||$ dependent on γ. The Gibbs sampling distributions are then given as follows, with θ' used to denote all the parameters not on the left of the conditioning bar:

$$\mathbf{w}_\gamma^* | \theta', \mathbf{X}_\gamma^*, \mathbf{Z}, \mathbf{y} \sim \mathcal{N}(\mathbf{w}_\gamma^* | \mathbf{V}_{\mathbf{w}_\gamma^*} \mathbf{X}_\gamma^{*\top}(\mathbf{y} - \mathbf{Z}\mathbf{b}) + \mathbf{V}_{\mathbf{w}_\gamma^*} \boldsymbol{\Sigma}_{\mathbf{w}_\gamma^*}^{-1} \mathbf{m}_\gamma, \sigma_\varepsilon^2 \mathbf{V}_{\mathbf{w}_\gamma^*}) \tag{12}$$

$$\mathbf{b} | \theta', \mathbf{X}_\gamma^*, \mathbf{Z}, \mathbf{y} \sim \mathcal{N}(\mathbf{b} | \tfrac{1}{\sigma_\varepsilon^2} \mathbf{V}_\mathbf{b} \mathbf{Z}^\top(\mathbf{y} - \mathbf{X}_\gamma^* \mathbf{w}_\gamma^*), \mathbf{V}_\mathbf{b}) \tag{13}$$

$$\sigma_{b,g}^2 | \theta', \mathbf{X}_\gamma^*, \mathbf{Z}, \mathbf{y} \sim \mathcal{IG}(\sigma_{b,g}^2 | \, ||\mathbf{b}_g||/2 + \alpha_{b,g}, \beta_{b,g} + \tfrac{1}{2}\mathbf{b}_g^\top \mathbf{b}_g) \tag{14}$$

$$\mu_{w,h} | \theta', \mathbf{X}_\gamma^*, \mathbf{Z}, \mathbf{y} \sim \mathcal{N}(\mu_{w,h} | V_{\mu_\gamma,h}^{-1}(\Sigma(\mathbf{w}_{\gamma,h})/\sigma_{w,h}^2 + \mu_{0,h}/\sigma_{0,h}^2), \sigma_\varepsilon^2 V_{\mu_\gamma,h}) \tag{15}$$

$$\sigma_{w,h}^2 | \theta', \mathbf{X}_\gamma^*, \mathbf{Z}, \mathbf{y} \sim \tag{16}$$
$$\mathcal{IG}(\sigma_{w,h}^2 | \, ||\mathbf{w}_{\gamma,h}||/2 + \alpha_{w,h}, \beta_{w,h} + \tfrac{1}{2\sigma_\varepsilon^2}(\mathbf{w}_{\gamma,h} - \mathbf{1}\mu_{w,h})^\top(\mathbf{w}_{\gamma,h} - \mathbf{1}\mu_{w,h}))$$

$$\sigma_\varepsilon^2 | \theta', \mathbf{X}_\gamma^*, \mathbf{Z}, \mathbf{y} \sim \mathcal{IG}(\sigma_\varepsilon^2 | (N + ||\mathbf{w}_\gamma^*|| + H)/2 + \alpha_\varepsilon, \beta_\varepsilon + \tfrac{1}{2} R_{\sigma_\varepsilon^2}) \tag{17}$$

$$\pi | \theta', \mathbf{X}_\gamma^*, \mathbf{Z}, \mathbf{y} \sim \mathcal{B}(\pi | \alpha_\pi + ||\gamma||, \beta_\pi + J - ||\gamma||) \tag{18}$$

where we sample $\sigma_{b,g}^2$, $\mu_{w,h}$ and $\sigma_{w,h}^2$ for each g and h respectively. We also define $\mathbf{V}_{\mathbf{w}_\gamma^*} = (\mathbf{X}_\gamma^{*\top} \mathbf{X}_\gamma^* + \boldsymbol{\Sigma}_{\mathbf{w}_\gamma^*}^{-1})^{-1}$, $\mathbf{V}_\mathbf{b} = (\tfrac{1}{\sigma_\varepsilon^2} \mathbf{Z}^\top \mathbf{Z} + \boldsymbol{\Sigma}_\mathbf{b}^{-1})^{-1}$, $V_{\mu_\gamma,h} = ((||\mathbf{w}_{\gamma,h}||/\sigma_{w,h}^2)^{-1} + (\sigma_{0,h}^2)^{-1})^{-1}$ and $R_{\sigma_\varepsilon^2} = (\mathbf{y} - \mathbf{X}_\gamma^* \mathbf{w}_\gamma^* - \mathbf{Z}\mathbf{b})^\top(\mathbf{y} - \mathbf{X}_\gamma^* \mathbf{w}_\gamma^* - \mathbf{Z}\mathbf{b}) + (\mathbf{w}_\gamma^* - \mathbf{m}_\gamma)^\top \boldsymbol{\Sigma}_{\mathbf{w}_\gamma^*}^{-1} (\mathbf{w}_\gamma^* - \mathbf{m}_\gamma) + \sum_{h=1}^H (\mu_{w,h} - \mu_{0,h})^2/\sigma_{0,h}^2$ for notational simplicity.

References

Andrieu, C., Doucet, A.: Joint Bayesian model selection and estimation of noisy sinusoids via reversible jump MCMC. IEEE Trans. Sig. Process. **47**(10), 2667–2676 (1999)

Bates, D., Maechler, M., Bolker, B.: lme4: linear mixed-effects models using S4 classes (2013)

Davies, V., Reeve, R., Harvey, W., Maree, F., Husmeier, D.: Sparse Bayesian variable selection for the identification of antigenic variability in the Foot-and-Mouth Disease Virus. J. Mach. Learn. Res. Workshop Conf. Proc. (AISTATS) **33**, 149–158 (2014)

Gelman, A., Carlin, J.B., Stern, H.S., Dunson, D.B., Ventari, A., Rubin, D.B.: Bayesian Data Analysis, 3rd edn. Chapman & Hall, London (2013)

Gelman, A., Rubin, D.: Inference from iterative simulation using multiple sequences. Stat. Sci. **7**, 457–511 (1992)

George, E.I., McCulloch, R.E.: Variable selection via Gibbs sampling. J. Am. Stat. Assoc. **88**(423), 881–889 (1993)

Grazioli, S., Moretti, M., Barbieri, I., Crosatti, M., Brocchi, E.: Use of monoclonal antibodies to identify and map new antigenic determinants involved in neutralisation on FMD viruses type SAT 1 and SAT 2. In: Report of the Session of the Research Group of the Standing Technical Committee of the European Commission for the Control of Foot-and-Mouth Disease, pp. 287–297, Appendix 43 (2006)

Harvey, W.T., Gregory, V., Benton, D.J., Hall, J.P., Daniels, R.S., Bedford, T., Haydon, D.T., Hay, A.J., McCauley, J.W., Reeve, R.: Identifying the genetic basis of antigenic change in influenza A (H1N1). arXiv preprint arXiv:1404.4197 (2015)

Hastie, T., Tibshirani, R., Friedman, J.: The Elements of Statistical Learning. Springer, New York (2009)

Hastings, W.: Monte Carlo sampling methods using Markov chains and their applications. Biometrika **57**(1), 97–109 (1970)

Jow, H., Boys, R.J., Wilkinson, D.J.: Bayesian identification of protein differential expression in multi-group isobaric labelled mass spectrometry data. Stat. Appl. Genet. Mol. Biol. **13**(5), 531–551 (2014)

Maree, F.F., Borley, D.W., Reeve, R., Upadhyaya, S., Lukhwareni, A., Mlingo, T., Esterhuysen, J.J., Harvey, W.T., Fry, E.E., Parida, S., Paton, D.J., Mahapatra, M.: Tracking the antigenic evolution of Foot-and-Mouth Disease Virus (2015, in submission)

Metropolis, N., Rosenbluth, A., Rosenbluth, M., Teller, A., Teller, E.: Equations of state calculations by fast computing machines. J. Chem. Phys. **21**(6), 1087–1092 (1953)

Mitchell, T., Beauchamp, J.: Bayesian variable selection in linear regression. J. Am. Stat. Assoc. **83**(404), 1023–1032 (1988)

Mohamed, S., Heller, K., Ghahramani, Z.: Bayesian and l_1 approaches for sparse unsupervised learning. In: Proceedings of the 29th International Conference on Machine Learning (ICML 2012), pp. 751–758 (2012)

Pinheiro, J.C., Bates, D.: Mixed-Effects Models in S and S-PLUS. Springer, New York (2000)

Reeve, R., Blignaut, B., Esterhuysen, J.J., Opperman, P., Matthews, L., Fry, E.E., de Beer, T.A.P., Theron, J., Rieder, E., Vosloo, W., O'Neill, H.G., Haydon, D.T., Maree, F.F.: Sequence-based prediction for vaccine strain selection and identification of antigenic variability in Foot-and-Mouth Disease Virus. PLoS Comput. Biol. **6**(12), e1001027 (2010)

Schelldorfer, J., Bühlmann, P., van de Geer, S.: Estimation for high-dimensional linear mixed-effects models using ℓ1-penalization. Scand. J. Stat. **38**(2), 197–214 (2011)

Tibshirani, R.: Regression shrinkage and selection via the lasso. J. Roy. Stat. Soc. B **58**, 267–288 (1996)

Watanabe, S.: Asymptotic equivalence of Bayes cross validation and widely applicable information criterion in singular learning theory. J. Mach. Learn. Res. **11**, 3571–3594 (2010)

Comparison of Gene Expression Signature Using Rank Based Statistical Inference

Kumar Parijat Tripathi[1]([✉]), Sonali Gopichand Chavan[2],
Seetharaman Parashuraman[2], Marina Piccirillo[1], Sara Magliocca[1],
and Mario Rosario Guarracino[1]

[1] Laboratory for Genomics, Transcriptomics and Proteomics (LAB-GTP),
High Performance Computing and Networking Institute (ICAR),
National Research Council (CNR), Via Pietro Castellino 111, 80131 Naples, Italy
parijat.tripathi@icar.cnr.it
[2] Institute of Protein Biochemistry, National Research Council of Italy (CNR),
Via Pietro Castellino 111, 80131 Napoli, Italy

Abstract. To understand the unique characteristics of biological state or phenotype, such as disease or cellular homeostasis, it is of vital importance to analyze the behavior of global gene expression. In the field of transcriptomics, gene expression patterns under the corresponding phenotypic state could be used as a proxy to determine the physiological and chemical response from the cellular system of an organism. Studying these kinds of patterns helps us to unveil the response of molecular machinery of cell, and predict regulation of a particular metabolic pathway. To understand the biological implication of these gene expression signatures is still an open question. In the present work, we are studying the behaviour of gene expression signatures of 22 knock down (perturbed) genes involved in secretory pathways, in distinct human cancer cell lines at different time scales, with the help of rank based statistical approach. The aim of our work is to compare the consequence of these gene perturbations at the transcriptional level, independently from the specific cell line effects, and categorize these perturbations to understand the inter connected network with in these perturbed genes and their shared influence on the regulation of sectretory pathway. To achieve this goal, we compared the gene expression signatures with respect to each perturbation per cell lines, using three different approaches implying non parametric rank based statistics. In the first approach, we generated prototype rank lists (PRLs) from gene expression data from given perturbation experiments, and calculated distance between expression signatures using pattern matching similarity based on kolmogorov-Smirnov statistic. In second approach, we implemented rank-rank hyper-geometric overlap maps (RRHO) for the identification of statistically significant overlapping genes between gene-expression signatures with respect to 22 genes perturbation experiments. Finally, we carried out gene set enrichment analysis (GSEA) on the previously obtained PRLs for the respective perturbations, and identify statistically significant KEGG pathways for which expression signatures of these 22 pertubations are enriched. Based on the comparative study of gene expression signature,

© Springer International Publishing Switzerland 2016
C. Angelini et al. (Eds.): CIBB 2015, LNBI 9874, pp. 28–41, 2016.
DOI: 10.1007/978-3-319-44332-4_3

22 pertubations are clustered into 4 groups. Our results show that the transcriptional response with respect to each perturbation does not have an independent behavior, but perturbations with in each cluster share a common transcriptional response. Sister perturbations in each cluster have a cumulative role in shaping up the behaviour of cellular system.

Keywords: Expression signature · Perturbation · Prototype rank list · RRHO · Secretory pathway

1   Introduction

The secretory pathway is responsible for the delivery of a large variety of proteins to their proper cellular location and it is essential for cellular function and multicellular development. It is composed of a series of compartment that includes endoplasmic reticulum (ER), golgi apparatus, trans golgi network (TGN), through which the cargo (protein or lipid) is transported in an orderly fashion starting from the ER where the biosynthesis of cargoes is initiated. The processing of proteins is carried out by the addition of glycan groups through golgi apparatus and then sorted to their appropriate sites by the Trans Golgi Network [1].

At each step of protein transport, including forward and recycling pathways are controlled by regulatory modulator genes that maintain the homeostasis of the system [2] (see Fig. 1).

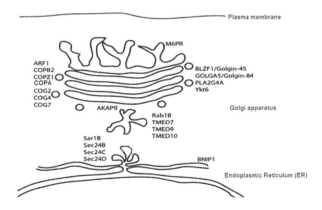

Fig. 1. Modulators genes in secretory pathway.

With the online availability of huge experimental data, it is now possible to extensively predict novel functions and interactions of genes involved in secretory pathway [3,5]. In our research work, we are keenly interested in studying the pathways and functional components regulated in the secretory pathway. In the present era of high-tech experimental opportunity, expression data provide an easy and large-scale analysis platform for understanding biological mechanism at the cellular level. In the present work, we put our focus on the list of genes, which are localized in a secretory pathway and share gene ontology terms related to protein transport machinery (Fig. 2).

Fig. 2. Word cloud for important gene ontology terms represented by modulators genes in secretory pathway.

Our aim is to highlight the role of those genes which are involved in secretory pathways, by studying the consequences of their perturbations at the transcriptional level independently from the specific cell line effects. It certainly provide an insight into their influence on the biological system. We employed rank based statistical approach to compare their gene expression profiles to predict their global effects with respects to pathways and functions, which might be directly or indirectly linked with the gene perturbed and thus the secretory pathway.

2   Materials and Methods

2.1   Data Retrieval

The gene expression profile data is a collection of 22 genes perturbation (knockdown) experiments in 12 cancer cell lines at different time points such as 96 and 120 h (Fig. 3). For each perturbation experiment, we collected expression data for all the biological and technical replicates in different cell lines at 96 and 120 h time points and merged them into single file to create the expression profile. The data is downloaded from LINCS (http://www.lincscloud.org/perturbagens/), NIH program that funds the generation of perturbation profiles across multiple cells and perturbation types(genetic and chemical). The expression data comprises of signature profiles of genes obtained through micro-array technology, which utilized directly measured 1000 landmark transcripts to detect the over all gene expression of all the genes. It has been normalized using invariant set scaling followed by quantile normalization [4]. In these profiles expression data are represented in terms of fold change across the distinctive cell line at different time points.

Gene perturbation	A375	A549	ASC	HA1E	HCC515	HEPG2	HT29	MCF7	NPC	PC3	SKL	VCAP
						Cancer Cell Lines						
ARF1	✓	✓		✓		✓	✓	✓		✓		✓
COG2	✓	✓	✓	✓	✓	✓	✓	✓	✓	✓	✓	✓
COG4	✓	✓	✓	✓	✓	✓	✓	✓	✓	✓	✓	✓
COG7	✓	✓	✓	✓	✓	✓	✓	✓	✓	✓	✓	✓
COPA	✓	✓	✓	✓	✓	✓	✓	✓		✓		✓
COPB2	✓	✓	✓	✓	✓	✓	✓	✓	✓	✓	✓	✓
COPZ1	✓	✓	✓	✓	✓	✓	✓	✓	✓	✓	✓	✓
GOLGA5												✓
M6PR	✓	✓	✓	✓	✓	✓	✓			✓		✓
PLA2G4A	✓	✓	✓	✓	✓	✓	✓	✓		✓		✓
RAB1B	✓	✓	✓	✓	✓	✓	✓	✓	✓	✓	✓	✓
SAR1B	✓	✓	✓	✓	✓	✓	✓	✓		✓		✓
SEC24B	✓	✓	✓	✓	✓	✓	✓	✓		✓	✓	✓
SEC24C	✓	✓	✓	✓	✓	✓	✓	✓	✓	✓	✓	✓
SEC24D	✓	✓	✓	✓	✓	✓	✓	✓		✓		✓
TMED10	✓	✓	✓	✓	✓	✓	✓	✓	✓	✓	✓	✓
TMED7	✓	✓	✓	✓	✓	✓	✓	✓		✓		✓
TMED9	✓	✓	✓	✓	✓	✓	✓	✓		✓		✓
YKT6	✓	✓	✓	✓	✓	✓	✓	✓	✓	✓	✓	✓
BLZF1	✓	✓	✓	✓		✓	✓	✓		✓		✓
AKAP9	✓	✓	✓	✓		✓	✓	✓		✓		✓
BNPI1	✓	✓		✓	✓	✓	✓	✓		✓		✓

Fig. 3. Perturbed genes in corresponding cell lines for which gene expression profiles are downloaded from LINCS.

2.2 Comparison of Gene Expression Signature Using Prototype Rank List

To obtain gene expression signature and compute distances between pre-processed gene expression profiles for the selected genes, we used "Gene Expression Signature Package" from Bioconductor in R [6]. This package provides the implementation of a methodology to determine the gene expression signature for the perturbation data and calculate distance between them. Gene expression signature is represented as a list of genes whose expression is correlated with a biological state of interest. The distance between the gene expression signature is calculated non parametrically using rank based pattern matching similarity based on Kolmogorov-Smirnov statistic [7]. There are four basic steps involved;

1. The gene expression profiles were sorted according to the differential expression values with respect to the controls, and each gene in the profile is ranked accordingly to its expression value within the sorted list. A matrix is generated, which is composed of ranked list representing the corresponding gene expression profile in each cell line for a given gene perturbation. The graded lists of profiles known as PRLs (prototype ranked list).
2. The PRLs in all the corresponding cell lines are aggregated by a rank merging procedure to negate the effects of specific cell lines. We used built-in "krubor" function of "Gene Expression Signature Package" to carry out the rank merging process. It comprises of two sub steps;
   - a distance is measured between two ranked list using Spearman's footrule [8] and two or more ranked lists are merged using Borda Merging method.
   - a single ranked list is obtained in a hierarchical way using Kruskal algorithm [9].
3. A signature of optimum length such as 250, 500, 800 and 1000 of the most up-regulated genes (near the top of the list) and the most down-regulated genes (near the bottom of the list) is used for the distance calculation between all

the PRLs representing the individual perturbation experiment, with the help of ScorePGSEA and ScoreGSEA functions in "Gene expression signature" in R package. We considered this signature of genes as a general cellular response to the perturbation. In other words, we obtained sets of genes which shows variable expression change in response to the perturbation across different experimental conditions (e.g., different cell lines, different dosages). The signature of perturbation x, $p = p1, p2...pn$ (up-regulated) and $q = q1, q2....qn$ (down-regulated), we defined as the distance between perturbation x and perturbation y the Inverse Total Enrichment Score (TES) of the perturbation x signature p, q, with respect to the PRL of perturbation y, as follows: $TES(x, y) = 1 - (ES_y^p - ES_y^q)/2$, here $ES_y^r (r \in p, q)$ is the Enrichment-Score of the optimal signature with respect to the PRL of y. ES_y^r ranges in $[-1,1]$, it is a measure based on the Kolmogorov-Smirnov statistics, and it quantifies how much a set of genes is at the top of a ranked list [12]. The closer that this measure is to 1, the more the genes are at the top of the list, whereas the closer to -1, the more the genes are at the bottom of the list. TES(x,y) ranges in $[0,2]$, it takes into account two sets of genes, and it checks how much the genes in the first set (p) are placed at the top of the y PRL and how much the genes in the second set (q) are placed at the bottom. The more these two statements are true, the more the value of TES(x,y) is close to 0.

We take into consideration two distance measurements between PRLs:

- Average Enrichment Score Distance $D_{avg} = (TES_{x,y} + TES_{y,x})/2$;
- Maximum Enrichment Score Distance $D_{max} = Min(TES_{x,y}, TES_{y,x})/2$.

4. Affinity propagation clustering (AP) [10] is used to group these PRLs representing different perturbation experiments, inferring distance measures among respective gene expression signatures. AP iteratively searches for optimal clustering by maximizing an objective function called net similarity.

2.3   Rank-Rank Hypergeometric Overlap Test Analysis

To highlight the correlation strength between two expression profiles, we carried out the rank-rank hypergeometric overlap test analysis using "RRHO package" from Bioconductor in R [11]. This algorithm compares two gene expression profiles ranked by the degree of differential expression. It is used to infer the amount of agreement between two sorted lists (PRL's) by computing the number of overlapping elements in the first $i * stepsize$ and $j * stepsize$ elements of each list, where stepsize represents the number of genes selected from the complete ranked gene list i and j, and return the observed significance of this overlap using a hyper geometric test (Fisher exact test). The output is returned as a list of matrices including: the overlap in the first $i * stepsize, j * stepsize$ elements and the significance of this overlap.

2.4   Gene Set Enrichment Analysis

The PRLs generated were further processed by Gene Set Enrichment Analysis (GSEA) [12] (http://www.broadinstitute.org/gsea/downloads.jsp) using the Molecular Signature Database (MsigDB). GSEA is a computational method that determines whether an initially defined set of genes shows statistical significant differences between two biological states. MsigDB has a collection of annotated gene sets (curated gene set, motif gene set, GO gene set, oncogene signature, immunologic signature, etc.) for use with GSEA software. In order to study the enriched pathways, we used KEGG (Kyoto Encyclopedia of Genes and Genomes) pathway gene set that includes genes contributing to each of the pathways listed in this gene set. Subsequently, using GSEA, enriched KEGG pathways and thus enriched genes, were predicted for all the 22 PRLs and sorted for further analysis.

3   Results

3.1   Distance Calculation and Clustering of Gene Expression Profiles

In order to calculate the pair wise distances among samples (PRLs representing expression profiles in response to gene perturbation experiments), gene's lists are ranked accordingly to the gene expression ratio (fold change). We include 250 top up-regulated genes (near the top of the list) and the most down-regulated genes (near the bottom of the list) for the distance calculation (Fig. 4). We also tested signature length for 500, 800, 1000 genes to obtain the best possible average distance and cluster the gene expression profile through affinity propagation approach. Using scoreGSEA function, we obtained Average Enrichment Score Distance D_{avg} as a measure of pairwise distance between PRLs. The matrix table is generated using pairwise distance between all 22 PRLs versus each other (Fig. 4).

	ARF1	BLZF1	BNPI1	COG2	COG4	COG7	COPA	COPB2	COPZ1	GOLGA	M6PR	PLA2G4	AKAP9	RAB1B	SAR1B	SEC24B	SEC24C	SEC24D	TMED7	TMED9	TMED10	YKT6
ARF1	0	0.9	1.03	1.01	0.99	0.92	0.9	1.02	0.86	1.06	1.07	1.06	0.93	0.96	1.05	1.07	0.92	0.91	1.1	1.03	0.96	1.04
BLZF1	0.9	0	1.03	0.92	1.13	0.94	1.05	1.14	0.87	1.06	1.16	0.98	1.05	0.83	0.96	0.9	1.08	0.91	1.11	0.96	0.85	0.91
BNPI1	1.03	1.03	0	0.95	1.01	0.93	1.03	0.98	1.1	1.17	1	1.1	0.91	0.98	0.92	1.04	1.03	0.94	0.93	1.03	1.01	0.95
COG2	1.01	0.92	0.95	0	0.99	1.03	0.93	1.03	0.83	0.94	0.99	0.88	0.97	1.07	1.01	0.91	1	1	1.09	0.9	0.94	0.97
COG4	0.99	1.13	1.01	0.99	0	0.88	0.85	0.81	1	0.87	0.97	0.91	0.86	1.06	0.92	1.16	0.94	0.96	0.88	0.92	1.09	1.09
COG7	0.92	0.94	0.93	1.03	0.88	0	0.82	0.85	0.86	0.93	1.07	1.13	0.94	0.96	1.1	1	0.88	0.92	1.04	1.09	0.83	0.89
COPA	0.9	1.05	1.03	0.93	0.85	0.82	0	0.89	0.78	0.95	1.12	1.05	1.07	0.97	1	1.06	0.85	0.91	0.85	0.92	1	1.05
COPB2	1.02	1.14	0.98	1.03	0.81	0.85	0.89	0	1.12	0.95	0.89	1.02	0.85	0.89	0.92	1.11	0.87	0.98	0.92	1.01	1	0.95
COPZ1	0.86	0.87	1.1	0.83	1	0.86	0.78	1.12	0	1.07	1.07	1	1.01	1.07	0.96	0.92	0.85	0.93	1.07	1.04	0.88	0.92
GOLGA	1.06	1.06	1.17	0.94	0.87	0.93	0.95	0.95	1.07	0	0.92	0.88	1.1	1.1	1.19	0.9	1.03	0.91	1.05	0.85	1.03	0.99
M6PR	1.07	1.16	1	0.99	0.97	1.07	1.12	0.89	1.07	0.92	0	0.85	0.9	1	0.97	1.08	0.93	1.07	0.9	1.01	1.09	1.01
PLA2G4	1.06	0.98	1.1	0.88	0.91	1.13	1.05	1.02	1	0.88	0.85	0	0.9	0.94	1.04	1.08	1.04	0.99	1.1	1.02	0.97	1.11
AKAP9	0.93	1.05	0.91	0.97	0.86	0.94	1.07	0.85	1.01	1.1	0.89	0.9	0	1.04	0.9	1.13	0.98	1.03	1.08	1.06	1.01	1.01
RAB1B	0.96	0.83	0.98	1.07	1.06	0.96	0.97	0.89	1.07	1.1	1.12	0.94	1.04	0	0.94	1.07	1	1.05	0.97	0.9	1.02	0.92
SAR1B	1.05	0.96	0.92	1.01	0.92	1.1	1	0.92	0.96	1.19	0.97	1.04	0.9	0.94	0	0.91	1	0.93	0.76	0.93	1.11	0.91
SEC24B	1.07	0.9	1.04	0.91	1.16	1	1.06	1.11	0.92	0.9	1.08	1.08	1.13	1.07	0.91	0	1.03	1.07	0.97	1.08	1.05	0.93
SEC24C	0.92	1.08	1.03	1	0.94	0.88	0.85	0.87	0.85	1.03	0.93	1.04	0.98	1	1.03	1.07	0	0.9	1.02	1.06	1.11	0.94
SEC24D	0.91	0.91	0.94	1	0.96	0.92	0.91	0.98	0.93	0.91	1.07	0.99	1.03	1.05	0.93	1.07	0.9	0	1.08	1.14	1	0.89
TMED7	1.1	1.11	0.93	1.09	0.88	1.04	0.85	0.92	1.07	1.05	0.9	1.1	1.08	0.97	0.76	0.97	1.02	1.08	0	0.87	1.11	0.89
TMED9	1.03	0.96	1.03	0.9	0.92	1.09	0.92	1.01	1.04	0.85	1.01	1.02	1.06	0.9	0.93	1.08	1.06	1.14	0.87	0	1.02	0.87
TMED10	0.96	0.85	1.01	0.94	1.09	0.83	1	1	0.88	1.03	1.09	0.97	1.01	1.02	1.11	1.05	1.11	1	1.11	1.02	0	0.95
YKT6	1.04	0.91	0.95	0.97	1.09	0.89	1.05	0.95	0.92	0.99	1.01	1.11	1.01	0.92	0.91	0.93	0.94	0.89	0.89	0.87	0.95	0

Fig. 4. An average enrichment score distance calculation between PRLs corresponding to each perturbation.

The closeness of expression profiles, was further verified with the affinity propagation clustering method. For the reproducibility of the results, we carried out comparison between the clusters obtained for 250, 500, 800 and 1000 signature length and voted out the genes which are always grouped together irrespective of signature length (Fig. 5).

APResult object				
Gene Signature length	250	500	800	1000
Number of samples	22	22	22	22
Number of iterations	152	147	160	170
Input preference	-0.07108233	-0.06518968	-0.05895802	-0.05682894
Sum of similarities	2.661744	2.270408	2.04184	1.928065
Sum of preferences	-0.2843293	-0.1955691	-0.1768741	-0.1704868
Net similarity	2.377415	2.074839	1.864965	1.757578
Number of clusters	4	3	3	3
Cluster 1	COG4 COG7 COPB2 M6PR AKAP9 RAB1B	ARF1 BLZF1 COG2 COG7 COPA COPZ1 SEC24C TMED10	ARF1 BLZF1 COG2 COG7 COPA COPZ1 SEC24B SEC24C SEC24D TMED10	ARF1 BLZF1 COG2 COG7 COPA COPZ1 SEC24B SEC24C SEC24D TMED10
Cluster 2	ARF1 BLZF1 COG2 COPA COPZ1 SEC24C TMED10	COG4 GOLGA5 M6PR PLA2G4A SEC24D TMED9	BNPI1 COG4 COPB2 M6PR PLA2G4A AKAP9 SAR1B	BNPI1 COG4 COPB2 M6PR PLA2G4A AKAP9 SAR1B
Cluster 3	GOLGA5 PLA2G4A TMED9	BNPI1 COPB2 AKAP9 RAB1B SAR1B SEC24B TMED7 YKT6	GOLGA5 RAB1B TMED7 TMED9 YKT6	GOLGA5 RAB1B TMED7 TMED9 YKT6
Cluster 4	BNPI1 SAR1B SEC24B SEC24D TMED7 YKT6	NA	NA	NA

Fig. 5. Clustering of gene perturbation on the basis of affinity propagation. It shows the respective clusters obtained for signature length 250, 500, 800 and 1000.

The AP clusters obtained for different signature length are in concordance with each other. The clusters for 250 and 500 signature length are more similar in comparison to 800 and 1000 signature led AP clusters. By voting for gene togetherness in each cluster for the different signature length, we determined 4 set of genes, which mostly ($>=2$ of 4 times) group together irrespective of signature length. The first set of genes includes ARF1, BLZF1, COG2, COG7, COPA, COPZ1, SEC24C, (SEC24D, SEC24B only in 800 and 1000 signature length) and TMED10. This set of genes are also making important biological sense as all these genes are involved in either COPI coating of golgi vesicles or COPII vesicle coating and involved in ER to Golgi vesicle mediated transport of proteins with in the cell. The second set of genes includes COG4, M6PR, COPB2, PLA2G4A and AKAP9. These genes are involved in signal transduction, endocytosis, mitotic cell cycle, retrograde vesicle-mediated transport and regulation of membrane repolarization. It is important to observe that signal transduction and membrane repolarization activity in this set of genes are associate to retrograde vesicle mediated transport and mitotic cell cycle. It is possible that by pertubing genes which are involved in retrograde vesicle mediated transport, mitotic cycle of the cell experience a substantial effect. The third group of genes includes YKT6, TMED9 and GOLGA5, in the case of 250 and 500 signature length, PLA2G4A also group together with these genes. These genes generally involved in protein transport and Golgi organization. The fourth and final set includes BNP1, TMED7, RAB1B and SARIB. The fourth group does not show strong evidence of grouping all-together with respect to the applied voting criteria, yet it is observed that BNP1, SAR1B as one sub set, while RAB1B, TMED7 as another sub set follow the group voting criteria with respect to the given signature length. These genes are mostly involved in membrane transport, and post-translational protein modification. Initially, we carried out hierarchical clustering approach to cluster genes (results not shown), but due to the fact that hierarchical clustering starts with every data point as its own cluster and then recursively merges pairs of clusters which led this approach to makes hard decisions that can cause it to get stuck in local minima. While, on the other hand, affinity propagation is a clustering algorithm that requires each cluster to vote for a good exemplar from within its data points, and thus it provides better clustering of perturbed genes in this study as far as biological inference is concerned. For example, hierarchical clustering shows M6PR (signal receptor activity) closer to GOLGA5 (Golgi organization) in spite of TMED9 (Golgi-organization), which is known to be closer to GOLGA5 with respect to gene ontology. These results further reveal although the gene expression signature was obtained independently concerning distinct gene perturbation experiments in different cancer cell lines, they still share the same biological connectivity as far as functional regulation of cellular activity is concerned within each respective cluster. For example, in set 1, the gene expression profiles obtained by the perturbation of ARF1, BLZF1, COG2, COG7, COPA, COPZ1, SEC24C, (SEC24D, SEC24B only in 800 and 1000 signature length) and TMED10 genes in separate experiments within different cancer cell lines, shares a common transcriptional response. It will help us to understand the basic biology in which these genes are involved. It also helps us to address and characterize the biological properties of less-known genes in comparison with its elated sister genes within the cluster, as far as their biological activity and molecular functionality in the cellular system and metabolic pathways are concerned.

3.2 RRHO Analysis

In addition to "Gene Expression Signature" analysis, we also carried out rank-rank hyper-geometric overlap test to measure the statistical significance of the number of overlapping genes between two expression profiles. Although the gene expression signature is a robust methodology there is a drawback existed in the form of "signature length". In the previous analysis, we provide the result based on pre-defined signature length such as 250 which means that we only selected the top 250 and bottom 250 genes in our study. Though it is possible to change this criterion, and we also obtained results for 500, 800 and 1000 preselected genes by altering the signature length significantly in our work. Nevertheless, previous methodology does not consider the whole 22000 genes in gene expression profile all together while calculating the average distance between profiles. To over come this bottleneck and consider all the genes simultaneously within the study, we used RRHO analysis. In RRHO analysis, we select a window of 2000 (around 10 %) genes from the complete list of genes in a expression profile, which moves throughout all the gene's identifiers listed on the ranked list. The output is returned as a list of matrices including: the overlap in the first $i * stepsize, j * stepsize$ elements and the significance of this overlap.

The significance of the overlap between two lists is calculated as function aggregating information from the whole overlap matrix into one summary statistic, typically the min p-value, or max on $-\log(\text{pval})$ scale. We also calculated this summary statistic between each pair of knock down genes (table not shown). We carried out network analysis for all the 22 perturbed genes. Taking cut-off value > 70 ($-\log(\text{pval})$ scale from 0–160), a gene network is obtained.

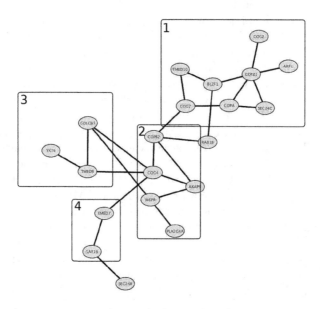

Fig. 6. Network representation of perturbed genes based on sum static of $-\log$ P value of hypergeometric overlap between them greater than 70.

In this network (Fig. 6), we obtained 4 subnetwork representing the perturbed genes cluster, representing perturbed genes which shares the common pattern of gene expression. The cluster obtained from the network analysis shows satisfactory coherence with the genes set obtained through by comparing and voting approach for AP clusters. As in AP approach, we clearly obtained perturbations seperated in different groups, while in case of RRHO, RAR1B, SEC24B, and BNP1 does not behave in the similar manner. This may be due to the fact that in RRHO analysis, we consider the complete set of probes in an experiment, instead of fixed set of signature genes. Rank-rank hypergeometric overlap maps provides two level of information. First information is about the level of intensity of an overlap as well as position of an overlap between the two ranked list.

While the sub networks obtained from the network based on sum static of $-\log$ P value of hypergeometric overlap between them greater than 70 also shows modest coherence with the previous obtained clusters through affinity propagation approach in PRL analysis. We have selected sister perturbed genes in cluster/set 1 and cluster/set 2 respectively (see Fig. 7). Taking these sister perturbation with in each cluster and generating RRHO heat maps reveals common

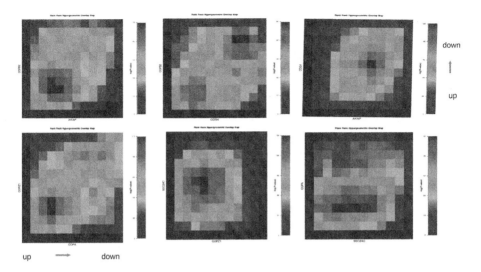

Fig. 7. RRHO map between genes perturbation profiles clustered through affinity propagation and representing sum static of −log P value of hypergeometric overlap greater than 70.

pattern of differential behaviour of gene expression among them. In simple terms, from these results, it is obvious that these perturbation do not have an independent effect, instead these perturbations behave in a complex interconnected manner to pursue a global effect on the transcriptional response during perturbation of any of the sister genes with in each cluster respectively. This commonalities in expression behaviour is more prominent in highly up and down-regulated genes between sister perturbations within each cluster.

3.3   Gene Set Enrichment Analysis

The PRLs generated were further analyzed with Gene Set Enrichment Analysis (GSEA) using a java desktop application available [13] at Molecular Signature Database (MsigDB). GSEA is a computational method that determines whether a priori defined set of genes shows statistically significant, differences between two biological states. In order to study the enriched pathways, we used KEGG (Kyoto Encyclopedia of Genes and Genomes) pathway gene sets that includes genes contributing to each of the pathways listed in this gene set. KEGG pathway is a collection of manually drawn pathway maps representing our knowledge on the molecular interaction and reaction networks for metabolism, genetic information processing, cellular process, etc. [14]. Thus using GSEA, enriched KEGG pathways and thus enriched genes were predicted for all the 22 PRLs and sorted for further analysis. GSEA assigns an enrichment score (ES) along with the false discovery rate (FDR) to each of the enriched predicted pathways. ES (Enrichment Score) is calculated by walking down the ranked list of genes increasing a running-sum statistic when a gene is in the gene set and decreasing it when it is not. Based on the ES obtained, an ES matrix was generated. The number of pathways enriched were narrowed down with FDR cut-off of 5 %. Of the 22 PRLs analyzed; only 8 showed enriched pathways within the 5 % FDR cut-off (result not shown). Further, these set of enriched pathways at 5 % FDR and 25 % FDR respectively, were clustered into groups based on shared common genes (>50 %) among the pathways. Thus for analysis at 5 % and 25 % FDR, 4 and 11 groups were obtained respectively (Figs. 8 and 9).

Group	KEGG pathways	Gene perturbation
1	KEGG_STEROID_BIOSYNTHESIS	ARF1
	KEGG_STEROID_HORMONE_BIOSYNTHESIS	GOLGA5
2	KEGG_RIBOSOME	TMED9
	KEGG_RIBOSOME	GOLGA5
3	KEGG_GLYCOSAMINOGLYCAN_BIOSYNTHESIS_CHONDROI TIN_SULFATE	SEC24C
	KEGG_GLYCOSAMINOGLYCAN_BIOSYNTHESIS_CHONDROI TIN_SULFATE	SAR1B
4	KEGG_PARKINSONS_DISEASE	GOLGA5
	KEGG_OXIDATIVE_PHOSPHORYLATION	GOLGA5

Fig. 8. Groups generated for enriched pathways at 5 % FDR analysis

In Fig. 8, based on 5 % FDR, four groups are obtained. Each group represent the sister perturbation for which KEGG Pathways are enriched. It shows that TMED9 and GOLGA5 in group 2, while SAR1B and SEC24C within group 3 share the common enriched pathways, though previously from both PRL and RRHO analysis SAR1B and SEC24C does not group together. From biological point of view, it is important that SAR1B and SEC24C are involved in N-linked glycosylation process. The results obtained from GSEA analysis at 5 % FDR are not complete and loose pathway enrichments for some of the perturbations which does not fit into such stringent statistical cutoff. So as a trade off for biological inference of our results over statistical significance, we lower down the stringency from 5 % to 25 %, and obtained 11 groups of gene perturbations sharing common KEGG pathways (See Fig. 9). The results obtained show that gene perturbation such as GOLGA5 and TMED9 form one group, while perturbations such as COPZ1, COG7 and BLZF1 form another group. These results are not in complete coherence with the previously obtained results, due to the fact that these results represents more biological interpretation instead of providing statistical strength to the study.

4   Conclusion

In the present work, we analyze the perturbation profiles of 22 knock-down genes in distinctive cell lines involved in secretory pathways using three different

Group	KEGG pathways	Gene perturbation
1	KEGG_RIBOSOME	TMED9
	KEGG_RIBOSOME	GOLGA5
2	KEGG_AMINOACYL_TRNA_BIOSYNTHESIS	COPA
	KEGG_AMINOACYL_TRNA_BIOSYNTHESIS	BLZF1
	KEGG_AMINOACYL_TRNA_BIOSYNTHESIS	COPZ1
3	KEGG_MISMATCH_REPAIR	COPZ1
	KEGG_MISMATCH_REPAIR	COG4
	KEGG_DNA_REPLICATION	COPZ1
	KEGG_DNA_REPLICATION	COG7
4	KEGG_PANTOTHENATE_AND_COA_BIOSYNTHESIS	COG7
	KEGG_PANTOTHENATE_AND_COA_BIOSYNTHESIS	COPZ1
	KEGG_PANTOTHENATE_AND_COA_BIOSYNTHESIS	RAB1B
5	KEGG_GLYCOSAMINOGLYCAN_BIOSYNTHESIS_CHONDROITIN_SULFATE	COPZ1
	KEGG_GLYCOSAMINOGLYCAN_BIOSYNTHESIS_CHONDROITIN_SULFATE	COPA
	KEGG_GLYCOSAMINOGLYCAN_BIOSYNTHESIS_CHONDROITIN_SULFATE	SEC24C
	KEGG_GLYCOSAMINOGLYCAN_BIOSYNTHESIS_CHONDROITIN_SULFATE	SAR1B
6	KEGG_ALZHEIMERS_DISEASE	GOLGA5
	KEGG_HUNTINGTONS_DISEASE	GOLGA5
	KEGG_OXIDATIVE_PHOSPHORYLATION	GOLGA5
	KEGG_PARKINSONS_DISEASE	GOLGA5
7	KEGG_ARACHIDONIC_ACID_METABOLISM	GOLGA5
	KEGG_ETHER_LIPID_METABOLISM	GOLGA5
	KEGG_LINOLEIC_ACID_METABOLISM	GOLGA5
	KEGG_ARACHIDONIC_ACID_METABOLISM	SEC24D
8	KEGG_STEROID_BIOSYNTHESIS	BLZF1
	KEGG_STEROID_BIOSYNTHESIS	ARF1
	KEGG_STEROID_HORMONE_BIOSYNTHESIS	GOLGA5
	KEGG_METABOLISM_OF_XENOBIOTICS_BY_CYTOCHROME_P450	GOLGA5
	KEGG_METABOLISM_OF_XENOBIOTICS_BY_CYTOCHROME_P450	SEC24D
	KEGG_RETINOL_METABOLISM	SEC24D
	KEGG_RETINOL_METABOLISM	GOLGA5
	KEGG_DRUG_METABOLISM_OTHER_ENZYMES	SEC24D
	KEGG_DRUG_METABOLISM_OTHER_ENZYMES	GOLGA5
9	KEGG_PROTEIN_EXPORT	COPA
	KEGG_PROTEIN_EXPORT	GOLGA5
	KEGG_PROTEIN_EXPORT	TMED9
10	KEGG_PHENYLALANINE_METABOLISM	GOLGA5
	KEGG_PHENYLALANINE_METABOLISM	TMED9
	KEGG_PHENYLALANINE_METABOLISM	YKT6
	KEGG_TYROSINE_METABOLISM	GOLGA5
	KEGG_HISTIDINE_METABOLISM	GOLGA5
	KEGG_HISTIDINE_METABOLISM	RAB1B
11	KEGG_STARCH_AND_SUCROSE_METABOLISM	RAB1B
	KEGG_STARCH_AND_SUCROSE_METABOLISM	TMED7
	KEGG_STARCH_AND_SUCROSE_METABOLISM	GOLGA5

Fig. 9. Groups generated for enriched pathways at 25 % FDR analysis

methodologies based on non parametric rank statistics. The effect of perturbations across the different cell lines are taken into consideration and based on their differential regulation of genes, a comparison study between their gene expression signature is carried out. The gene expression profiles with respect to each perturbation are checked for their conservation across the cell lines to get an insight into novel functions. In this work, we shows that the transcriptional response with respect to each perturbation does not have an independent behavior, but somehow these perturbations shares a common gene expression response. It regulates the secretory pathway with the help of complicated gene interactions network. To study the perturbation of these 22 genes in secretory pathways, we employed rank based statistical approach to compare their gene expression profile to predict their global effects with respects to pathways and functions, which might be directly or indirectly linked with the gene perturbed and thus the secretory pathway. Based on expression signature, these 22 knock-down genes are categorized into four different sets and sister perturbation in each cluster have a cumulative role in shaping up the behaviour of cellular system. Considering the facts, that these 22 genes are involved in numerous biological functions from golgi organization, vesicle coating, protein transport, membrane, signal transduction to mitotic cycle with in the cell. It is not possible to group these genes, using one methodology, due to the several facts such as: (i) genes do performs various other biologically distant activities apart from sharing common biological functions, which are not easy to relate; (ii) transcription rate of genes associated to different biological activities are also distinct, for example, genes involved in cellular component organization have higher expression compare to a genes involved in signal transduction. In the case of PRL analysis, we only consider the predefined highly up or down regulated genes for the analysis. It means that in this approach, we could not take into account for those genes which shows subtle expression change during perturbation, though these genes may have higher impact on the cellular complexity during perturbation exper-iments. In the other approach like RRHO, we do not consider pre-defined gene signature, instead we took the complete gene set according to their rank. In this way, we take into account of all the genes, irrespective of the level of change in their expression. In the GSEA method, we only focus on KEGG pathway to determine gene-gene association with respect to their common role in a given biological pathway. These three approaches complement each other, while obtain-ing novel informations, which is required to cluster the sister perturbations in a distinct group. We obtained the outcome from three different methods (PRLs analysis, RRHO and GSEA) and analyze them to retrieve significant information for the systematic understanding of all these perturbation with in the secretory pathway. This approach in future is also very helpful to characterize the novel perturbation. We are working on the standardization of the appropriate method for studying large-scale perturbation data.

Acknowledgments. We would like to thank the INTEROMICS flagship project, PON02-00612-3461281 and PON02-00619-3470457 for the funding support.

References

1. Kelly, R.B.: Pathways of protein secretion in eukaryotes. Science **230**(4721), 25–32 (1985)
2. Luini, A., Mavelli, G., Jung, J., Cancino, J.: Control systems and coordination protocols of the secretory pathway. F1000prime reports, vol. 6 (2014)
3. Butte, A.: The use and analysis of microarray data. Nat. Rev. Drug Discovery **1**(12), 951–960 (2002)
4. Bolstad, B.M., Irizarry, R.A., Strand, M., Speed, T.P.: A comparison of normal-ization methods for high density oligonucleotide array data based on variance and bias. Bioinformatics **19**(2), 185–193 (2003)
5. Werner, T.: Bioinformatics applications for pathway analysis of microarray data. Curr. Opin. Biotechnol. **19**(1), 50–54 (2008)

6. Li, F., Cao, Y., Han, L., Cui, X., Xie, D., Wang, S., Bo, X.: GeneExpressionSignature: an R package for discovering functional connections using gene expression signatures. OMICS **17**(2), 116–118 (2013)
7. Smirnov, N.: Table for estimating the goodness of fit of empirical distributions. Ann. Math. Stat. **19**, 279–281 (1948)
8. Diaconis, P., Graham, R.L.: Spearman's footrule as a measure of disarray. J. Roy. Stat. Soc. B **39**(2), 262–268 (1977)
9. Kruskal, J.B.: On the shortest spanning subtree of a graph and the Traveling Salesman problem. Proc. Am. Math. Soc. **7**, 48–50 (1956)
10. Bodenhofer, U., Kothmeier, A., Hochreiter, S.: APCluster: an R package for affinity propagation clustering. Bioinformatics **27**(17), 2463–2464 (2011)
11. Plaisier, S.B., Taschereau, R., Wong, J.A., Graeber, T.G.: Rank-rank Hypergeometric overlap: identification of statistically significant overlap between gene-expression signatures. Nucleic Acids Res. **38**(17), e169 (2010)
12. Subramanian, A., et al.: Gene set enrichment analysis: a knowledge-based approach for interpreting genome-wide expression profiles. PNAS **102**(43), 15545–15550 (2005)
13. http://www.broadinstitute.org/gsea/downloads.jsp
14. http://www.genome.jp/kegg/pathway.html

Managing NGS Differential Expression Uncertainty with Fuzzy Sets

Arianna Consiglio[1,2][(✉)], Corrado Mencar[2], Giorgio Grillo[1], and Sabino Liuni[1]

[1] CNR, Institute for Biomedical Technologies, Bari, Italy
{arianna.consiglio,giorgio.grillo,sabino.liuni}@ba.itb.cnr.it
[2] Department of Informatics, University of Bari A. Moro, Bari, Italy
corrado.mencar@uniba.it
http://www.ba.itb.cnr.it
http://www.di.uniba.it

Abstract. High-performance Next-Generation Sequencing (NGS) has become a widely used technology to characterize case-control comparison studies for RNA transcripts, such as mRNAs and small non-coding RNAs. The first step in the analysis strategies is mapping NGS reads against a reference database and a critical issue emerges in this phase: the problem of multireads. In this paper we present a novel approach to represent and quantify read mapping ambiguities through the use of fuzzy sets and possibility theory. The aim of this work is to obtain a list of candidate differential expression events, providing a description of the uncertainty of the results due to multiread presence. In a preliminary experiment on HeLa cells, the method correctly detected the possibility of false positiveness, while on a case-control study of human endobronchial biopsies, the method identified 11 genes with possible different expression, four of them with an uncertain fold change. This last result was confirmed by FDR adjusted Fisher's test, while DESeq2 did not provide significant differences between case and control.

Keywords: RNA-Seq · Differential expression · Multireads · Fuzzy sets · Possibility measure

1 Scientific Background

NGS technology is continuously improving and the produced reads are increasingly numerous. When working with alignment-based methods, a confounding factor is the presence of gene duplication, repetitive regions and overlapping genes. These events induce the problem of *multireads* in the NGS mapping procedure when a significant proportion of reads map to more than one location. This issue can lead to mistakes and imprecision in differential expression or alternative splicing analysis based on counts of reads mapping to some reference databases.

When multireads are sporadic, usually such reads are discarded from the analysis, but this option leads to an underestimation of the read counts.

© Springer International Publishing Switzerland 2016
C. Angelini et al. (Eds.): CIBB 2015, LNBI 9874, pp. 42–53, 2016.
DOI: 10.1007/978-3-319-44332-4_4

In the last years, alternative strategies have been developed for the estimation of read counts in presence of multireads. The simplest choice is to randomly assign multireads to references (as in best-match mapping) or proportionally to the expression of uniquely mapped reads [1]. More complex techniques compute an estimation of the read counts using probabilistic models, based on some assumptions on the distribution of data [2–4].

The estimated expressions are given as input to the tools for the analysis of differential expression [5,6]. Such tools scale the counts in order to make the expression values comparable, then they compute the fold change and a p-value with a statistical test, and eventually select a list of candidate differentially expressed genes. These results may contain many false positives and must be validated with further laboratory assays [7].

In this paper we propose a novel method, based on fuzzy sets and possibility theory, that deals with the inherent uncertainty of multiread mapping. The adopted approach is compliant with the work of Zadeh [8], who proposed possibility distributions as suitable interpretations of fuzzy sets. The possibility measure is used in this paper following the notation introduced by Pedrycz [9].

The aim of this work is to obtain a list of candidate differential expression events ordered by significance, while also providing a description of the uncertainty of the results due to the multiread issue, for an easier detection of false positives. The proposed approach is based on the idea of representing and quantifying read mapping ambiguities without heavy simplifications or stringent probabilistic assumptions.

2   Materials and Methods

In this section, we first introduce the representation of gene expression in terms of fuzzy sets. Then, we outline the general idea of differential expression analysis with fuzzy gene expressions through a graphical evaluation. Finally, we introduce a complete workflow for differential expression analysis, which requires a preliminary definition of fuzzy fold change.

2.1   Fuzzy Representation of Gene Expression

The uncertainty of multireads is modeled through a possibility distribution describing the possibility that each gene has a given read count. Such a possibility distribution is naturally represented in terms of a fuzzy sets in the domain of read counts.

When a read is mapped against a reference database, we have one of the following results:

(i) the read does not map to any reference sequence;
(ii) the read maps to only one reference sequence (unique mapping);
(iii) the read maps to more than one reference sequence with equal or different mapping quality (multiple mapping).

In all cases, a read is the actual expression of a single gene, therefore multiple mapping introduce uncertainty on the real expression quantification. Furthermore, usually the mapping algorithms assign a score to each match, in order to rank the mapping results, with the biologically plausible assumption that the higher the score, the more likely the read is an expression of that gene. Since our aim is to transfer the uncertainty in mapping to a description of the uncertainty in read count computation, we define a possibility degree to each mapping by scaling the score between 0 and 1 (see Table 1 for a simple example).

Table 1. Example of mapping results, with scaled mapping scores.

Read ID	Gene ID	Mapping score
read-1	gene-1	1.0
read-2	gene-1	1.0
read-2	gene-2	0.8
read-3	gene-1	1.0
read-3	gene-2	1.0

For each gene, we can now compute the possibility degree that a given count x is the real number of reads mapping to the gene. Roughly speaking, the possibility of having a read count equal to x for a gene is the maximal possibility of having x reads as true matches and all the others as false matches. Based on this definition, four main values have a critical importance:

A = number of uniquely mapping reads;
B = number of reads having the gene as unique best match (i.e. other genes may match, but with lower quality);
C = number of reads having the gene as best match, although not unique (i.e. other genes may match with the same quality);
D = number of reads having the gene as match, even if not best.

According to their definition, $A \leq B \leq C \leq D$. Of course, it is impossible that a gene is mapped to less than A reads and more than D reads. Also, it is maximally possible that the number of reads is between B and C because for each x in this interval, there are at least x reads mapping to the gene with maximal quality. On the other hand, the possibility degree increases between A and B and decreases between C and D, thus reflecting the relation between possibility and mapping quality. According to the example in Table 1, for gene-1: $A = 1$, $B = 2$, $C = 3$, $D = 3$, while for gene-2: $A = B = 0$, $C = 1$, $D = 2$.

The possibility distribution of the gene expression, based on read counts, can be approximated by a trapezoidal fuzzy set characterized by the values A, B, C and D. A fuzzy set represents a collection of elements with possibly partial membership; it models an ill-defined variable where the possibility degree that the variable assumes a value x corresponds to the membership degree of x to the fuzzy set. Formally, a fuzzy set is a function $X \mapsto [0,1]$ being X the universe of discourse; in our work we use trapezoidal fuzzy sets, defined on real numbers, with the following form:

$$\mathrm{Tr}\,[A', B, C, D']\,(x) = \begin{cases} 0, & x \leq A' \vee x \geq D' \\ \frac{x-A'}{B-A'}, & A' < x \leq B \\ 1, & B < x < C \\ \frac{x-D'}{C-D'}, & C \leq x < D' \end{cases} \tag{1}$$

where[1] $A' = A - 1$ and $D' = D + 1$ to give non-null possibility to counts A and D respectively. Linear increase/decrease can be assumed for simplicity from A' to B and from C to D'. The width of the fuzzy set (1) (defined as $D' - A'$) quantifies the uncertainty in the evaluation of the expression value, which in turn generates uncertainty in differential expression evaluation.

2.2   Graphical Evaluation of Uncertainty in Differential Expression

For a qualitative evaluation of differential expression of genes in a case-control study, a graphical method can be proposed as a first approach.

Two trapezoidal fuzzy sets, representing the expression of the same gene in different samples, can be plotted on a 3-dimensional graph, which is useful to fully understand the use of fuzzy sets and the related possibility distributions. The count values for the two experimental samples are drawn on the x axis and y axis respectively, while the possibility degrees are represented on the z axis. As shown in Fig. 1 (left subfigure), the Cartesian product of two trapezoidal fuzzy sets, representing the expression of the same gene in different samples, yields a truncated pyramid of possibility degrees (3D plot). The z-value of the pyramid is the possibility degree that the first sample has x reads *and* the second sample has y reads for the gene under consideration.

The projection on the xy plane highlights two rectangles that bound the possibility: the innermost covers the area with highest possibility, while the outermost limits the area with non-null possibility. Larger rectangles represent wider uncertainty, small rectangles (possibly degenerating to a single point) represent more definite results. The position of the rectangle with respect to the bisector line describes the differential expression result in the case-control comparison.

As an example, depicted in the right part of Fig. 1, in the case that the outermost rectangle does not intersect the bisector line (as in the case of the red rectangles), then it is fully possible that a gene is over-expressed or under-expressed, depending on the relative position of the rectangle w.r.t. the bisector line, but it is impossible that there is no differential expression. On the other hand, if the bisector intersects the outermost rectangle but not the innermost (yellow case in the figure), then there is full possibility of over-expression or under-expression, but non-null possibility of false positiveness (i.e. no differential expression). Finally, if the bisector intersects the innermost rectangle (green

---

[1] $A' = A - 1$ if $A > 0$, otherwise $A' = A = 0$.

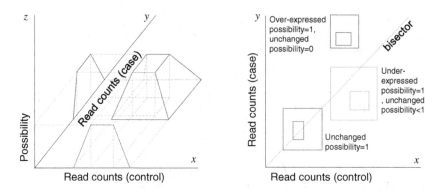

Fig. 1. Graphical interpretation of the fuzzy sets and their comparison for differential expression evaluation. Each couple of rectangles of the same color in the right figure corresponds to a 2-D projection of a fuzzy relation, as depicted in the left figure. (Color figure online)

case), then there is full possibility that the mapping results do not show any differential expression between case and control.

2.3   Fuzzy Fold Change Computation

The proposed quantitative method for the evaluation of differential expression extends the fold change metric, usually adopted for differential expression, by integrating fuzzy sets representing uncertain read counts. In particular, given a control sample with fuzzy expression $\mathrm{Tr}\,[A_1', B_1, C_1, D_1']$ and a control case with fuzzy expression $\mathrm{Tr}\,[A_2', B_2, C_2, D_2']$, we extend the usual fold change metric to the following fuzzy fold change metric:

$$\mathrm{Tr}\left[\log_2 \frac{A_1'}{D_2'}, \log_2 \frac{B_1}{C_2}, \log_2 \frac{C_1}{B_2}, \log_2 \frac{D_1'}{A_2'}\right] \tag{2}$$

This fuzzy set follows from the application of the extension principle to the standard fold change metric (limited to the points A', B, C, D'), eventually simplified to a trapezoidal fuzzy set for ease of computation[2].

The fuzzy fold change is very useful to highlight potential false positives when the value of 0 (corresponding to null variation between case and control) belongs to fuzzy fold change with high possibility degree. However, we should take into account that its values depend on the amount of expression of the genes, since it is a ratio among read counts.

---

[2] A standard application of the extension principle to the fold change results in a fuzzy set with a complex membership function, which requires complex computations without any real benefits.

2.4   Fuzzy Representation of Data and Differential Expression

For a complete differential expression analysis, all the genes of both the case and the control samples must be taken into account. The last approach we propose ranks the genes of both samples in order of possibility that their expression in the case and control is significantly different. This approach combines the fuzzy fold change metric with a fuzzy representation of the dataset of genes.

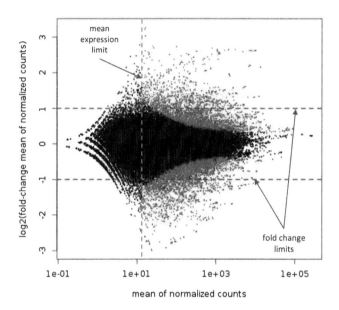

Fig. 2. MA-plot is a compact representation of the results of a differential expression analysis. Existing tools usually select varied genes combining several metrics: a statistical test of significant variation (red points selected by DESeq2), a user-defined threshold for the fold change values (horizontal lines) and, optionally, a minimum value of mean expression (vertical line). This figure is based on a plot published on www. genomatix.de. (Color figure online)

In order to analyze the trend of the logarithmic fold change, differential expression analysis results are usually represented with an MA-plot[3]. Figure 2 shows an example of how the results are managed by an existing tool for differential expression analysis, DESeq2. Genes are considered as differentially expressed if they pass a statistical test based on some user-defined thresholds. In our approach we take into account the uncertainty of read mappings and we will not fix any arbitrary threshold for detecting differential expression, because we rely on

---

[3] Given two expressions e_1 and e_2 of a gene in two samples, the MA-plot places the gene on a plane (M, A) where $M = \log_2(e_1/e_2)$ (the fold change) and $A = (1/2)\log_2(e_1 e_2)$ (average intensity).

ranking genes according to their possibility of being differentially expressed as
well as the possibility degree of false positiveness.

We represent expression data in the MA-plot, as in Fig. 3 (main plot). For
simplicity, each gene is represented as a point and its expression value is the centroid of its trapezoid[4]. The plot clearly shows that the variability of fold change
decreases as the mean expression value increases. The genes that are far from
the main rhomboidal figure are the best candidates for differential expression.

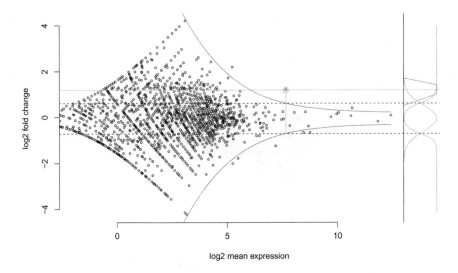

Fig. 3. Computation of the differential expression possibility of a gene. The MA-plot of
estimated data shows the variability of log2 fold change for low expression values (main
plot). The blue boundaries enclose the genes with insignificant expression variation. By
projecting the mean expression of one gene (in yellow), its intersection with the blue
curves can be used to define the system of three fuzzy sets of differential expression
possibility for that gene (green plot on the right) and intersected with the fuzzy fold
change (in red). (Color figure online)

In order to compute the possibility of a gene to be differentially expressed,
we estimate the boundaries of the rhomboidal distribution of points (blue curves
in Fig. 3)[5]. These curves serve as fuzzy boundaries for differential expression
events (we are not interested in the left part of the rhomboidal distribution,
because it represents genes with too small expressions). The points enclosed
in the boundaries are those genes with a higher possibility of having expression
variations which are unrelated to the experimental conditions. The genes lying on
the boundaries can be associated to a possibility of being differentially expressed

---

[4] The centroid is computed with the constraint of falling inside the interval $[B, C]$.

[5] The boundaries are estimated as hyperbolas, with their parameters fitted on the
dataset; their horizontal asymptotes represent a limit fold change value under which
differential expression loses significance.

equal to 0.5; on the other hand, genes above the upper boundary or below the lower boundary have higher possibility of being differentially expressed.

Thereby, for each value of mean expression, a system of three fuzzy sets is used to describe and to compute a differential expression possibility value for each gene, according to the following procedure.

1. Given a gene, its corresponding mean expression is computed.
   (a) If it is located in the left part of the rhomboid, it is excluded from further analysis because its expression is not significant.
   (b) Otherwise, in correspondence of its abscissa (yellow vertical line in Fig. 3), the ordinates y^+, y^- of the two enclosing boundaries are produced (dotted lines in Fig. 3).
2. Three fuzzy sets are then defined in the domain of fuzzy fold change values, given y^+, y^- and a mean expression value (see Fig. 3, projections on the right). The first fuzzy set represents over-expression, the second fuzzy set represent insignificant variation, and the third fuzzy set represent under-expression.
   (a) The fuzzy sets representing under-expression and over-expression are defined in terms of a sigmoidal membership function, while the fuzzy set representing insignificant variation is defined as a Gaussian fuzzy set.
   (b) The fuzzy set representing under-expression (resp. over-expression) intersects the fuzzy set representing insignificant variation at y^- (resp. y^+), with membership degree equal to 0.5.
3. The three fuzzy sets are used to evaluate the possibility of the gene to be differentially expressed. More specifically, the possibility measure is computed between the fuzzy fold change and the under-expression fuzzy set, the insignificant variation fuzzy set, and the over-expression fuzzy set[6].
   (a) The possibility measure between the fuzzy fold change and the sigmoidal fuzzy set representing over-expression (resp. under-expression), quantifies the possibility that the gene is over-expressed (resp. under-expressed) in the control sample.
   (b) The possibility measure between the fuzzy fold change and the Gaussian fuzzy set representing insignificant variation evaluates the possibility of false positiveness.

By repeating the procedure for all the genes, a ranked list is eventually produced with genes sorted according to their possibility of being differentially expressed, and accompanied with an additional information of possible false-positiveness.

3   Results

The proposed method was tested using two datasets downloaded from NCBI-SRA archive: DRP000527 and SRP014005. The datasets were mapped against

---

[6] The possibility measure between two fuzzy sets F_1 and F_2 is defined as $\Pi(F_1, F_2) = \max_x \min\{F_1(x), F_2(x)\}$.

Table 2. Example of differential expression analysis results for dataset SRP014005. For each gene, the table lists the Fisher's p-value adjusted with False Discovery Rate computed on centroids (FDR), the Log2 Fold Change computed on centroids (Log2fc), the fuzzy fold change computed on fuzzy read counts, the possibility of differential expression as the maximum between over- or under-expression possibility (DE-p) and of false positive risk (FP-p). The results are sorted by FDR.

n	Gene ID	FDR	Log2fc	Fuzzy fold change	DE-p	FP-p
1	RP5-857K21.6	2.58e−11	0.45	Tr[0.24, 0.45, 0.45, 0.57]	0.91	0.44
2	MUC5AC	1.39e−10	1.24	Tr[1.13, 1.24, 1.24, 1.32]	1.00	0.06
3	EEF1A1	6.91e−6	−1.21	Tr[−5.01, −1.89, −0.57, 1.94]	1.00	0.88
4	POSTN	9.53e−5	2.21	Tr[2.08, 2.21, 2.21, 2.35]	0.94	0.20
5	CH507-513H4.3	9.77e−4	−0.63	Tr[−10.32, −10.31, 9.75, 9.76]	1.00	1.00
6	CH507-513H4.4	9.77e−4	−0.63	Tr[−10.32, −10.31, 9.75, 9.76]	1.00	1.00
7	CH507-513H4.6	9.77e−4	−0.63	Tr[−10.32, −10.31, 9.75, 9.76]	1.00	1.00
8	SCGB1A1	0.00489	−1.96	Tr[−3.07, −2.01, −1.92, −0.23]	0.91	0.58
9	LYZ	0.00942	−1.74	Tr[−2.14, −1.86, −1.62, −1.46]	0.91	0.37
10	RP11-380M21.2	0.0162	1.75	Tr[−0.14, 1.33, 2.25, 3.50]	0.93	0.75
11	CLCA1	0.0221	4.00	Tr[3.88, 4.00, 4.00, 4.11]	0.75	0.36
12	MTRNR2L12	0.0292	0.13	Tr[−0.28, 0.03, 0.23, 0.55]	0.71	0.99
13	PDE4DIP	0.0292	2.42	Tr[−0.17, 2.25, 2.70, 5.90]	0.90	0.69
14	ENO2	0.1087	3.75	Tr[3.61, 3.75, 3.75, 3.88]	0.58	0.48
15	WAC	0.1272	−4.00	Tr[−4.07, −4.00, −4.00, −3.92]	0.75	0.36
16	TSPAN3	0.1488	2.58	Tr[2.19, 2.58, 2.58, 3.06]	0.59	0.55
17	AFF4	0.1750	−2.41	Tr[−2.89, −2.41, −2.41, −2.02]	0.64	0.52
18	GSTA2	0.1881	−3.92	Tr[−4.07, −3.92, −3.92, −3.75]	0.70	0.40
19	MALAT1	0.1906	−0.15	Tr[−0.16, −0.15, −0.15, −0.14]	0.13	0.83
20	FAM213A	0.1906	3.61	Tr[3.46, 3.61, 3.61, 4.00]	0.51	0.54

Vega transcript database [10], while DESeq2 [5], TopHat-Cuffdiff [6] and Fisher's Exact test p-value (adjusted with False Discovery Rate, FDR) were used to evaluate differential expression.

The first dataset contains two samples coming from the HeLa cells. There is no difference between the two samples, except for one gene: in one sample the U2AF1 gene is suppressed, therefore only this gene should result differentially expressed in the comparison of the two samples. The U2AF1 gene is 14632 bases long, it is present in the human genome on the chromosome 21, but on the same chromosome there is another gene similar to the 98 %, that we call U2AF1'. The presence of this gene is a clear source of multireads.

The Illumina reads were mapped using Bowtie2. The 30 % of mapped reads are multireads. The mapping identified 19816 genes, and 78 % of them are influenced by multireads.

By applying the proposed method, we obtain that both U2AF1 and U2AF1' have a possibility of being differentially expressed = 1, but also a possibility of

Table 3. Example of fuzzy read counts for dataset SRP014005. For each gene, the centroid and the fuzzy trapezoidal set for Control (Ctrl) and Case are listed. Genes are ordered as in Table 2.

n	Gene ID	Ctrl	Ctrl fuzzy read count	Case	Case fuzzy read count
1	RP5-857K21.6	1167	Tr[1076, 1166, 1168, 1176]	1593	Tr[1389, 1592, 1594, 1599]
2	MUC5AC	103	Tr[101, 103, 103, 108]	245	Tr[236, 245, 245, 252]
3	EEF1A1	174	Tr[34, 152, 226, 314]	75	Tr[9, 60, 102, 134]
4	POSTN	11	Tr[10, 11, 11, 12]	54	Tr[52, 54, 54, 55]
5	CH507-513H4.3	361	Tr[0, 0, 1268, 1282]	233	Tr[0, 0, 860, 868]
6	CH507-513H4.4	361	Tr[0, 0, 1268, 1282]	233	Tr[0, 0, 860, 868]
7	CH507-513H4.6	361	Tr[0, 0, 1268, 1282]	233	Tr[0, 0, 860, 868]
8	SCGB1A1	52	Tr[19, 50, 53, 60]	12	Tr[6, 12, 12, 16]
9	LYZ	59	Tr[53, 56, 62, 64]	17	Tr[14, 16, 18, 19]
10	RP11-380M21.2	12	Tr[8, 9, 16, 40]	44	Tr[36, 41, 48, 95]
11	CLCA1	0	Tr[0, 0, 0, 0]	15	Tr[14, 15, 15, 16]
12	MTRNR2L12	4459	Tr[3635, 4316, 4612, 4883]	4879	Tr[4015, 4724, 5049, 5339]
13	PDE4DIP	5	Tr[0, 4, 5, 14]	29	Tr[12, 28, 32, 59]
14	ENO2	0	Tr[0, 0, 0, 0]	12	Tr[11, 12, 12, 14]
15	WAC	15	Tr[14, 15, 15, 16]	0	Tr[0, 0, 0, 0]
16	TSPAN3	2	Tr[2, 2, 2, 3]	20	Tr[19, 20, 20, 21]
17	AFF4	24	Tr[23, 24, 24, 25]	4	Tr[2, 4, 4, 5]
18	GSTA2	14	Tr[12, 14, 14, 16]	0	Tr[0, 0, 0, 0]
19	MALAT1	3249	Tr[3235, 3249, 3249, 3262]	2934	Tr[2912, 2934, 2934, 2939]
20	FAM213A	0	Tr[0, 0, 0, 0]	11	Tr[10, 11, 11, 15]

insignificant variation (risk of false positive) greater than 0.9. This means that our method is able to detect the suppression of U2AF1, but, since almost all the reads mapping to U2AF1 are multireads (because they also map to U2AF1'), even its similar gene U2AF1' results as differentially expressed. Because of the multireads, both the results are labeled with a high risk of false positiveness. In fact, just one of the two genes was really suppressed. DESeq2 and Fisher's test, applied on the centroids of the fuzzy read counts, confirm the result on U2AF1 and U2AF1', with a Log2 fold change greater than 4. The same analysis, performed with TopHat and Cuffdiff, does not output any significant differential expression events: when we run these tools without multireads management option, all the reads mapping to U2AF1 are discarded because they all map also on U2AF1'; on the other hand, when we run the tools with multireads management (and Cuffdiff's rescue estimation), some reads are mapped to U2AF1, producing a log2 fold change of only 0.76 between case and control (but with a non significant p-value greater than 0.05).

The second dataset contains a case-control study of the Asthma disease, performed through 454 Roche sequencing of human endobronchial biopsies. The reads were mapped using BLAST, with 97 % of identity required, and 16 % of mapped reads are multireads. The mapping identified 14802 genes, 11 genes have a possibility of being differentially expressed greater than 0.9 and four of them have a degree of false positive risk greater than 0.9. Both centroid data and uniquely mapping read counts have been processed with DESeq2, but the tool only warns about the absence of replicates and outputs no significant differences in the two samples. The adjusted p-value of Fisher's test selects 13 differentially expressed genes: 11 of them are the same highlighted with fuzzy possibility sets, while the other 2 show a possibility of being differentially expressed between 0.71 and 0.75 and one of them has also a 0.99 risk of false positiveness. TopHat and Cuffdiff cannot be run on 454 data.

Tables 2 and 3 show the results of the computation for the first 20 genes, ordered by Fisher's p-value adjusted with FDR. The 1st and 10th genes are ribosomal proteins that showed over-expression, but one of them with an high FP-p (0.75), due to multireads. EEF1A1 gene shows a quite high possibility of insignificant variation (FP-p) in the result, due to multireads. In fact, even if the centroids generate a relevant fold change, the fuzzy read counts are partially overlapping and the fuzzy fold change includes 0. The 5th, 6th and 7th genes are lincRNA with identical sequence and each read mapping to one of these three genes maps at least on the other two. From the mapping accuracy point of view, it is impossible to distinguish the real source of the obtained reads. This means that it is possible that all of the three genes have really changed their expression, but it is also possible that only one did it, influencing the other with the ambiguous mapping. From the functional point of view, since the genes are identical, it is important to highlight the overall variation in their expression, because identical genes have the same function in the cell. For MTRNR2L12, the uncertainty is due both to multireads and to low values of log2 fold change. The results with high DE-p (maximum possibility of over or under-expression) and low FP-p are the most relevant. It is also useful to look at the obtained read counts, showed in Table 3. In this example, if we want to select the three most relevant and less uncertain differential expression events, they are the over-expression of MUC5AC and POSTN and the under-expression of LYZ. In general, DE-p and FP-p could be used to help the biologist in the selection of candidate differential expression events: genes with high DE-p are the most promising candidates, provided that they have a small FP-p value; otherwise they are possible false positive.

4   Conclusion

The described method exploits fuzzy sets to manage the uncertainty due to multireads, in particular during the evaluation of differential expression analysis with NGS RNA-Seq data. The model has been tested on case-control transcriptomic data produced by Roche 454 and Illumina sequencers.

Gene expressions are represented with trapezoidal fuzzy sets, which represent the ambiguities resulting from read mapping. Genes are ranked through a possibility measure of differential expression, accompanied with information about the uncertainty that could be present in the results, caused by multireads.

The uncertainty representation can also be used just to add information to the results obtained with other differential expression tools, in order to highlight the risk of false positives in the results.

The method can also be applied to different types of data, like genomic and metagenomic reads, and it will be extended to cope with biological replicates and different types of sample comparison (e.g. with more than two conditions or time series data).

In order to test the method, some scripts have been developed using R and Phyton. All the scripts and the datasets used for the experimentations are available at http://bioinformatics.ba.itb.cnr.it/multidea.

Acknowledgments. We thank Dr. Flavio Licciulli, Dr. Mariano Caratozzolo and Dr. Flaviana Marzano for their suggestions and help with NGS data elaboration. A.C. is supported by Progetto MICROMAP PON01_02589.

References

1. Faulkner, G.J., Forrest, A.R., Chalk, A.M., Schroder, K., Hayashizaki, Y., Carninci, P., HUme, D.A., Grimmond, S.M.: A rescue strategy for multimapping short sequence tags refines surveys of transcriptional activity by CAGE. Genomics **91**(3), 281–288 (2008)
2. Jiang, H., Wong, W.H.: Statistical inferences for isoform expression in RNA-Seq. Bioinformatics **25**(8), 1026–1032 (2009)
3. Li, B., Ruotti, V., Stewart, R.M., Thomson, J.A., Dewey, C.N.: RNA-Seq gene expression estimation with read mapping uncertainty. Bioinformatics **26**(4), 493–500 (2010)
4. Li, B., Dewey, C.N.: RSEM: accurate transcript quantification from RNA-Seq data with or without a reference genome. BMC Bioinf. **12**(1), 323 (2011)
5. Love, M.I., Huber, W., Anders, S.: Moderated estimation of fold change and dispersion for RNA-Seq data with DESeq2. Genome Biol. **15**(12), 550 (2014)
6. Trapnell, C., Roberts, A., Goff, L., Pertea, G., Kim, D., Kelley, D.R., Pimentel, H., Salzberg, S.L., Rinn, J.L., Pachter, L.: Differential gene and transcript expression analysis of RNA-Seq experiments with TopHat and Cufflinks. Nat. Protoc. **7**(3), 562–578 (2012)
7. Glaus, P., Honkela, A., Rattray, M.: Identifying differentially expressed transcripts from RNA-Seq data with biological variation. Bioinformatics **28**(13), 1721–1728 (2012)
8. Negoita, C., Zadeh, L.A., Zimmermann, H.J.: Fuzzy sets as a basis for a theory of possibility. Fuzzy Sets Syst. **1**, 3–28 (1978)
9. Pedrycz, W., Gomide, F.: An Introduction to Fuzzy Sets: Analysis and Design. MIT Press, Cambridge (1998)
10. Wilming, L.G., Gilbert, J.G.R., Howe, K., Trevanion, S., Hubbard, T., Harrow, J.L.: The vertebrate genome annotation (Vega) database. Nucleic Acids Res. **36**(suppl 1), D753–D760 (2008)

Module Detection in Dynamic Networks by Temporal Edge Weight Clustering

Paola Lecca[1,2(✉)] and Angela Re[3]

[1] Department of Mathematics, University of Trento,
Via Sommarive 14, 38123 Povo, Trento, Italy
paola.lecca@unitn.it
[2] Association for Computing Machinery, New York, USA
[3] Laboratory of Translational Genomics, Centre for Integrative Biology,
Univesrity of Trento, Via Sommarive 9, 38123 Povo, Trento, Italy
angela.re@unitn.it

Abstract. While computational systems biology provides a rich array of methods for network clustering, most of them are not suitable to capture cellular network dynamics. In the most common setting, computational algorithms seek to integrate the static information embedded in near-global interaction networks with the temporal information provided by time series experiments. We present a novel technique for temporally informed network module detection, named TD-WGcluster (Time Delay Weighted Graph CLUSTERing). TD-WGcluster utilizes four steps: (i) time-lagged correlations are calculated between any couple of interacting nodes in the network; (ii) an unsupervised version of k-means algorithm detects sub-graphs with similar time-lagged correlation; (iii) a fast-greedy optimization algorithm identify connected components by sub-graph; (iv) a geometric entropy is computed for each connected component as a measure of its complexity. TD-WGcluster notable feature is the attempt to account for temporal delays in the formation of regulatory modules during signal propagation in a network.

Keywords: Graph clustering · Network analysis · Time lagged correlation · Geometric entropy · Protein network

1 Introduction

Interaction networks can reveal the overall landscape of biological systems. The common tendency in network mapping studies is to regard interaction networks as static information. Almost all interaction maps, to date, have been generated under a single standard laboratory condition, in the absence of information on the physiological context, either in space or in time, where such interactions may or may not occur. In fact biological systems often undergo highly dynamic processes such as during cell cycle and circadian rhythm, along development, age and disease progression or in response to a host of environmental and genetic perturbations. It is clear that these dynamical, either reactive or programmed,

© Springer International Publishing Switzerland 2016
C. Angelini et al. (Eds.): CIBB 2015, LNBI 9874, pp. 54–70, 2016.
DOI: 10.1007/978-3-319-44332-4_5

processes effect or are affected by changes in the underlying interaction networks. The variation of the interactions over time guides the temporal aspects of the information flow through a network [22]. Since interaction network dynamics cannot be directly gauged by most currently available near-global interaction maps, the integration of static network data with time series data is decisive to our ability to infer the interaction network dynamics. The richest resources of temporal information at relatively reduced costs are provided by time series gene expression (e.g. transcriptomics, proteomics) data. Computational approaches based on the integration of network data with temporal data are largely proposed in the following settings.

- Identification of expression-activated sub-networks. A multitude of algorithms detect sub-networks on the basis of statistically significant expression changes of their nodes over some time intervals. Such approaches are typified by jActiveModules [11], ExprEssence [36], TimeXNet [25]. Fewer algorithms detect subnetworks by resorting to the responsiveness of the edges connecting the nodes in the sub-networks [10]. Another sort of approaches relies on the detection of time-sequenced modules, and applies static module detection methods for each time point in the time series experiment [17,24,35]. Recently, TS-OCD has been proposed based on the concept of overlapping temporal protein complexes in order to track the evolution (growth or shrinking) of modules across different time points [23]. However, in general, these approaches disregard the connections between the networks at consecutive time points.
- Identification of expression-activated pathways by algorithms which conduct enrichment analysis of differentially expressed genes in annotated pathways such as PathExpress [9], GSEA [34]; Pathway-Express [5] and SPIA [32] account also for the magnitude of gene expression changes and pathway topological properties.
- Identification of dynamical regulatory events on global scales by probabilistic algorithms such as DREM and SMARTS [28,29,37], regression-based algorithms such as Inferelator [1] and data decomposition algorithms like NCA [18]. These modules aim at providing several hypotheses regarding possibly causal relationships between nodes.

However, the methods deriving either static time-wise modules or dynamic regulatory modules often miss important temporal aspects. Since many causal events happen in a sequential manner, correlation-based analysis may miss key regulatory events whenever the relationships between regulators and regulated nodes are time-lagged. Lagged correlations refer to the correlation between two time series shifted in time relative to one another, and are characteristic of the dynamical behaviour of biological systems, which rely not only on impulse responses but also on responses delayed over time. The simple correlation coefficient between the two series is inadequate to characterize the relationship in such common situations in the biological systems.

In this paper, we introduce an advanced computational method, Time Delayed Weighted Edge Clustering (TD-WGcluster), which integrates static interaction networks with time series data in order to detect modules of genes between which the information flows at similar time delays and intensities. For this purpose, TD-WGcluster does not modularize the time series but the time-lagged correlation profiles between any two nodes connected in the static network. Since the basic units of the modules are time-lagged correlation curves, a module detected by TD-WGcluster may assemble pairs of nodes characterized by diverse temporal relationships, provided that these relationships are maximally intense at similar temporal delays. Therefore, TD-WGcluster does not set stringent constraints on node temporal profiles which finally may turn out to be unrealistic. Oppositely, TD-WGcluster permits heterogeneity in the temporal information flow between node pairs within a module and, in this way, it improves the adherence of the detected modules in the networks to the functional modules in biological systems, which indeed may rely on combinations of diverse types of causal (e.g. regulatory) relationships between module constituents. Additional TD-WGcluster advantages include the quantification of the complexity of the temporal evolution of the detected modules.

2   TD-WGcluster: The Algorithm

The TD-WGcluster algorithm is implemented in R (http://wwww.r-project.org) and takes as input network edges in Simple Interaction File (SIF Cytoscape) format and the time series of the abundance of each system's component (represented as a node of the network) in tabular text format. The algorithm sequentially executes four computational modules.

1. First it calculates the time-lagged correlation (TLC) between each couple of linked nodes in the network, i.e. the correlation between the nodes' time series shifted in time relative to one another. Then it analyzes the TLC curves to estimate the features describing their shape. i.e. the lag corresponding to the maximum of the curve, the trend index, the seasonality index, the autocorrelation test statistics, the non-linearity test statistics, the skewness index, the kurtosis index, and the Lyapunov coefficient [19,30].
2. A K-means algorithm detects sub-graphs by clustering the TLC curves of each couple of linked nodes according to the features describing the shape of the curves. Relying on the shape of the TLC curves, such clustering permits the identification of sub-graphs of nodes between which the information propagates from the source node to the target node at similar time lags. The similarity of the shapes of the TLC curves does not imply that those nodes have similar dynamics (i.e. similar time series), but only that the synchronization between the activities of directly linked nodes occurs at similar time lag. Therefore, each sub-graph can be characterized by its own time lag τ_r, representative of the time delay at which the correlation between the dynamics of the directly linked nodes reach its maximum value.
3. Fast-greedy modularity optimization procedure finds (if any) the connected components in each sub-graph.

4. Finally, the geometric entropy E_r of each connected component in the subgraph characterized by representative time delay τ_r is calculated as the negative natural logarithm of the determinant of the covariance matrix of the time series of the nodes in the component. The geometric entropy is used as a measure of the complexity of the connected component since it is indicative of its size (i.e. number of nodes and number of edges) and of the multiformity of the dynamics of its nodes. Indeed, if a component consists of D nodes, the geometric entropy is a global measure of the D-dimensional variance of the time series corresponding to the D nodes. The larger is the volume occupied by the node time series in the D-dimensional space, the larger is the geometric entropy. The most interesting cases are identified by extreme low/high values of the geometric entropy, since the former case can suggest purely stochastic or purely deterministic dynamics, whereas the latter case can suggest hybrid deterministic/stochastic dynamics. Furthermore, the distribution of the geometric entropy on the graph reflects the complexity of the graph itself. Indeed, the more the entropies of the network connected components differ, the more the dynamics vary across the network nodes, and the higher is the complexity of the network.

Finally, note that matrices with the same determinant (thus geometric entropy) can show different sets of eigenvalues. For this reason, TD-WGcluster performs the comparative analysis of the connected components by complementing the usage of the geometric entropy with the analysis of the eigenvectors and eigenvalues of the covariance matrix of each connected component. In particular, TD-WGcluster decomposes the contribution of each node to the variance of the dynamics in a connected component through variable factor analysis.

Computational modules of TD-WGcluster are now described in detail.

2.1   Analysis of Time Lagged Correlation

Lagged correlation refers to the correlation between two time series shifted in time relative to one another. Lagged correlation is important in studying the relationship between time series since one series may have a delayed response to the other series, or a delayed response to a common stimulus that affects both series.

The lagged correlation is estimated by the cross-correlation function (CCF). The CCF of two time series is the product-moment correlation as a function of lag between the series. It is helpful to begin defining the CCF by the cross-covariance function (CCFV). Consider N pairs of observations on two time series, $x(t)$ and $y(t)$, the sample CCFV is:

$$c_{xy} = \frac{1}{N} \sum_{t=1}^{N-\tau} (x(t) - \overline{x(t)})(y(t+k) - \overline{y(t)}), \quad k = 0, 1, \cdots, (N-1) \quad (1)$$

and, similarly, $c_{xy} = \frac{1}{N} \sum_{t=1-\tau}^{N} (x(t) - \overline{x(t)})(y(t+k) - \overline{y(t)})$, $k = -1, -2, \cdots, -(N-1)$ where $\overline{x(t)}$ and $\overline{y(t)}$ are the sample means, and τ is the

lag. The sample CCF is the CCFV scaled by the variances $c_{xx}(0)$ and $c_{yy}(0)$ of
the two series: $r_{xy}(\tau) = \frac{c_{xy}(\tau)}{\sqrt{c_{xx}(0)c_{yy}(0)}}$.

The CCFV and CCF are asymmetrical functions. The asymmetry is specified
in the definition given in Eq. (1). In this equation, the cross-correlation function
is described in terms of *lead* and *lag* relationships. The first part of the equation
applies to $y(t)$ shifted forward relative to $y(t)$. With this direction of shift, $x(t)$ is
said to *lead* $y(t)$. This is equivalent to saying that $y(t)$ *lags* $x(t)$. The second part
of Eq. (1) describes the reverse situation, and summarizes lagged correlations
when $y(t)$ *leads* $x(t)$ ($x(t)$ *lags* $y(t)$).

The analysis of the CCF permits to detect the lag at which the two time
series are maximally correlated and to determine if this correlation is significant.
The correlation is significant if its values does not belong to the confidence interval of the cross-correlation. This confidence interval relies on several simplifying
assumptions and can be computed from the sample size alone. For a two-tailed
test, the approximate γ confidence interval is $CI = 0 \pm z_\gamma \frac{1}{\sqrt{N}}$. The value z_γ is
the γ probability-point of the cumulative distribution function of the normal distribution. This confidence interval relies on assumptions that (1) the processes
generating $x(t)$ and $y(t)$ are uncorrelated, (2) the processes are not autocorre-
lated, (3) the populations are normally distributed, and (4) the sample size is
large. Under those assumptions, the sample cross-correlations are $\sim \mathcal{N}(0, 1/N)$,
or normally distributed with mean zero and variance $1/N$.

2.2 Detection of Sub-graphs

The number of optimal sub-graphs which partition the input graph is estimated by minimizing the total within-clusters sum of squares (WCSS) obtained
with a K-means procedure. K-means clustering is applied to the feature
vectors $\mathbf{tlc} = (tlc_1, tlc_2, \ldots, tlc_E)$, describing the shape of the time-lagged correlations between the time series of each couple of linked nodes. E is the number of edges in the input graph. The components of the feature vector for
the i-th TLC curve $C_i(\tau)$ ($i = 1, 2, \ldots, E$) are: (lag of maximal correlation)$_i$,
(seasonality index)$_i$, (trend index)$_i$, (non-linearity index)$_i$, (kurtosis index)$_i$,
(skewness index)$_i$, (Liapunov coefficient)$_i$, (autocorrelation statistics)$_i$.

The elements of $\mathbf{tlc}_i$ are typically used in time series analysis [19, 30] but,
within TD-WGcluster, are used to capture the main behavioural characteristics
of the TCL curves [16]. In particular, the lag of maximal correlation is the lag
at which the cross-correlation (i.e. the TLC curve) is maximum. Hereafter, this
lag will be denoted by τ_r and we will refer to it as to the *representative* of the
interactions in a sub-graph.

Applying K-means algorithm to the feature vector set $\{\mathbf{tlc}_i\}$, ($i = 1, 2, \ldots, E$), with an increasing putative number of clusters (sub-graphs) at each
run, we obtained the values of WCSS. An elbow in the curve interpolating the
WCSS curves points ($n_{\text{sub-graphs}}$, WCSS), where $n_{\text{sub-graphs}}$ is the increasing
putative number of sub-graphs, suggests the appropriate number of sub-graphs
n_{optimal}. n_{optimal} is estimated as the minimum value of n_{clusters} at which the

first derivative of WCSS w.r.t. $n_{\text{sub-graphs}}$ is null within a tolerance $0 < \epsilon \ll 1$, i.e. $\left| \frac{d\ WCSS}{dn_{\text{sub-graphs}}} \right| \leq \epsilon$. The first derivative of the curve $(n_{\text{sub-graphs}}, WCSS)$ is calculated by the Stineman algorithm [12]. The problem of WCSS minimization is known to be NP-hard. Furthermore, if the input data do not have a strong clustering structure, the procedure may not converge. For this reason, TD-WGcluster adopts the Lloyd's algorithm whose complexity is linear in the number of edges and number of sub-graphs, and is recommended in case of data poorly clustered [6].

2.3   Detection of Connected Components and Geometric Entropy

Each sub-graph $S_v(v = 1, \ldots, K)$ returned by the K-means clustering is decomposed into connected components $C_l^{(v)}$ (with $l = 1, 2, \ldots, L_v$, where L_v is the number of connected components in sub-graph S_v) via a fast-greedy optimization procedure [4]. The geometric entropy of $C_l^{(v)}(\tau_r)$, indicating the entropy value at the lag τ_r, is calculated as:

$$E_l^{(v)}(\tau_r) \equiv -\ln|\det M_l^{(v)}| \tag{2}$$

where $M_l^{(v)}(\theta_{l,v})$ is the matrix of covariances of the time series of the nodes in the connected component l of sub-graph v; its entries are

$$M_{(l,(a,b))}^{(v)} = \mathbf{E}(x_a(t)x_b(t + \tau_r)), \quad a, b = 1, 2, \ldots N_l$$

where $x_a(t)$, and $x_b(t)$ denote the time series of nodes "a" and "b", respectively. The entropy as defined in Eq. (2) estimates the N_l-dimensional variance of the dynamics of the nodes in the connected component.

3   Results

In this section we will introduce the TD-WGcluster features by an in silico experiment and we will present an application of it in a real case study of biological interest.

3.1   In Silico Study

As a running example to illustrate the features of TD-WGcluster, we constructed an in silico random network consisting of 100 nodes and 500 edges, where lags in correlation were induced by generating a bimodal distribution of distinct kinetic rates, characterizing slow and fast reactions (Fig. 1). Kinetic rates from this distribution have been randomly assigned to the edges of the network. The time series (consisting of 1000 time points between initial time 0 and final time 4 in arbitrary units) associated with each node were generated by simulating the deterministic dynamics of the network with Dizzy software (http://magnet.systemsbiology.net/software/Dizzy/).

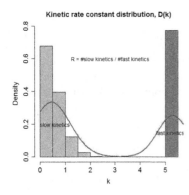

Fig. 1. Bimodal distribution $D(k)$, of the kinetic rates k of the benchmark network. $D(k)$ has been obtained as a mixture of two lognormal distributions with parameters $\mu_1 = 0, \sigma_1 = \ln(2)$ and $\mu_2 = \ln(200), \sigma_2 = 0$. The index R measures the disproportion between the number of slow and fast kinetics which are characterized by kinetic rate constant under the two peaks in the bimodal distribution.

The K-means clustering resulted in the identification of eight sub-graphs characterized by distinct feature sets, as shown in Fig. 2.

The fast-greedy modularity optimization algorithm resulted in the identification of 18 connected components (number of edges greater than 1). Three examples of connected components extracted from three distinct subgraphs and characterized by TLC curves of different shape are shown in Fig. 3. Interestingly, in more than half of the connected components (61 %), the representative time lag of maximal correlation was different from zero. Figure 4 displays the connected components in three coordinate spaces: (i) representative entropy $E_l^{(v)}(\tau_r)$ and number of nodes, (ii) representative entropy E_r and the representative maximum of the TLC defined by the average of the maxima of the TLC curves in a connected component, and (iii) representative entropy $E_l^{(v)}(\tau_r)$ and the standard deviation of the maxima of the TLC curves.

TD-WGcluster provides an effective approach in any applicative context where the assumption of absence of temporal delay, which is implicit in classical correlation-based clustering approaches, can represent a limitation. As mentioned, TD-WGcluster assigns a temporal delay characteristic of the interactions between the nodes in each identified connected component; in this way, it allows the user an easy investigation of the identified connected components in the light of the associated time lags, as shown in Fig. 7.

Finally, the aforementioned synthetic dataset was used to undertake the assessment of TD-WGcluster performances. This choice was motivated by the insufficiency of time series datasets of adequate quality and of clustering methods accounting for lagged-correlation of time series. To establish the benchmark clustering solution in the synthetic setting, we computed the expected optimal number of clusters of time-lagged correlation curves applying the Calinski criterion [2] and the Bayesian Information criterion [21] to the convolution functions of the time series associated with the nodes in the synthetic network (see TDWG-Cluster tutorial at [16] for more details on this regard). Indeed, time series convolution (as well as time-lagged correlation between time series) is determined principally by the distribution of kinetic rates and, secondly, by network topology. In this case, the topology is the simplest as possible as it was defined by an Erdös-Renyi random graph with no scale free property, and therefore, it is likely to scarcely influence the clustering structure of the time-lagged correlation curves. Both the Calinski and Bayesian Information criteria agreed

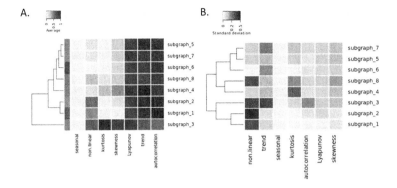

Fig. 2. Features of the sub-graphs identified by TD-WGcluster in the in silico network. (A) Heat map displaying the average value of each feature reported along the x-axis for each sub-graph reported along the y-axis. The colour bar along the y-axis displays distinct time lag in distinct colours. Average is computed across the TLC curves assembled in each sub-graph. (B) Heat map displaying the standard deviation value of each feature reported along the x-axis for each sub-graph reported along the y-axis. Standard deviation is computed across the TLC curves assembled in each sub-graph (Color figure online)

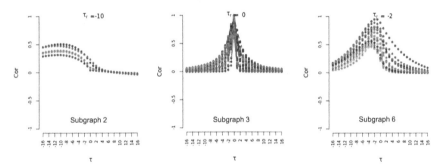

Fig. 3. The TLC curves derived from the time series of couples of nodes belonging to a sub-graph have similar shapes. Each sub-graph, and consequently each connected components in it, performs its own TLC curve.

on estimating 10 clusters as the optimal number of clusters. Therefore, the convolution curves were classified in 10 clusters with the recent dynamical time warping fuzzy clustering algorithm [7] implemented in the dtwclust R package [27]. We assessed the validity of the sub-graphs detected by TD-WGcluster by comparing both the number and the content of the TD-WGcluster sub-graphs with the expected number and content of the clusters inferred by the dynamic time warping algorithm. We compared clusters' content by counting the edges in common between each sub-graph detected by TD-WGcluster and each cluster detected by fuzzy clustering.

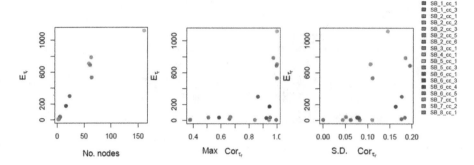

Fig. 4. Connected components are visualized as coloured points in the space of entropy at the representative time lag and number of nodes (first plot); in the space of entropy at representative time lag $(E_l^{(v)}(\tau_r))$, here simply indicated as E_r) and maximum value of time-lagged correlation at the representative time lag, here indicated as $\max Cor_{tau_r}$ (second plot); and in the space of entropy at representative time lag and standard deviation of the maximum of time-lagged correlation (third plot). The plots show that connected components of larger size (i.e. number of nodes) have also a larger geometric entropy. Regardless of the number of nodes, connected components represented by larger value of the maximum of TLC (averaged on the TLC curve of the connected component) have also a larger entropy. Regardless of the number of nodes and the value of the average maximum of the TLC curves, the entropy seems to not have a functional dependency on the standard deviation of this maximum.

TD-WGcluster sub-graphs showed reasonable agreement with the clusters detected by fuzzy clustering in terms of clusters' number and content, as shown in Fig. 5A. TD-WGcluster performances were found to be sensitive to the proportion of interactions of slow (low k) and fast (high k) kinetics, as shown in Figs. 5B and 5C. In particular, the presence of a larger number of slow (fast) kinetics jeopardizes the accuracy of the tool, as slow (fast) kinetics tend to result in flat (stiff) dynamics causing classification errors in algorithms relying on the shape of a curve. In order to understand how the performance of TD-WGcluster depends on the ratio R between slow and fast kinetics, and thus to establish the conditions for an optimal performance, we generated different sets of kinetic rate distributions characterized by a different value of R (see Fig. 6). We found that for an optimal performance of the tool R is recommended to be in the interval [21 %, 72 %]. Nonetheless, we also noticed that even for values of R outside of this range, the performance does not decrease below 50 %.

3.2   Time-Resolved Proteomics Case Study

We applied TD-WGcluster to a time-resolved mass-spectrometry-based quantitative proteomic profiling of mouse embryonic stem cell (ESC) differentiation [20] in combination with the protein-protein interaction network provided by the IntAct database (www.ebi.ac.uk/intact/). This application is of particular

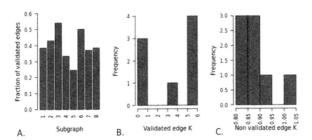

Fig. 5. A. Performance of TD-WGcluster in clustering a network of 100 nodes and 500 interactions whose kinetic rates follow a bimodal distribution obtained as a mixture of two log-normal distributions. **B.** Distribution of the kinetic rate constants of the interactions common to TD-WGcluster sub-graphs and the corresponding best matching benchmark clusters. **C.** Distribution of the kinetic rate constants of the interactions which are present in a TD-WGcluster sub-graph and which are absent from the best matching benchmark cluster.

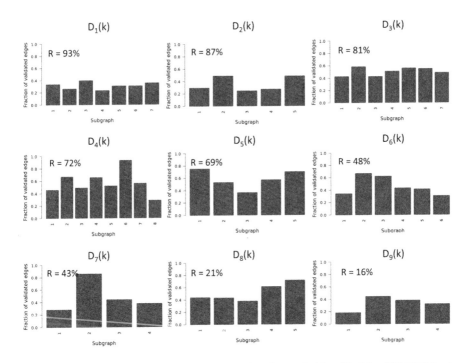

Fig. 6. Percentage of validated edges for each sub-graph identified by TD-WGcluster in graphs of equal topology but different distribution of kinetic rate constants. Nine bimodal distributions, $D_i(k), i = 1, \ldots, 9$, with a different ratio R between the number of interactions with low and high kinetic rates. TD-WGcluster has best performance for $R \in [21\%, 72\%]$.

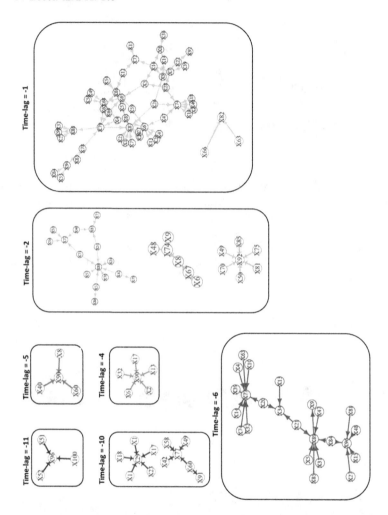

Fig. 7. The connected components identified by TD-WGcluster in the in silico case study are displayed in order of representative time lag. Connected components whose representative time lag is zero are not displayed.

interest since ESCs represent a valuable model for systematically studying mammalian embryonic development.

Applying TD-WGcluster to this case study resulted in the identification of a total of 86 connected components which were arranged in 7 sub-graphs of distinct correlation profiles (Fig. 8). Figure 9 displays the eigenvectors of the covariance matrices for the connected components with a number of nodes greater than 3. In the majority of cases the first axis (PCA 1) extracted more than 90 % of the variation in the time series data. An increment of the entropy E_r (and of the number of nodes) was found to correspond to a decrement in the norm of the majority of the eigenvectors.

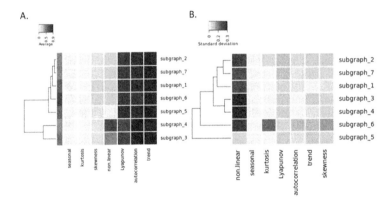

Fig. 8. Features of the subgraphs identified by TD-WGcluster in Intact protein network (see Fig. 2 for a detailed explanation of this graphical representation).

We performed gene set enrichment analysis for each connected component by using the Gene Ontology categories in order to explore the biological interest raised by the connected components identified by TD-WGcluster (Fisher exact test, false discovery test <0.05). The connected components characterized by synchronized time series (time lag = 0) consisted of nodes annotated to plasma membrane organization, cell junction organization, to developmental maturation, and cytoskeleton-dependent intracellular transport (Fig. 10). This observation is interesting since genes functioning in the organization of the embryonic microenvironment have been shown to directly or indirectly modulate the critical balance between ESC self-renewal and differentiation [14,15,31]. Furthermore, human ESC differentiation into neuro-ectodermal spheres has been shown to be associated with a marked reorganization of the cellular cytoskeleton [3]. Therefore, the connected components of temporally synchronized nodes are in agreement with previous observations about the ESC differentiation process.

In general, the connected components where the time lag of maximal correlation is different from zero can be of equal interest since they can capture regulatory events delayed over time. In the present study, such connected components were found to consist of genes linkable to the ESC fate determination. In addition to genes involved in the crucial process of ESC environment modification, the connected components were found to contain genes involved in energy production and cell cycle-related processes. This observation elicited our attention since bioenergetics shifts fuel cell state transitions [33]. ESCs are derived from a relatively hypoxic environment and, accordingly, mainly rely on glycolytic ATP generation regardless of oxygen availability [38], and a glycolytic engagement was shown to mobilize the induction of pluripotency [8,13]. Conversely, the conversion of ESCs into differentiated phenotypes involves a glycolytic to oxidative metabolic transition [26] accompanied by a mitochondrial reorganization. In summary, these results make it likely that the connected components of time-lagged correlation profiles include interactions worth of attention in the study of ESC differentiation.

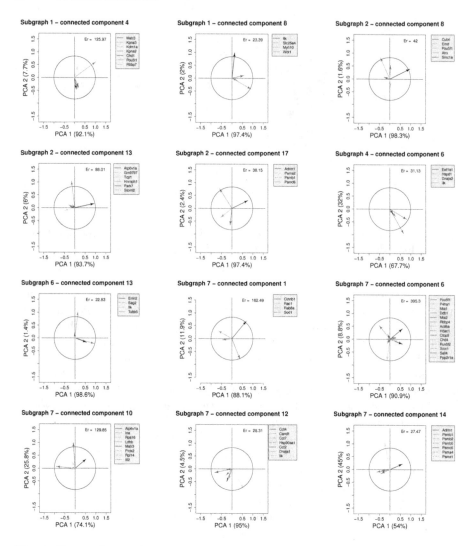

Fig. 9. First and second components of the eigenvector of covariance matrix for the connected components with a number of nodes greater than 3. In the majority of cases the first axis (PCA 1) explains the 90 % of the variation in time series dataset. We note as an increment of the entropy E_r corresponds to a decrement of the norm of the vectors.

4   Discussion

A central aim in cell biology is to describe the molecular programmes that drive cellular functions. Although these are encoded in the genome, they are executed primarily by networks of interacting proteins. The strategies used to map interaction networks are insensitive to the physiological contexts within which

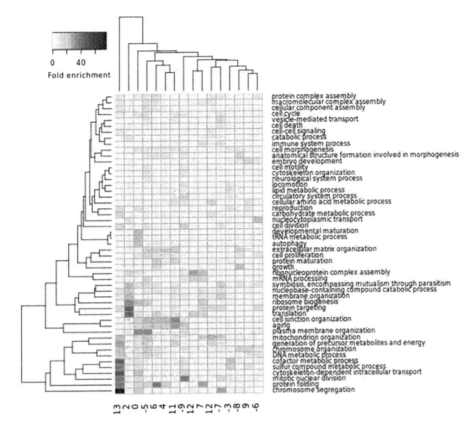

Fig. 10. Functional characterization of the connected components identified by TD-WGcluster in the time-resolved proteomics case study. The heat map shows the fold enrichment of Gene Ontology categories (Biological Process domain) by connected component.

interactions can occur. Nevertheless, analysing the temporal aspects underlying the interactions is crucial to understand cellular function. Given this, a common approach is to combine static protein interaction networks with temporal data from time series experiments.

Here, we introduce the TD-WGcluster method that utilizes interaction networks and time series data in order to execute temporally-informed network clustering analysis. The temporal information is embedded in the time-lagged correlation profiles, which are associated with the edges in the network. The requirement of similar time-lagged correlations coupled to the interactions between nodes is the cornerstone of the TD-WGcluster approach since it drives the identification of the sub-networks and of the connected components therein. The modules identified by our approach can be viewed as a framework for hypothesis generation and for design and interpretation of focused studies that address the dynamic underpinnings of interactions of interest.

5 Software

TD-WGcluster is available at

https://sites.google.com/site/paolaleccapersonalpage/software

along with a short tutorial and a technical report describing the theoretical foundations of the tool and script (R and Dizzy) for the generation of the synthetic case study presented in the paper.

References

1. Bonneau, R., Reiss, D., Shannon, P., Facciotti, M., Hood, L., Baliga, N., Thorsson, V.: The inferelator: an algorithm for learning parsimonious regulatory networks from systems-biology data sets de novo. Genome Biol. **7**(5), R36 (2006)
2. Calinski, T., Harabasz, J.: A dendrite method for cluster analysis. Commun. Stat. **3**, 1–27 (1974)
3. Chae, J., Kim, J., Woo, S., Han, H., Cho, Y., Oh, K., Nam, K., Kang, Y.: Cytoskeleton-associated proteins are enriched in human embryonic-stem cell-derived neuroectodermal spheres. Proteomics **9**(5), 1128–41 (2009)
4. Clauset, A., Newman, M.E.J., Moore, C.: Finding community structure in very large networks. Phys. Rev. E **70**, 066111 (2004)
5. Draghici, S., Khatri, P., Tarca, A., Amin, K., Done, A., Voichita, C., Georgescu, C., Romero, R.: A systems biology approach for pathway level analysis. Genome Res. **17**(10), 1537–1545 (2007)
6. Du, Q., Emelianenkom, M., Ju, L.: Convergence of the Lloyd algorithm for computing centroidal Voronoi tesellation. SIAM J. Numer. Anal. **44**(1), 102–119 (2006). http://www.personal.psu.edu/qud2/Res/Pre/dej06sinum.pdf
7. DâĂŹUrso, P., Maharaj, E.A.: Autocorrelation-based fuzzy clustering of time series. Fuzzy Sets Syst. **160**(24), 3565–3589 (2009)
8. Folmes, C., Nelson, T., Martinez-Fernandez, A., Arrell, D., Lindor, J., Dzeja, P., Ikeda, Y., Perez-Terzic, C., Terzic, A.: Somatic oxidative bioenergetics transitions into pluripotency-dependent glycolysis to facilitate nuclear reprogramming. Cell Metab. **14**(2), 264–271 (2011)
9. Goffard, N., Weiller, G.: Pathexpress: a web-based tool to identify relevant pathways in gene expression data. Nucleic Acids Res. **35**, W176–W181 (2007). Web Server issue
10. Guo, Z., Wang, L., Li, Y., Gong, X., Yao, C., Ma, W., Wang, D., Li, Y., Zhu, J., Zhang, M., Yang, D., Rao, S., Wang, J.: Edge-based scoring and searching method for identifying condition-responsive protein-protein interaction sub-network. Bioinformatics (Oxford, England) **23**(16), 2121–2128 (2007)
11. Ideker, T., Ozier, O., Schwikowski, B., Siegel, A.: Discovering regulatory and signalling circuits in molecular interaction networks. Bioinformatics (Oxford, England) **18**(Suppl 1), S233–S240 (2002)
12. Johannesson, T., Bjornsson, H.: Stineman, a consistently well behaved method of interpolation (2012). http://rpackages.ianhowson.com/cran/stinepack/. Accessed 01 July 2015

13. Kida, Y., Kawamura, T., Wei, Z., Sogo, T., Jacinto, S., Shigeno, A., Kushige, H., Yoshihara, E., Liddle, C., Ecker, J., Yu, R., Atkins, A., Downes, M., Evans, R.: ERRs mediate a metabolic switch required for somatic cell reprogramming to pluripotency. Cell Stem Cell **16**(5), 547–555 (2015)
14. Kinney, M., Saeed, R., McDevitt, T.: Mesenchymal morphogenesis of embryonic stem cells dynamically modulates the biophysical microtissue niche. Sci. Rep. **4**, 4290 (2014)
15. Krieg, M., Arboleda-Estudillo, Y., Puech, P., KÃd'fer, J., Graner, F., MÃijller, D., Heisenberg, C.: Tensile forces govern germ-layer organization in zebrafish. Nat. Cell Biol. **10**(4), 429–436 (2008)
16. Lecca, P.: Software - TD-WGcluster Technical Report (2016). https://sites.google.com/site/paolaleccapersonalpage/software
17. Li, M., Wu, X., Wang, J., Pan, Y.: Towards the identification of protein complexes and functional modules by integrating PPI network and gene expression data. BMC Bioinformatics **13**, 109 (2012)
18. Liao, J., Boscolo, R., Yang, Y., Tran, L., Sabatti, C., Roychowdhury, V.: Network component analysis: reconstruction of regulatory signals in biological systems. Proc. Nat. Acad. Sci. U.S.A. **100**(26), 15522–15527 (2003)
19. Makridakis, S.G., Wheelwright, S.C., Hyndman, R.J.: Forecasting: Methods and Applications. Wiley, New York (1998)
20. Mulvey, C., Schröter, C., Gatto, L., Dikicioglu, D., Fidaner, I., Christoforou, A., Deery, M., Cho, L., Niakan, K., Martinez-Arias, A., Lilley, K.: Dynamic proteomic profiling of extra-embryonic endoderm differentiation in mouse embryonic stem cells. Stem Cells (Dayton, Ohio) **33**(9), 2712–2725 (2015)
21. Neath, A.A., Cavanaugh, J.E.: The Bayesian information criterion: background, derivation, and applications. Wiley Interdisc. Rev. Comput. Stat. **4**(2), 199–203 (2012). http://dx.doi.org/10.1002/wics.199
22. Nooren, I., Thornton, J.: Diversity of protein-protein interactions. EMBO J. **22**(14), 3486–3492 (2003)
23. Ou-Yang, L., Dai, D., Li, X., Wu, M., Zhang, X., Yang, P.: Detecting temporal protein complexes from dynamic protein-protein interaction networks. BMC Bioinformatics **15**, 335 (2014)
24. Park, Y., Bader, J.: How networks change with time. Bioinformatics (Oxford, England) **28**(12), i40–i48 (2012)
25. Patil, A., Nakai, K.: Timexnet: identifying active gene sub-networks using time-course gene expression profiles. BMC Syst. Biol. **8**(Suppl 4), S2 (2014)
26. Pereira, S., GrÃčos, M., Rodrigues, A., Anjo, S., Carvalho, R., Oliveira, P., Arenas, E., Ramalho-Santos, J.: Inhibition of mitochondrial complex III blocks neuronal differentiation and maintains embryonic stem cell pluripotency. PloS One **8**(12), e82095 (2013)
27. Sarda-Espinosa, A.: Time series clustering along with optimizations for the dynamic time warping distance (2016). http://rpackages.ianhowson.com/cran/dtwclust/
28. Schulz, M., Devanny, W., Gitter, A., Zhong, S., Ernst, J., Bar-Joseph, Z.: Drem 2.0: improved reconstruction of dynamic regulatory networks from time-series expression data. BMC Syst. Biol. **6**, 104 (2012)
29. Segal, E., Shapira, M., Regev, A., Pe'er, D., Botstein, D., Koller, D., Friedman, N.: Module networks: identifying regulatory modules and their condition-specific regulators from gene expression data. Nat. Genet. **34**(2), 166–176 (2003)
30. Shumway, R.H., Stoffer, D.S.: Time Series Analysis and its Applications: With R Examples. Springer, New York (2011)

31. Suh, H., Han, H.: Collagen I regulates the self-renewal of mouse embryonic stem cells through α2β1 integrin- and DDR1-dependent BMI-1. J. Cell. Physiol. **226**(12), 3422–3432 (2011)

32. Tarca, A., Draghici, S., Khatri, P., Hassan, S., Mittal, P., Kim, J., Kim, C., Kusanovic, J., Romero, R.: A novel signaling pathway impact analysis. Bioinformatics (Oxford, England) **25**(1), 75–82 (2009)

33. Teslaa, T., Teitell, M.: Pluripotent stem cell energy metabolism: an update. EMBO J. (Oxford, England) **34**(2), 138–153 (2015)

34. Tian, L., Greenberg, S., Kong, S., Altschuler, J., Kohane, I., Park, P.: Discovering statistically significant pathways in expression profiling studies. Proc. Nat. Acad. Sci. U.S.A. **102**(38), 13544–13549 (2005)

35. Wang, J., Peng, X., Li, M., Pan, Y.: Construction and application of dynamic protein interaction network based on time course gene expression data. Proteomics **13**(2), 301–312 (2013)

36. Warsow, G., Greber, B., Falk, S., Harder, C., Siatkowski, M., Schordan, S., Som, A., Endlich, N., SchÄűler, H., Repsilber, D., Endlich, K., Fuellen, G.: Expressence-revealing the essence of differential experimental data in the context of an interaction/regulation network. BMC Syst. Biol. **4**, 164 (2010)

37. Wise, A., Bar-Joseph, Z.: Smarts: reconstructing disease response networks from multiple individuals using time series gene expression data. Bioinformatics **31**(8), 1250–1257 (2015)

38. Yoshida, Y., Takahashi, K., Okita, K., Ichisaka, T., Yamanaka, S.: Hypoxia enhances the generation of induced pluripotent stem cells. Cell Stem Cell **5**(3), 237–241 (2009)

A Novel Technique for Reduction of False Positives in Predicted Gene Regulatory Networks

Abhinandan Khan[1](✉), Goutam Saha[2], and Rajat Kumar Pal[1]

[1] Department of Computer Science and Engineering,
Acharya Prafulla Chandra Roy Siksha Prangan, University of Calcutta,
JD-2, Sector - III, Saltlake, Kolkata 700106, India
khan.abhinandan@gmail.com, pal.rajatk@gmail.com
[2] Department of Information Technology,
North Eastern Hill University, Shillong 793022, Meghalaya, India
dr.goutamsaha@gmail.com

Abstract. In this paper, we have proposed a novel method for the reduction of the number of inferred false positives in gene regulatory networks, constructed from time-series microarray genetic expression datasets. We have implemented a hybrid statistical/swarm intelligence technique for the purpose of reverse engineering genetic networks from temporal expression data. The theory of combination has been used to reduce the search space of network topologies effectively. Recurrent neural networks have been employed to obtain the underlying dynamics of the expression data accurately. Two swarm intelligence techniques, namely, Particle Swarm Optimisation and a Bat Algorithm inspired variant of the same, have been used to train the corresponding model parameters. Subsequently, we have identified and used their common portions to construct a final network where the incorrect predictions have been filtered out. We have done preliminary investigations on experimental (*in vivo*) data sets of the real-world SOS DNA repair network in *Escherichia coli*. Furthermore, we have implemented our proposed algorithm on medium-scale networks, consisting of 10 and 20 genes. Experimental results are quite encouraging, and they suggest that the proposed methodology is capable of reducing the number of false positives, thus, increasing the overall accuracy and the biological plausibility of the predicted genetic networks.

Keywords: Bat algorithm · Gene regulatory network · Particle swarm optimisation · Recurrent neural network · Time-series microarray data

1 Scientific Background

To fully comprehend the critical cellular activities of living beings [1], it is imperative that we learn the exact nature of the genetic relationships using the knowledge of genetic expression patterns. Investigations on gene regulatory networks

© Springer International Publishing Switzerland 2016
C. Angelini et al. (Eds.): CIBB 2015, LNBI 9874, pp. 71–83, 2016.
DOI: 10.1007/978-3-319-44332-4_6

(GRNs), thus, have enticed the research fraternity considerably. Thus, the development of suitable methodologies to completely understand the causal relationships between genes has ensued.

Transcriptional regulation of genes involves DNA, RNA, protein and other molecules. A GRN represents the inter-genetic relationships, and they are indirect, i.e. genes do not interact with each other directly. The interactions are essentially realised via proteins (aka transcription factors) although other factors such as small and long non coding RNA (e.g., miRNA, lncRNA, etc.) are also involved. For example, if a gene g_1 is said to be regulated by another gene g_2, it indicates that the transcription factors encoded by g_2 are responsible for controlling the expression/transcription of g_1 (also known as the target gene). There are other underlying activities involved in the regulatory process, viz. mRNA splicing, structural modifications of proteins post-transcription, etc. Nevertheless, contemporary research is concentrated on the transcription stage (mRNA) only, because of a lack of information concerning the other processes (advancements are being constantly made in this regard). Regardless of these restrictions, the simplicity allows large scale simultaneous measurements of genetic expression values in the form of microarrays. Although, microarray technology is slowly being replaced with more accurate techniques line RNA sequencing, next generation sequencing, etc. expression datasets from such methods are still not freely and easily available.

A GRN represents these complex, inter-genetic regulatory relationships. Depending upon the nature of regulation, the relationships may be of two types: (i) activation: the expression of the target gene increases and (ii) repression: there is a decrease in the expression of the target gene.

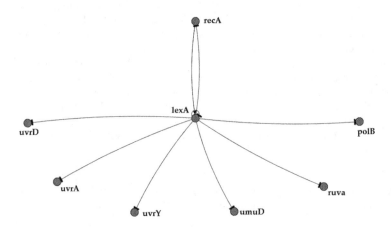

Fig. 1. The original structure of the SOS DNA repair transcriptional network of *E. coli*. Arrowheads represent activation; T-heads represent repression. (Courtesy: http://wws.weizmann.ac.il/mcb/UriAlon/sites/mcb.UriAlon).

The simultaneous measurement of the genetic expression levels of several thousand genes has been made possible by the innovations of DNA microarray technology. However, the microarray data contains unwanted, experimental noise. Additionally, the number of genes investigated is two to three orders of magnitude higher compared to the number of time points. The problem is known as the curse of dimensionality and it severely undermines the potential of any applied methodology for the construction of GRNs from temporal microarray datasets. Several research endeavours have been made to solve this problem but researchers have achieved only partial success in this regard. Various methodologies for reverse engineering of GRNs from time-series expression data such as Boolean Networks [2], Recurrent Neural Networks (RNN) [3], S-systems [4], etc. have been investigated. A review of the various methods used for reverse engineering of GRNs from time-series expression datasets is given in [5]. The training of the corresponding model parameters is an optimisation problem. Thus, metaheuristic techniques like particle swarm optimisation (PSO) [6] are quite popular among researchers worldwide. In this investigation, apart from PSO, a bat algorithm (BA) [7] inspired variant of PSO has been introduced and implemented. Results were evaluated based on their combination.

Although metaheuristics are extensively used for model parameter training purposes, the number of parameters increases in a quadratic manner with respect to the number of genes in a GRN. Thus, for $N \sim 10^2$ or 10^3 numbers of genes, optimisation becomes computationally implausible. Researchers have proposed to solve this particular problem by decomposing the global optimisation problem (model parameter optimisation of all genes in a GRN) into several local optimisation problems (model parameter optimisation of a single gene) [8].

Also, extensive investigations on GRNs reveal that a GRN contains only a handful of regulators [9], i.e. GRNs are sparse in nature. This information points towards the possibility of some form of topological constraint being applied on the predicted GRNs. It, thus, becomes feasible to decouple the architectural and the dynamical features of this reverse engineering problem. This can be realised by decoupling the discrete network architecture search space from the continuous model parameter search space [8]. The continuous search supervises the discrete search.

Unfortunately, none of the above computational techniques [5] for the reconstruction of GRNs from time series microarray data is completely accurate, and all true positives have never been predicted reliably without any false predictions. This inaccuracy increases exponentially as the number of genes involved in the GRN increases. Thus, the very effectiveness of the endeavour of reconstruction of GRN using computational techniques has been challenged.

The main hindrance in these techniques is the inference of many false positives amongst the genes along with true positives. False positives are naturally unwanted entities in the entire domain of investigation into the computational reconstruction of GRNs. Here, we have undertaken an investigation as how to minimise false positive prediction in computational GRN reconstruction. These are initiated by reconstructing GRNs using the RNN technique where several types of optimization techniques have been adopted. Later on, all these techniques are combined to derive a single GRN, where it was found that the number of false positives has been decreased to a significant extent.

2   Materials and Methods

2.1   Materials

First, we have implemented the proposed approach on the *in vivo* time-series genetic expression datasets of the SOS DNA repair network of *E. coli* [10]. This DNA repair mechanism involves only eight genes as studied by Ronen *et al.* [10]. The original network comprising of these eight genes has been shown in Fig. 1, and involves a total of nine interactions. These datasets are mostly used as a benchmark for the comparison of the results obtained from different computational methodologies for reconstruction of GRNs. Ronen *et al.* [10] performed four such experiments, producing four microarray datasets. Each dataset contains the expression values of the eight genes for 50 time points at an interval of six minutes. The expression value at the first time point for each gene, in each dataset, is zero. Hence, we have ignored the particular time-point for our training purpose.

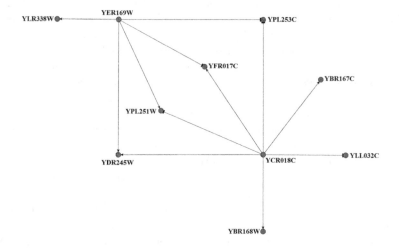

Fig. 2. The original structure of the GRN [8, 13].

Secondly, we have implemented the proposed algorithm on two medium-scale GRNs: one 10-gene and the other 20-gene. We have extracted the GRNs from GeneNetWeaver (GNW) [11] from the genome of yeast and *E. coli*, respectively. We have used DREAM4 [12] settings for the generation of the gene expression time-series. The generated time series contains 50 time points. The 10-gene GRN has been studied previously in [8, 13], and the original GRN has been shown in Fig. 2. The original structure of the 20-gene GRN has been shown in Fig. 3.

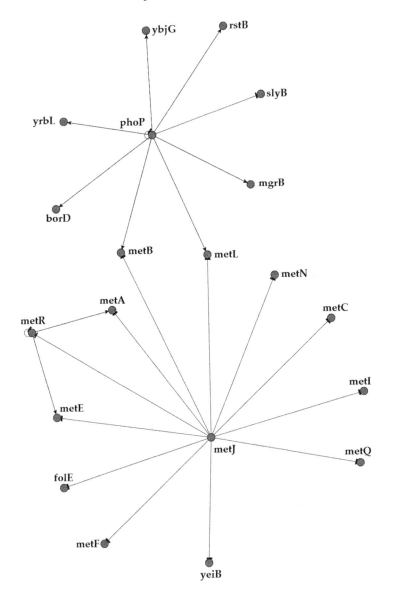

Fig. 3. The original structure of the 20-gene GRN extracted from GNW.

2.2   Methods

In this work, we have proposed to implement a statistical framework hybridised
with Particle Swarm Optimisation (PSO) [6], and another one, hybridised with a
Bat Algorithm [7] inspired Particle Swarm Optimisation (BAPSO) algorithm for
the construction of GRNs. Previous knowledge [14] allows us to reduce the search
space of network topologies by prefixing the maximum number of allowable

regulators for a particular gene. In this work, we have assumed $m = 4$, as the GRNs under consideration are small to medium-scale only [14]. Also, the theory of combination has been utilised to search the reduced network space, exhaustively. Thus, we have achieved a reduction of the search space from a maximum dimension of 2^N to $^NC_4$ or $\binom{N}{4}$ (i.e. the combination of N things taken 4 at a time).

To explain this, let us suppose that there are four genes (A, B, C, D) in a GRN and there are maximum two regulators allowed for each gene. Then the probable candidate solutions for gene A would be: [0011] i.e. A regulated by C and D; [0101] i.e. A regulated by B and D; [0110] i.e. A regulated by B and C; [1001] A regulated by A and D; [1010] A regulated by A and C; and [1100] i.e. A regulated by A and B. This, in essence, is the combination of four things (ones representing the presence of regulatory relationships) taken two at a time, i.e. $^4C_2 = 6$. Proceeding in this manner, if there is no restriction on the number of regulators, the number of solutions become: $^4C_1 + ^4C_2 + ^4C_3 + ^4C_4 = 2^4$. In other words, for a N-gene network, the search space is originally 2^N which is reduced to ^NC_m by the restriction on regulators. Due to the exhaustive search, the likelihood of inferring biologically plausible networks increases, along with a lesser chance of false predictions. The RNN technique has been employed for the purpose of modelling the underlying dynamics of the temporal genetic expression data [3], according to the following equation:

$$x_i\left(t + \Delta t\right) = \frac{\Delta t}{\tau_i \left[1 + exp\left(-\sum_{j=1}^{N}\left[w_{ij}x_j + \beta_i\right]\right)\right]} - \left(1 - \frac{\Delta t}{\tau_i}\right)x_i(t) \quad (1)$$

In the above equation, x_i denotes the expression level of the i-th gene. The equation predicts the expression value of a gene, at a particular time point, based on the expression values of the all the genes at the previous time point only. The term β_i denotes an external bias, and τ_i is a constant parameter. The vector, $W = [w_{ij}]$ contains the regulatory relationships among the genes. A positive value of w_{ij} indicates activation of gene i by gene j, a negative value of w_{ij} indicates repression of gene i by gene j, and $w_{ij} = 0$ indicates no relationship among genes i and j. The efficient training of the parameters: w_{ij}, β_i, and τ_i from the given temporal expression data is an optimization problem. And, like any optimization problem, an objective or a fitness function is required which can be defined by the mean squared error (MSE), in this work:

$$MSE = \frac{1}{NT}\sum_{1}^{N}\sum_{1}^{T}\left(x_i(t) - \tilde{x}_i(t)\right)^2 \quad (2)$$

Here, N is the total number genes in the GRN, T is the number of time points, $x_i(t)$ is the original expression data and $\tilde{x}_i(t)$ is the simulated data at any point of time t. The MSE determines the fitness of a solution. Traditional PSO has been used for the first implementation of the proposed methodology, described by Eqs. (3) and (4).

$$v_i' = w \cdot v + r_1 c_1 \cdot (p_i^b - p_i) + r_2 c_2 \cdot (g^b - p_i) \qquad (3)$$

$$p_i' = p_i + v_i' \qquad (4)$$

Here, v' and v denote the particle velocities for the next and the current generations, respectively; p' and p are the particle positions in the next and the current generations, respectively; p^b is the best solution achieved by a particle; g^b is the best solution achieved by the swarm; w is the inertia weight term that controls the efficient balance between exploration and exploitation undertaken by a particle; r_1, r_2 are random numbers in the range $[0, 1]$; and c_1, c_2 are taken as 2.

In the proposed BAPSO, the update of w is inspired by the frequency update of BA. We have uniformly draw a random value of w from $[w_{min}, w_{max}]$. We have assigned $w_{min} = 0$ and $w_{max} = 1$, in this work. This, to a certain extent, counterbalances the problem of getting trapped at local minima thus, getting an upper hand over one the very few problems plaguing PSO.

Another modification introduced into the proposed BAPSO algorithm, is the initialisation of the particle velocity to zero, inspired by virtual bats in BA. This might help in preventing particles from acquiring an initial unguided velocity that may sidetrack it from a potential optimal solution in the search space.

Last but not the least, for this investigation, GRNs need to be represented computationally, and that is most easily achieved via directed graphs. A directed graph, $G = (V, E)$ can represent a GRN, where V is the set of all genes (nodes or vertices) and E is the set of all interactions between the elements of V (edges). The set E contains an edge e_{ij}, if and only if a causal relationship is present between the vertices (genes), i and j. This structure can be represented with the help of an adjacency matrix, $G = [g_{ij}]_{N \times N}$ where $N =$ the number of nodes (genes) in the graph. The element g_{ij} has a value 0 or 1 depending on the absence or presence of any regulatory interaction from gene j to gene i, respectively.

The novelty of this work lies in the final network construction strategy based on the results of the two formalisms above. We have compared, rather, superimposed the two GRNs constructed using the two swarm intelligence techniques. Subsequently, we have used only the edges, common to both the structures, to assemble the final inferred GRN. This filters out the false positives while keeping the true positives intact.

This can be explained based on the fact that GRNs are sparse. Thus, false positives can be scattered throughout the entire search space and the possibility of identifying the same false positive, by different metaheuristic techniques, is low. As a result, the incorrect predictions vary in their positions in the inferred GRNs for different methodologies and get filtered out. On the other hand, the true positives are unwavering in their positions and thus, survive the filtering process.

3 Experimental Results

All the simulations have been run on 64-bit Matlab 2014a, running on a 64-bit version of MS Windows 7. The simulations have been run on a desktop computer with an Intel Core $i7$ Processor running at 3.4 GHz, with 8 GB of RAM.

In this work, we have employed a collaborative learning scheme due to the stochastic quality of the methodologies involved (as a consequence the GRNs vary in their structures for each independent experiment). Subsequently, we assign a plausibility score ps_{ij} for each edge e_{ij} for its inclusion in the final predicted network, as:

$$ps_{ij} = \frac{1}{L} \sum_{1}^{L} g_{ij} \tag{5}$$

In the above equation, $g_{ij} \in G$, $ps_{ij} \in [0, 1]$, and $L = $ numbers of independent experiments conducted (inferred GRNs) corresponding to each methodology. After the evaluation of ps_{ij} for all i and j, we have stored the final inferred GRN as a matrix $G^F = [g_{ij}^f]_{N \times N}$. The value of g_{ij}^f can be either 0 or 1 and we have adopted the following technique to assign a value to $g^f ij$:

$$g_{ij}^f = \begin{cases} 0, & \text{if } ps_{ij} \geq \alpha \\ 1, & \text{otherwise} \end{cases} \tag{6}$$

The parameter α is a threshold of ps_{ij} i.e. the plausibility score. It regulates the inclusion of a particular edge in the final GRN. In order to evaluate the accuracy of the implemented methodology, we compare this final GRN, G_F with G_O, the original GRN. The experimental results have been validated in this manner. An inferred edge is categorised into four types:

True Positive (TP): if $g_{ij}^o = 1$ and $g_{ij}^f = 1$
True Negative (TN): if $g_{ij}^o = 0$ and $g_{ij}^f = 0$
False Positive (FP): if $g_{ij}^o = 0$ and $g_{ij}^f = 1$
False Negative (FN): if $g_{ij}^o = 1$ and $g_{ij}^f = 0$

The following metrics help in the quantitative comparison of the proposed methodology with those in the contemporary literature:

$$Sensitivity(S_n) = \frac{TP}{TP + FN} \tag{7}$$

$$Specificity(S_p) = \frac{TN}{TN + FP} \tag{8}$$

$$PositivePredictiveValue(PPV)/Precision = \frac{TP}{TP + FP} \tag{9}$$

$$Accuracy(ACC) = \frac{TP + TN}{TP + FP + TN + FN} \tag{10}$$

$$F\text{-}score = \frac{2TP}{2TP + FN + FP} \tag{11}$$

Here, in this work, we have conducted $L = 10$ independent experiments for each of the GRNs studied. The experiment on the SOS DNA repair network of *E. coli* have been conducted with a swarm population of $n = \binom{8}{4}$, and a maximum number of 5000 iterations. For the 10-gene GRN, we have used a swarm population of $n = \binom{10}{4}$, and a maximum number of 10000 iterations. Finally, for the 20-gene GRN, we have used a swarm population of $n = \binom{20}{4}$, and a maximum number of 10000 iterations.

Table 1. Experimental results for the *E. coli* SOS DNA repair network involving eight genes.

Dataset	Technique	TP	FP	S_n	S_p	PPV	ACC	F-score	Graph edges
1	eDSF [8]	3	10	0.33	0.82	0.23	0.75	0.27	13
	PSO	5	09	0.56	0.84	0.36	0.80	0.43	14
	BAPSO	7	09	0.78	0.84	0.44	0.83	0.56	16
	Proposed	**7**	**09**	**0.78**	**0.84**	**0.44**	**0.83**	**0.56**	**16**
2	eDSF [8]	8	05	0.89	0.91	0.62	0.91	0.73	13
	PSO	4	10	0.44	0.73	0.29	0.77	0.35	14
	BAPSO	7	15	0.78	0.82	0.32	0.73	0.45	22
	Proposed	**7**	**12**	**0.78**	**0.78**	**0.37**	**0.78**	**0.50**	**19**
3	eDSF [8]	3	10	0.33	0.82	0.23	0.75	0.27	13
	PSO	4	09	0.44	0.84	0.31	0.78	0.36	13
	BAPSO	4	09	0.44	0.84	0.31	0.78	0.36	13
	Proposed	**7**	**09**	**0.78**	**0.84**	**0.44**	**0.83**	**0.56**	**16**
4	eDSF [8]	0	09	0.00	0.84	0.00	0.72	0.00	09
	PSO	3	08	0.33	0.78	0.27	0.78	0.30	11
	BAPSO	4	12	0.44	0.85	0.25	0.73	0.32	16
	Proposed	**3**	**00**	**0.44**	**1.00**	**1.00**	**0.92**	**0.62**	**04**

Table 1 shows the comparison of the inferred network structures with those inferred in [8]. The proposed framework is consistent with respect to the number of correct (true positive) and incorrect (false positive) predictions for each dataset unlike the unevenness in the results of [8]. The obtained prediction error, MSE is $\sim 10^{-2}$

Moreover, the novel scheme of constructing the final inferred topology has succeeded in the reduction of false positives, as seen in Table 1. In general, BAPSO identifies a greater number of edges compared to PSO. As a result, there are more true positives identified, but at the same time, more false positives creep into the final inferred topology as well. Constructing a GRN with only those edges that are common to both the methodologies, helps in filtering out the noisy information i.e. false positives from the GRNs at the same time preserving the increased number of true predictions. The results show the extent

Table 2. Variation of results with α for the *E. Coli* SOS DNA repair network.

TP	FP	S_n	S_p	ACC	TP	FP	S_n	S_p	ACC
α = 0.60									
Dataset 1					**Dataset 2**				
7	12	0.78	0.78	0.78	7	15	0.78	0.73	0.73
Dataset 3					**Dataset 4**				
9	12	1.00	0.78	0.81	4	03	0.44	0.95	0.88
α = 0.70									
Dataset 1					**Dataset 2**				
7	12	0.78	0.78	0.78	7	13	0.78	0.76	0.77
Dataset 3					**Dataset 4**				
9	10	1.00	0.82	0.84	4	00	0.44	1.00	0.92
α = 0.80									
Dataset 1					**Dataset 2**				
7	10	0.78	0.82	0.81	7	12	0.78	0.78	0.78
Dataset 3					**Dataset 4**				
7	10	0.78	0.82	0.81	3	00	0.33	1.00	0.91
α = 0.90									
Dataset 1					**Dataset 2**				
5	08	0.56	0.85	0.81	4	10	0.44	0.82	0.77
Dataset 3					**Dataset 4**				
5	08	0.56	0.85	0.81	1	00	0.11	1.00	0.88
α = 1.00									
Dataset 1					**Dataset 2**				
2	06	0.22	0.89	0.8	1	10	0.11	0.82	0.72
Dataset 3					**Dataset 4**				
2	07	0.22	0.87	0.78	1	00	0.11	1.00	0.88

Table 3. Experimental results for the 10-gene GRN extracted from GNW.

Technique	TP	FP	S_n	S_p	PPV	ACC	F-score	Graph edges
eDSF [8]	5	11	0.42	0.88	0.31	0.82	0.36	16
PSO	5	11	0.42	0.88	0.31	0.82	0.36	16
BAPSO	6	12	0.50	0.86	0.33	0.82	0.40	18
Proposed	**3**	**09**	**0.25**	**0.90**	**0.25**	**0.82**	**0.25**	**12**

of such filtering: in the first three datasets, the false positives reduce but for the final dataset, the false positives vanish altogether. The variation of the results with increasing threshold α has been shown in Table 2.

Table 4. Variation of results with α for the 10-gene GRN extracted from GNW.

α	TP	FP	S_n	S_p	PPV	ACC	F-score	Graph edges
0.6	3	13	0.25	0.85	0.19	0.78	0.21	16
0.7	3	09	0.20	0.89	0.25	0.79	0.22	12
0.8	3	09	0.20	0.89	0.25	0.79	0.22	12
0.9	3	07	0.20	0.92	0.30	0.81	0.24	10
1.0	3	06	0.20	0.93	0.33	0.82	0.25	09

Table 5. Experimental results for the 20-gene GRN extracted from GNW.

Technique	TP	FP	S_n	S_p	PPV	ACC	F-score	Graph edges
BAPSO	3	26	0.13	0.93	0.10	0.88	0.11	29
PSO	2	32	0.08	0.91	0.06	0.87	0.07	34
Proposed	**2**	**16**	**0.08**	**0.96**	**0.11**	**0.91**	**0.10**	**18**

The results for the 10-gene network have been shown in Table 3. The MSE is $\sim 3 * 10^{-3}$. The results have been compared with [8], and it indicates that the proposed algorithm is able to reduce the number of false positives in comparison to [8]. The variation in the false positive reduction with the threshold has been shown in Table 4.

The results of the 20-gene network have been shown in Table 5, and in Table 6, we have shown the variation in the results with increasing threshold α. The MSE is $\sim 10^{-3}$. The results indicate that there is a marked reduction in the number of false predictions when the GRNs inferred using PSO and BAPSO are combined. The results are very encouraging when we consider that there is no biological information used in the inference process. In the contemporary literature, inference results for medium-scale and large-scale networks are not at all satisfactory, and researchers have resorted to including significant biological information in their formalisms [15]. The results obtained in this work, however, show that the accuracy of the inferred networks can be increased, by reducing false predictions, even for medium-scale networks without any biological information included.

Table 6. Variation of results with α for the 20-gene GRN extracted from GNW.

α	TP	FP	S_n	S_p	PPV	ACC	F-score	Graph edges
0.6	3	27	0.13	0.93	0.10	0.88	0.11	30
0.7	3	21	0.13	0.94	0.13	0.90	0.13	24
0.8	2	16	0.08	0.96	0.11	0.91	0.10	18
0.9	2	16	0.08	0.96	0.11	0.91	0.01	18
1.0	1	11	0.04	0.97	0.08	0.92	0.06	12

4   Conclusion

In this work, we have explored the construction of GRNs from temporal expression datasets where a decoupled methodology based on the ideas of combining swarm intelligence algorithms with RNN has been implemented. Results show that the proposed methodology is capable of achieving a remarkable reduction in the number of false positives without sacrificing true positives. In the present context, the identification and reduction of the number of false predictions require conscious efforts that may have been slightly neglected in the endeavour of predicting more and more true positives. Also, the proposed methodology, in essence, is based upon the concept of ensemble learning [16]. Although, it is quite an established concept, ensemble learning algorithms, like boosting, Bayes optimal classifier, etc. have not been implemented in the do-main of reverse engineering of GRNs yet and their application in this domain is also not yet clearly understood. This provides further scope of future research in the domain of this work.

References

1. McLachlan, G., Do, K.A., Ambroise, C.: Analysing Microarray Gene Expression Data. Wiley, New York (2005)
2. Kauffman, S.A.: Metabolic stability and epigenesis in randomly constructed genetic nets. J. Theor. Biol. **22**(3), 437–467 (1969)
3. Vohradsky, J.: Neural model of the genetic network. J. Biol. Chem. **276**(39), 36168–36173 (2001)
4. Voit, E.O.: Computational Analysis of Biochemical Systems: A Practical Guide for Biochemists and Molecular Biologists. Cambridge University Press, Cambridge (2000)
5. Hache, H., Lehrach, H., Herwig, R.: Reverse engineering of gene regulatory networks: a comparative study. EURASIP J. Bioinform. Syst. Biol. **1**, 1–12 (2009)
6. Eberhart, R.C., Kennedy, J.: A new optimizer using particle swarm theory. In: Proceedings of the Sixth International Symposium on Micro Machine and Human Science, vol. 1, pp. 39–43 (1995)
7. Yang, X.S.: A new metaheuristic bat-inspired algorithm. In: González, J.R., Pelta, D.A., Cruz, C., Terrazas, G., Krasnogor, N. (eds.) Nature Inspired Cooperative Strategies for Optimization (NICSO), vol. 284, pp. 65–74. Springer, Heidelberg (2010)
8. Kentzoglanakis, K., Poole, M.: A swarm intelligence framework for reconstructing gene networks: searching for biologically plausible architectures. IEEE/ACM Trans. Comput. Biol. Bioinf. **9**(2), 358–371 (2012)
9. Someren, E.V., Wessels, L.F.A., Backer, E., Reinders, M.J.T.: Genetic network modeling. Pharmacogenomics **3**(4), 507–525 (2002)
10. Ronen, M., Rosenberg, R., Shraiman, B.I., Alon, U.: Assigning numbers to the arrows: parameterizing a gene regulation network by using accurate expression kinetics. Proc. Nat. Acad. Sci. **99**(16), 10555–10560 (2002)
11. Schaffter, T., Marbach, D., Floreano, D.: GeneNetWeaver: in silico benchmark generation and performance profiling of network inference methods. Bioinformatics **27**(16), 2263–2270 (2011)

12. Greenfield, A., Madar, A., Ostrer, H., Bonneau, R.: DREAM4: combining genetic and dynamic information to identify biological networks and dynamical models. PloS One **5**(10), e13397 (2010)
13. Khan, A., Mandal, S., Pal, R.K., Saha, G.: Construction of gene regulatory networks using recurrent neural networks and swarm intelligence. Scientifica 2016(1060843), 14 pages (2016)
14. Bolouri, H., Davidson, E.H.: Modeling transcriptional regulatory networks. BioEssays **24**(12), 1118–1129 (2002)
15. Chowdhury, A.R., Chetty, M.: Network decomposition based large-scale reverse engineering of gene regulatory network. Neurocomputing **160**, 213–227 (2015)
16. https://en.wikipedia.org/wiki/Ensemble_learning

Unsupervised Trajectory Inference
Using Graph Mining

Leen De Baets[1(✉)], Sofie Van Gassen[1,2], Tom Dhaene[1], and Yvan Saeys[2,3]

[1] Internet Based Communication Networks and Services (IBCN),
Ghent University - iMinds, Technologiepark-Zwijnaarde 15, 9052 Ghent, Belgium
`leen.debaets@ugent.be`
[2] Data Mining and Modelling for Biomedicine (DaMBi),
VIB Inflammation Research Center, Ghent, Belgium
[3] Department of Internal Medicine, Ghent University, Ghent, Belgium

Abstract. Cell differentiation is a complex dynamic process and although the main cellular states are well studied, the intermediate stages are often still unknown. Single cell data (such as obtained by flow cytometry) is typically analysed by clustering the cells into distinct cell types, which does not model these gradual changes. Alternative approaches that explicitly model such gradual changes using seriation methods seems promising, but are only able to model a single differentiation pathway. In this paper, we introduce a new, graph-based approach that is able to model multiple branching differentiation pathways as continuous trajectories. Results on synthetic and real data show that this is a promising approach which is moreover robust to parameter changes.

1   Scientific Background

Flow cytometry offers a high-throughput platform for single cell analysis. Typically, 10 to 20 different cell surface proteins (markers) are stained with fluorochromes, enabling their detection using laser-based systems on millions of individual cells during an experiment. Traditional analysis of flow cytometry data is done by examining two-dimensional scatter plots in order to identify distinct cell populations.

However, the amount of possible scatter plots increases exponentially with the number of proteins measured, which makes it infeasible to examine them all. To alleviate this problem, alternative visualisation techniques such as SPADE [1], Visne [2] and FlowSOM [3] have been proposed. These map the multidimensional data to two-dimensional plots, incorporating the similarities of all marker dimensions. Figure 1 shows the result of a FlowSOM analysis on flow cytometry data concerning hematopoietic stem cells differentiating into common myeloid and lymphoid progenitors. While the different cell types are represented in different branches and the corresponding median fluorescence intensities for all markers are indicated, we do not infer any information about their developmental trajectory.

© Springer International Publishing Switzerland 2016
C. Angelini et al. (Eds.): CIBB 2015, LNBI 9874, pp. 84–97, 2016.
DOI: 10.1007/978-3-319-44332-4_7

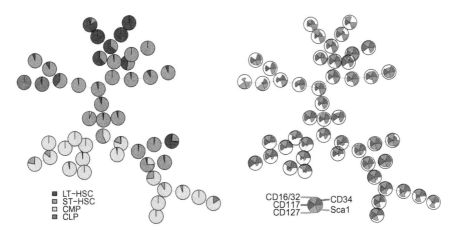

Fig. 1. FlowSOM representation of selected bone marrow cells of a wildtype mouse. Each circle represents a group of similar cells. On the left, the manual gating annotations are visualized in pie charts. The different branches in the tree correspond roughly to the different cell types. On the right, the median fluorescence intensities are indicated. Slight variations in marker intensities for a single cell type are discernable, but one cannot infer the known developmental process: the long-term hematopoietic stem cells (LT-HSC) differentiate into short-term hematopoietic stem cells (ST-HSC) and these can in turn differentiate into either common myeloid progenitor cells (CMP) or common lymphoid progenitor cells (CLP). (Color figure online)

A schematic example of cell differentiation is given in Fig. 2(a) where cells start in an immature state (state 1) and evolve through an intermediate state (state 2) to finally result in two distinct mature cell types (either state 3 or 4). While transitioning from one state to another, certain aspects of the cells change, for example increasing values of two markers when the immature cells differentiate to the intermediate state (Fig. 2(b)). As cells evolve in a continuous manner with many cells being in different states, a single snapshot of the developmental system can be presented in Fig. 2(c), showing only two markers for simplicity. As there is still a lot of uncertainty about the developmental trajectories that cells follow, it would be interesting to infer such trajectories automatically from the data without using any external information about the intermediate stages. This results not only in information about the differentiation state of the cells, but can also present a concise overview of marker behaviour during the differentiation, thereby providing novel hypotheses about cell differentiation and possibly revealing new intermediate cell stages.

Inferring a temporal ordering is not limited to deduce the developmental chronology of cells, and occurs in many other situations, such as reconstruction of temporal ordering of biological samples using microarray data [4]. Other related fields are seriation and ordination, which also try to find an ordering without explicitly using the order-defining property (such as time). Seriation was developed to chronologically order archaeological artefacts from numerous

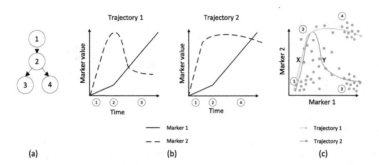

Fig. 2. This figure illustrates the concept of trajectory modelling where (a) shows a schematic overview of the developmental differentiation, (b) the marker changes during development, and (c) a snapshot of the cells following trajectories which are indicated with the coloured lines. (Color figure online)

sites belonging to the same culture [5] and ordination orders objects so that similar objects are near each other and dissimilar objects are farther from each other, often applied on the field of community ecology [6].

The inference of differentiation pathways is a relatively new research. On the one hand, there are methods working on expression data such as Monocle [7] that is an unsupervised algorithm that can infer temporal ordering in single cells based on their expression profiles. At the other hand, there are methods working on flow cytometry data such as Wanderlust [8] that will be explained in the following section. This is the first paper that proposes a method for identifying multiple differentiation pathways in flow cytometry data.

2   Materials and Methods

In this work, we present a new computational method to identify multiple trajectories given a flow cytometry dataset. As input, we use the measured data plus a single cell selected to represent the most immature cell state and a cell selected for each mature cell state. Our algorithm forms a wrapper around the Wanderlust algorithm [8], which is capable of detecting a single trajectory where all cells must differentiate to the same mature cell type. Our method can be easily adapted to use any other approach that is capable of automatically detecting a single trajectory.

In Sect. 2.1, we explain shortly how Wanderlust constructs a single trajectory. In Sect. 2.2, we formulate our solution to find multiple trajectories in unordered data.

2.1   Inference of a Single Trajectory

We define a trajectory as an ordering of cells such that the cells are sorted according to their developmental order: immature cells should thus be sorted before mature cells. It is important to note that the developmental distance often does not correspond with Euclidean distance: cells that are developmentally close will typically be close in Euclidean distance as well, but the opposite is not necessarily true. An example is given in Fig. 2(c) by elements X and Y. Although they are close in Euclidean distance, X is still differentiating towards the intermediate state, while Y is much closer to the mature state. The Wanderlust algorithm solves this by representing the data as a k-nearest neighbour (knn) graph such that only developmentally close cells are connected. Due to noise, it is still possible that cells get connected even though they are developmentally far apart. To handle these short circuits, an ensemble is created consisting of m graphs where in each graph each node contains only $l(< k)$ edges that are randomly chosen from the k edges present in the knn graph, resulting in m l-knn graphs.

Wanderlust creates a trajectory by ordering all cells according to their similarity to the cell selected as the most immature (the start cell). This similarity is defined as the distance in the l-knn graph and is calculated for each graph in the ensemble, resulting in l trajectories. The final trajectory is obtained by averaging each cell's distance across all m l-knn graphs.

The running time of this algorithm is dominated by the creation of the knn graph which takes $O(N^2)$ if the amount of cells is N. Thus if the amount of cells increases linearly, then the running time increases quadratically.

The most restrictive assumption of this algorithm is that all cells must follow one and the same trajectory. This is a necessary assumption as the cells are ordered with respect to their distance to the starting cell of the trajectory. In the following section, we will propose an algorithm that can infer multiple trajectories.

2.2   Inference of Multiple Trajectories

An overview of our algorithm to infer multiple trajectories is given in Fig. 3. As input, we do not only use a start cell representing the most immature cell, but also one end cell for each mature state. Using this information, we infer a trajectory for each end cell. First, we assign each cell in the dataset to one or more trajectories. Once this is done, we apply the original Wanderlust algorithm on all cells assigned to one trajectory and aggregate the results in one final trajectory per mature state.

Assigning cells to a trajectory is done by clustering the data with k-means clustering using a large amount of clusters (overclustering). In these clusters, we find the ones to which the given end cells belong. This allows us to represent each mature state with not only one end cell, but with all the cells in the cluster: the representatives. In the next step, we determine for each representative the edges of the shortest path to the starting cell, resulting in a collection of edges for each mature state. Then for every cell, we determine the edges of the shortest path to the start cell and compare it to the edge collections of the mature states, resulting in the detection of the most similar collection. This comparison is done by determining the amount of overlap. Knowing the most similar collection, we assign the cell to the trajectory going from the starting cell to the selected

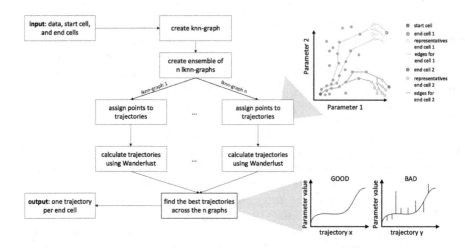

Fig. 3. This figure illustrates the proposed algorithm with a flow chart. The steps concerning the assignments of points and finding the best trajectories are elaborated in more detail with a figure.

end cell. When the overlap is the same for different edge collections, the cell is assigned to multiple trajectories.

Due to the ensemble of graphs, we have m possible trajectories for each mature state. To choose the best trajectory without using extra information, we add a regularisation procedure which assumes that changes between the states in the trajectory should be gradual. Due to this property, the curves representing the parameter values of the cells along the trajectory (Fig. 2(b)) must be smooth. More specific, if these curves contain large jumps (see Fig. 3), there is a high probability that cells are missing or that the ordering is wrong. Concretely, we calculate the smoothed curves using a median filter and calculate a score as the difference between these curves and their smoothed version. The difference between a curve c and his smoothed version $\hat{c}$ is defined as

$$\sum_i |c_i - \hat{c}_i|$$

The trajectory with the smallest score is chosen as it differs the least from its smoothed version and thus the assumption that changes occur gradual is satisfied. This results in one trajectory per mature state. Note that indeed no extra information is used.

The running time of this algorithm is also dominated by the creation of the knn graph which takes $O(N^2)$ if the amount of cells is N. Thus if the amount of cells increases linearly, then the running time increases quadratically.

The parameters for our algorithm are the amount of neighbours k for creating the knn graph, the amount of randomly selected edges l out of the k edges to reduce the effect of short circuits, and the amount of times m this sampling must happen, resulting in m l-knn graphs. For assigning points to a trajectory, the amount of clusters cl must be given.

3   Results

We used synthetic data to evaluate our algorithm and perform a sensitivity analysis for the different parameters. Additionaly, we evaluted our algorithm on real flow cytometry data.

3.1   Synthetic Data

The synthetic data was created using Bézier curves representing an artificial branching structure in a three-dimensional space. We have chosen six points A, B, C, D, E and $F \in \mathbb{R}^3$. The structure starts at point A going to B where it branches to C and endpoint D. From C, it branches then further to the endpoints E and F (Fig. 4(a)). All these connections are created using a quadratic Bézier curve:

$$\mathbf{B}(t) = (1 - t)\left[(1 - t)\mathbf{P}_0 + t\mathbf{P}_1\right] + t\left[(t - 1)\mathbf{P}_1 + t\mathbf{P}_2\right], t \in [0, 1] \qquad (1)$$

where P_0 and P_2 are two of the six chosen points that need to be connected in the branched structure (e.g. point B and C) and P_1 is a random point. We thus created a branched trajectory with three endpoints, as shown in Fig. 4(b). Note that for each endpoint we have a trajectory that starts in A, e.g. the trajectory for endpoint E traverses A, B, C and E. Another way to visualise this branched trajectory is shown in Fig. 4(c). Here, one plot is used for each trajectory and it visualises the parameter values (y-axis) of the points along the trajectory (x-axis), one curve for each feature. As we work in a three dimensional space (amount of parameters $= 3$), this results in three curves on one plot. The smoothness of these curves is used to define a good trajectory.

From each curve, we uniformly sample 100 points. Noise is added by perturbing each point,

$$\mathbf{P} = \mathbf{P} + int * \mathbf{n}, int \in [0, 1] \qquad (2)$$

where $\mathbf{n}$ is a noise vector ($\in \mathbb{R}^3$) containing numbers randomly sampled from a Gaussian distribution with $\mu = 0$ and $\sigma = 1$. The number int represents the intensity of the noise. It is not realistic to assume that during each stage in a trajectory the same amount of cells is available. Rather, we will have states that are very well represented and states that are rare because the cells evolve through them quickly. We model this with a random density functions f_{dens}. For each point, we generate x noisy variants, with x depending on f_{dens}.

$$x(\mathbf{P}) = f_{dens}(\mathbf{P}) \qquad (3)$$

For each curve in the branched structure, we define one f_{dens} by randomly choosing four points in the range $]0, 20]$ and by interpolating these with cubic splines. We make sure that at the connection points (B and C) the densities are the same for all relevant trajectories. An example of the density functions for each trajectory is given in Fig. 4(d) resulting in 3984 points representing the branched trajectory.

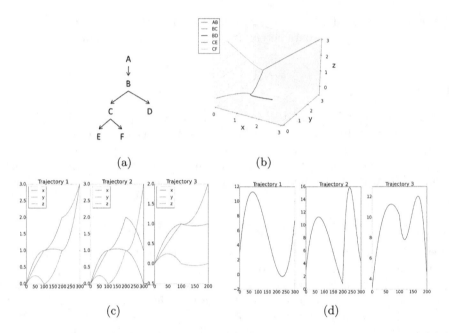

Fig. 4. This figure shows (a) a schematic branched trajectory with three endpoints, (b) the different functions $\mathbf{B}(t)$ following the branched structure in (a), (c) an alternative visualisation which shows a curve following the feature values (y-axis) of the points along the trajectory (x-axis) for each feature for each trajectory, and (d) the densities for each trajectory used for sample representation.

Evaluation. The main advantage of using synthetic data is that we know the exact underlying trajectories without noise. As such, a ground truth is available indicating which points belong to which trajectories and how they must be ordered. This underlying ground truth allows us to define two measures. First, we check if the points are assigned to the correct trajectories. From this comparison we extract the True Positive Rate (TPR or the sensitivity) and the False Positive Rate (FPR) for each trajectory, defined as:

$$TPR = \frac{TP}{TP + FN}, FPR = \frac{FP}{FP + FN} \tag{4}$$

where TP are the true positives, or points that are present in the calculated and ground truth assignments, FP the false positives or the points that are present in the calculated assignment but not in the ground truth assignment, FN the false negatives or the points that are not present in the calculated assignment but should be. The TPR must be as close to one as possible, and the FPR as close to zero.

A second criterion checks for each trajectory how well the TP are ordered according to the ground truth ordering, so we check if the points that are assigned correctly are ordered correctly. To this end, we calculate the Spearman rank correlation coefficient.

Parameter Sensitivity. As the described algorithm has a number of parameters, the robustness of the algorithms' output needs to be investigated by comparing it for different parameter settings. In the following, we change the amount of clusters cl, the amount of edges k and l, and the starting and ending points. We do not change the amount of graphs m as we assume that if this is sufficiently large, the best trajectory will be detected no matter how many graphs are created ($m = 50$ is sufficient). We perturbed the data with 15 % noise. Increasing the noise to 20 % does not pose a problem, but for higher values the results deteriorate.

Sensitivity to the Amount of Clusters. While changing the amount of clusters, the amount of representatives for the endpoints changes. The smaller the amount of clusters, the higher the amount of points assigned to each cluster, thus the higher the amount of points representing an endpoint. If there are a lot of representatives, it is likely that a point belonging to another trajectory is incorrectly assigned to the cluster resulting in wrong assignments of points to the trajectory going to this endpoint. As the change in the amount of clusters only affects the assignment of objects to trajectories it suffices to calculate the TPR and FPR.

In our test case, we set $k = 20$ and $l = 15$, and let the number of clusters vary between $\{4, 10, 20, 30, 40, 3950$ (amount of points in V)$\}$. The last case is equivalent to the case when there are no clusters. The TPR and FPR for each trajectory in function of the amount of clusters are shown in Fig. 5(a). These indicate that when using too few clusters, this leads to the scenario described above which is visible by the fact that when using 4 clusters all elements belonging to trajectory 3, are added to trajectory 3 (TPR $= 1$), but at the same time we notice that the FPR is high, meaning that points belonging to trajectory 1 and 2 are also added to trajectory 3. If we plot the trajectories in the 3-dimensional space (Fig. 5(b)), we indeed see that a lot of points are erroneously assigned to trajectory 3, and consequently a lot of points are missing in trajectory 1 and 2, resulting in a corresponding low TPR. When using enough clusters ($cl \geq 10$), we see that the result varies with the best being achieved when $cl = 20$. We can also see the big advantage of using clusters when comparing with the case without clusters ($cl = 3950$), namely for trajectory 1 and 2 the TPR stays approximately the same but the FPR decreases when no clusters are used. For trajectory 3 it remains approximately the same. When plotting the trajectories in the 3-dimensional space for both cases (Fig. 5(c) and (d)), we see this difference: trajectory 1 and 2 improve when clusters are used, as less points from respectively branch CF and CE are wrongly assigned to them, and for trajectory 3 points are wrongly assigned whether or not clusters are used, explaining the permanent high FPR. However, when clusters are used, these wrongly assigned points are more spread across other branches. In general, the FPR for trajectory 3 is high for every possible amount of clusters indicating this is a difficult branch to detect. For the following test cases we use $cl = 20$. The usage of clusters doesn't seem to be advantageous, but their use will be justified when using real data.

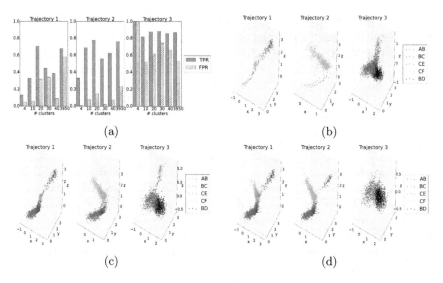

Fig. 5. This figure shows (a) the TPR and FPR concerning the assignments of points to the three different trajectories when $k = 20$, and $l = 15$, (b) the three trajectories in the 3-dimensional space when $k = 20, l = 15$, and $cl = 4$, (c) $k = 20, l = 15$, and $cl = 20$, and (d) $k = 20, l = 15$, and $cl = 3950$.

Sensitivity to k and l. When changing k and l, this has consequences on both the assignment of points to a trajectory and the execution of Wanderlust. In [8], it is stated that Wanderlust is robust against changes in the parameters k and l. We recheck this by using our second evaluation metric, and the robustness of our own assignment algorithm is checked using the first evaluation metric.

In our test case, we set $cl = 20$ and let k vary between $\{10, 20, 50, 100\}$ and l between $\{5, 10, 15, 20\}$ where we guarantee that $l < k$. The Spearman rank correlation coefficient, the TPR and FPR in function of k and l are shown in Fig. 6. When comparing for each trajectory, we see that a low TPR corresponds to a low Spearman rank correlation coefficient. This is logical as determining the real ordering using only a few points (low TPR) is hard, resulting in a low correlation coefficient. Otherwise, the Spearman rank correlation coefficient remains high indicating the robustness of the Wanderlust algorithm against changes in k and l.

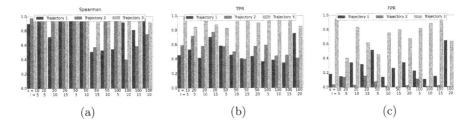

Fig. 6. This figure shows (a) the Spearman rank correlation coefficient, (b) the TPR and (c) the FPR when $cl = 20$.

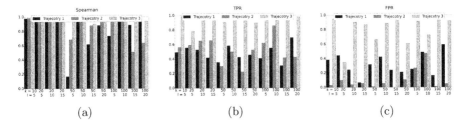

Fig. 7. This figure shows (a) the Spearman rank correlation coefficient and (b) the TPR and (c) FPR when $cl = 20$ and the start point and ending points are altered.

Sensitivity to Start and Endpoints. In the previous investigations, we assumed the real starting and endpoints to be given a priori. When working with synthetic data, this can be given without problem. But when working with real data, it is difficult to know exactly which points are the start and endpoints, i.e. know which points have no predecessor or successor in the trajectory. It is thus likely that a point with predecessors is given as starting point and points with successors as endpoints. To check if the algorithm can handle this scenario, we changed the real starting and end points. This is done by taking as start point, the point that is normally ordered as 30th, and as endpoints we pick the points that are normally ranked 30 places before the end.

The results are given in Fig. 7. If we compare these with the results obtained with the real endpoints (Fig. 6), we see that the Spearman rank correlation coefficient is as good, and that the TPR values range between the same intervals, but higher FPR values are more frequent. E.g. for the case when $k = 20, l = 15$ there is clearly an increase in the FPR in Fig. 7(b) when comparing to Fig. 6, and at the same time we also note that the corresponding TPR increases. This translates into the fact that more points are assigned to the trajectory, both good and wrong ones.

3.2 Real Data

Context. We evaluated our algorithm on the dataset visualized in Fig. 1, containing 4647 bone marrow cells. These are manually analysed and are known to differentiate from long-term hematopoietic stem cells (LT-HSC) into short-term hematopoietic stem cells (ST-HSC), which can in turn differentiate into either common myeloid progenitor cells (CMP) or common lymphoid progenitor cells (CLP). The cells can thus follow two trajectories, both of which are expected to include the hematopoietic stem cells. Our data set includes measurements for five surface markers: CD34, CD16/CD32, CD117, CD127, and Sca-1. Typically only a few of these marker values are used to define each of the distinct cell types, indicated in Fig. 8(a). By forming a trajectory from the unordered data, it can be checked if the known feature value changes correspond, and which other changes happen.

Evaluation. As the data is gated manually, we have cell annotations at our disposal. A simple way to visualise the quality of the resulting trajectories is to plot a density function which indicates the presence of a specific cell type in a specific place in the trajectory (using a sliding window approach). This leads to two plots, one for each trajecory, where the plots contain four curves representing the density of the four different cell types. This does not only show us if the cells are assigned to the correct trajectories, but also if the order in which cell types are assigned is as expected. Another way to visualise the resulting trajectories is to plot curves representing the marker values of the cells along the trajectory as done in Fig. 2(b). This again results in two plots, one for each trajectory, where each plot contains five curves, one for each marker value. A way to evaluate this result, is to compare these curves with the known marker presence shown in Fig. 8(a). For example, we can check if CD34 is low at first and then gradually increasing when the cells differentiate from LT-HSC to ST-HSC.

Results. As input, the algorithm takes one starting cell (a LT-HSC cell) and two end cells (a CLP and a CMP cell). These are chosen from the manual analysis, relying on the assumed values for a subset of markers. The start cell is the cell with the lowest value for CD34 and the highest value for CD117, the end cell representing CMP is the cell with highest CD34 and the end cell representing CLP is the cell with the highest CD117. As parameters we used $k = 100$, $l = 15$ and $cl = 10$. Changing the parameters did not have a big influence on the general trend of the results. The result of the algorithm is given in Fig. 8(b). On top, we see the density of the different cell types along the calculated trajectories. From this, we can conclude that the algorithm takes the correct cells for each trajectory. The trajectory ending in the CMP state contains LT-HSCs, ST-HSCs and CMPs but no CLPs, and these cell types are traversed in the expected order. The trajectory ending in the CLP state does contain a few CMPs, but is also mainly as expected from the assumed differentiation path in Fig. 8(a). When looking at the bottom of Fig. 8(b), we can also inspect the marker changes through differentiation. As expected, we see an increase in the CD34 intensity corresponding to the transition from LT-HSC to ST-HSC.

These results might help to gain new information as well. For example, there seems to be a subset in the CMPs which gains CD127, which is not used at all

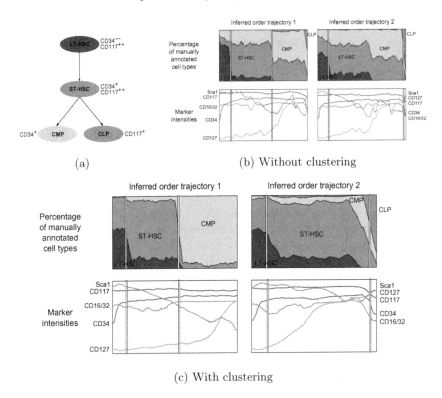

(a)   (b) Without clustering

(c) With clustering

Fig. 8. This figure shows (a) the theoretical branched trajectory, (b) the density of the different cell states and the feature values of the cells along the calculated branch going to CMP (on the left), and to CLP (on the right) when using $k = 100, l = 15$, and no clusters, and (c) the results when using 10 clusters. Only with clustering, the separation between the CMPs and CLPs is detected correctly.

in our manual gating. Often it is also very hard to know at which point a gate should be drawn if there is a gradual marker change. By looking at changes in other markers as well, a more informed decision might be taken.

The real data also emphasises the need for the usage of clusters for representing the endpoints. This is indicated in Fig. 8(c) where the density of the different cell states and the parameter values of the objects along the calculated trajectory is shown when no clusters are used. Comparing this with the previous result in Fig. 8(b), we see several mistakes are made, especially for trajectory 1 where many CMP cells are assigned in the CLP differentiation, resulting in a high FPR (Fig. 8(c) top left).

Note that for example in the top left of Fig. 8(b) the length of the interval where the CLP density is high, is very small in comparison to the length of the interval where the ST-HSC density is high. This has the simple reason that there are just far more ST-HSC-cells present in the data than CLP-cells (2192 versus 122). This phenomenon is also present in the bottom left of Fig. 8(b) where the feature values for CLP-cells are only shown in a short interval. A nice extension would thus be to automatically detect the different cell states and rescale the axis.

4   Conclusion

In this paper, we proposed a graph-based approach to infer multiple trajectories from unlabelled data. This is done by extending the Wanderlust algorithm [8] that was only able to model a single trajectory. Our algorithm extends Wanderlust by first assigning each cell to one or multiple trajectories. On each separate trajectory, Wanderlust is then executed. Results indicate that the assignment of the cells to the different trajectories and the ordering of the cells works well.

Future work includes finding the branch points and using this information to avoid wrong assignments. Knowing the branch points would also be beneficial for a better representation of the trajectories with a non-uniform density as this would allow rescaling of the trajectory. It is also important to take into account the fact that not all cells in the data might belong to the trajectories. While assigning the cells, such noisy data should be removed. This is a challenging extension that must be made as it will allow us to make use of data sources where not everything is known.

Acknowledgments. We would like to thank Lianne van de Laar and Bart Lambrecht for providing a biologically relevant dataset to test our algorithm. Sofie Van Gassen is funded by the Flanders Agency for Innovation by Science and Technology (IWT).

References

1. Qiu, P., Simonds, E.F., Bendall, S.C., Jr. Gibbs, K.D., Bruggner, R.V., Linderman, M.D., Sachs, K., Nolan, G.P., Plevritis, S.K.: Extracting a cellular hierarchy from high-dimensional cytometry data with SPADE. Nat. Biotechnol. **29**(10), 886–891 (2011)
2. Amir, E.D., Davis, K.L., Tadmor, M.D., Simonds, E.F., Levine, J.H., Bendall, S.C., Shenfeld, D.K., Krishnaswamy, S., Nolan, G.P., Pe'er, D.: viSNE enables visualization of high dimensional single-cell data and reveals phenotypic heterogeneity of leukemia. Nat. Biotechnol. **36**(6), 545–552 (2013)
3. Van Gassen, S., Callebaut, B., Van Helden, M.J., Lambrecht, B.N., Demeester, P., Dhaene, T., Saeys, Y.: FlowSOM: Using selforganizing maps for visualization and interpretation of cytometry data. Cytometry Part A (2015)
4. Magwene, P.M., Lizardi, P., Kim, J.: Reconstructing the temporal ordering of biological samples using microarray data. Bioinformatics **19**(7), 842–850 (2003)
5. Liiv, I.: Seriation and matrix reordering methods: an historical overview. Stat. Anal. Data Min. ASA Data Sci. J. **3**(2), 70–91 (2010)
6. Braak, T., Cajo, J.F., Van Tongeren, O.F.R.: Data Analysis in Community and Landscape Ecology. Cambridge University Press, Cambridge (1995)
7. Trapnell, C., et al.: The dynamics, regulators of cell fate decisions are revealed by pseudotemporal ordering of single cells. Nat. Biotechnol. **32**(4), 381–386 (2014)

8. Bendall, S.C., et al.: Single-cell trajectory detection uncovers progression andreg-ulatory coordination in human B cell development. Cell **157**(3), 714–725 (2014)
9. Trapnell, C., et al.: The dynamics and regulators of cell fate decisions are revealed by pseudotemporal ordering of single cells. Nat. Biotechnol. (2014)
10. Seita, J., Weissman, I.L.: Hematopoietic stem cell: selfrenewal versus differentia-tion. Wiley Interdisc. Rev. Syst. Biol. Med. **2**(6), 640–653 (2010)

Supervised Term Weights for Biomedical Text Classification: Improvements in Nearest Centroid Computation

Mounia Haddoud[1,2], Aïcha Mokhtari[2], Thierry Lecroq[1],
and Saïd Abdeddaïm[1(✉)]

[1] Laboratoire d'Informatique, du Traitement de l'Information et des
Systèmes (LITIS), Université de Rouen, 76821 Mont-Saint-Aignan Cedex, France
`mounia.haddoud@etu.univ-rouen.fr, said.abdeddaim@univ-rouen.fr`
[2] Recherche en Informatique Intelligente, Mathématiques et Applications (RIIMA),
USTHB, BP 32, El-Alia, Bab-ezzouar, 16111 Algiers, Algeria

Abstract. Maintaining accessibility of biomedical literature databases
has led to development of text classification systems to assist human
indexers by recommending thematic categories to biomedical articles.
These systems rely on using machine learning methods to learn the association between the document terms and predefined categories. The accuracy of a text classification method depends on the metric used in order
to assign a weight to each term. Weighting metrics can be classified as
supervised or unsupervised according to whether they use prior information on the number of documents belonging to each category. In this
paper, we propose two supervised weighting metrics (One-way Klosgen
and Loevinger) which both improve the quality of biomedical document
classification. We also show that by using moment generating function
centroids, an alternative to the traditional arithmetical average centroids,
a nearest centroid classifier with Loevinger metric performs significantly
better than SVM on a biomedical text classification task.

1 Scientific Background

Medical Subject Headings (MeSH), a controlled set of keywords, are used to
index all the article abstracts contained in the MEDLINE/PubMed database [1]
to facilitate search and retrieval. The increasing size of the MEDLINE/PubMed
needs efficient text classification tools to assist indexers in labeling document
texts with the predefined thematic categories of MeSH [2–4].

In the two last decades a huge number of machine learning techniques were
proposed to automatically classify text documents [5,6]. In text classifier systems, documents are preprocessed in order to be suitable as training data for a
learning algorithm. Traditionally, each text document is converted into a vector
where each dimension represents a term which value is the weight that will be
used in the learning process. As the weight reflects the importance of the term
in the document, an appropriate choice of the metric function used for weighting
terms is crucial for correct classification.

© Springer International Publishing Switzerland 2016
C. Angelini et al. (Eds.): CIBB 2015, LNBI 9874, pp. 98–113, 2016.
DOI: 10.1007/978-3-319-44332-4_8

Table 1. Abbreviations used in the present paper

Abbreviation	Definition
AAC	Arithmetical Average Centroid
BIN	Binary Term Weight
BNS	Bi-Normal Separation
CFC	Class Feature Centroid
GR	Gain Ratio
ICF	Inverse Class Frequency
IDF	Inverse Document Frequency
IG	Information Gain
IR	Information Retrieval
ITF	Inverse Term Frequency
JS	Jensen-Shannon divergence
KL	Kullback-Leibler divergence
LTF	Logarithm of the Term Frequency
MGFC	Moment Generating Function Centroid
OR	Odds Ratio
RF	Relevance Frequency
RTF	Raw Term Frequency
SVM	Support Vector Machine
TF	Term Frequency
TFW	Term Frequency Weight

Among text classification methods, the nearest centroid approach is one of the most popular supervised learning techniques due to its computational efficiency [7]. Nearest centroid method computes for each category a centroid prototype vector and classifies a document in the closest categories according to the similarity between their prototype vector and the document vector. When we consider the problem of term weighting in the context of nearest centroid classification, we have to determine how each term is weighted in each document vector representation (document term weighting), and how each term is weighted in the centroid prototype vector of each category (centroid term weighting). The performance of centroid-based classifiers is very sensitive to the metric adopted for weighting terms in both document and centroid prototype vectors. Our aim is to find a metric that is efficient for biomedical articles classification using the nearest centroid approach.

1.1 Document Term Weighting

Common term weighting metrics for text classification were unsupervised and generally borrowed from Information Retrieval (IR) field. The simplest IR metric is the binary representation BIN which assigns a weight of 1 if the term appears in the document and 0 otherwise. The term can be assigned a weight TF (Term Frequency) that reflect its frequency in the document. TFIDF is the most commonly used weighting metric in text classification. TFIDF is the product of TF and IDF, the Inverse Document Frequency which favors rare terms in the corpus over frequent ones[1]. However, there are some drawbacks on using unsupervised weighting functions, as the category information is omitted.

Previous studies proposed different supervised weighting metrics where the document frequency factor IDF of TFIDF is replaced by a factor that use prior information on the number of documents belonging to each category. Several classical metrics were tested in the literature, for instance, Chi-square (χ^2), Information Gain (IG), Gain Ratio (GR) and Odds Ratio (OR) [8]. These early studies get an improvement with TF.χ^2, TF.IG, TF.GR and TF.OR term weights trained with a Support Vector Machine (SVM). Accurate SVM text classification was obtained using Bi-Normal Separation (BNS) metric for supervised term weighting [9]. More recently, other specific metrics were proposed for the supervised term weighting problem. Lan et al. [10] utilizes a term weight TF.RF based of the Relevance Frequency (RF) metric. Altınçay and Erenel [11] combined RF metric with mutual information and the difference of term occurrence probabilities in the collection of the documents belonging to the category and in its complementary set.

1.2 Centroid Term Weighting

Given a set of documents belonging to the same category and their corresponding vector representations, the centroid prototype vector is traditionally defined as nothing more than the vector obtained by averaging the weights of the various terms present in the documents [12]. This vector is known as the arithmetical average centroid.

Guan et al. [7] proposed a Class Feature Centroid (CFC) term weighting that is not reducible to averaging or adding document term weights. The CFC weight of a given term belonging to a given category, depends on both inner-class term weight and inter-classes term weight. The inner-class weight measures the importance of a term within a category while the inter-classes weight measures is used to evaluate if a term is discriminative relatively to other categories. In this later work the document vectors are computed by unsupervised term weighting using the metric TFIDF. Recently, Ren et al. [13] compute arithmetical average cendroids based on the Inverse Class Frequency (ICF) metric. They also experimented ICS$_\delta$F a variant of ICF metric. ICF is exactly the inter-class part of the class feature centroid defined by Guan et al. [7] but Ren et al. used this metric as a term weight for document vectors. Nguyen et al. [14] proposed a centroid term weighting based on a moment generating function centroid which corresponds to the inner-class used by Guan et al. [7] in CFC. But instead of

---

[1] Table 1 contains the abbreviations used in this paper.

using the inter-class frequency, they use Kullback-Leibler divergence (KL) [15, 16] or Jensen-Shannon (JS) divergence measures [17].

1.3   Our Contribution

In this paper we propose two metrics, One-way Klosgen and Loevinger, as an alternative to the weight metrics described in the literature (see Table 2, Sect. 2.3). We use these metrics on a new term representation [18] where traditional bag-of-words document representation is extended by integrating term frequencies and term positions in the representation. We experimentaly study these metrics with two learning algorithms (SVM and nearest centroid) for biomedical text classification and show that both the metrics and the term representation can improve significantly the classification of Ohsumed biomedical documents [19]. In a first version of the present work [20] we used the traditional Arithmetical Average Centroids (AAC) and observed that SVM classification with One-way Klosgen weighting metric performs better than AAC with Loevinger metric. In this paper we improve the work by considering Moment Generating Function Centroids (MGFC) [14]. It appears from the experiments that using MGFC instead of AAC improves the performances of nearest centroid classifiers with Loevinger metric to the point that it makes centroid method performing significantly better than SVM on biomedical text classification task.

The rest of the paper is organized as follows. Section 2 describes the extended term representation we use, the supervised metrics we propose for term weighting and the corpus used as benchmark. Section 3 explains the way of computing the centroid prototype vectors. The experimental comparison on Ohsumed biomedical documents of our metrics with those proposed in the literature is presented in Sect. 4.

2   Materials and Methods

In the following we show how a text document can be seen as a vector where each dimension is a feature that represent either a term, a term frequency or a term position. Given this representation we describe 14 metrics for weighting these features, among them two (One-way Klosgen and Loevinger) are proposed by the authors of this paper.

2.1   Extended Term Representation

In this classical representation, terms are viewed as the dimensions of the learning space. A term may be a single word or a phrase (n-gram). In this work, we represent each dimension by a term together with its minimal frequency in the document. Let us consider for example, a particular term t such that 25 % of the documents where t appears are in category c. If 45 % of the documents where t appears at least 3 times are in category c, then the term t is probably more correlated with the category c when its frequency exceeds 2. Hence, we propose features of the form (t, n) in documents containing t with a term frequency at least n. If a document d contains ten times a term t, we must generate ten features (t, i) $(i = 1, 2, \ldots, 10)$, meaning that t occurs at least once, twice, $\ldots$, ten times. This could unnecessarily grow the number of features so we consider only n powers of 2. Then, if t occurs ten times, we will generate the features $(t, 1)$, $(t, 2)$, $(t, 4)$ and $(t, 8)$. The number of frequency features associated to a term t which appears n times in a document d will only be $\log_2 p$ in the worst case.

Most of the terms that are related to the main topics of a document occur at its beginning. In order to validate this assumption we propose features of the form (t, p), meaning that the first position of t in the document is lower or equal to p. The position being defined as the number of words preceding the term occurrence. As for term frequency features, we generate only features (t, p) with p powers of 2. For example, if a term t first appears at position 5 in a document of size 100 words, we generate the features $(t, 8)$, $(t, 16)$, $(t, 32)$ et $(t, 64)$, meaning that the first position of t is lower or equal than 8, 16, 32 and 64. The number of position features associated to a term t which appears in a document d at first position p will be $\log_2 |d|$ in the worst case, where $|d|$ is the size of d in number of words.

Table 2. Metrics used for supervised feature weighting. See Sect. 2.2 for the notations.

Metric	Mathematical form
Inverse Document Frequency (IDF)	$\log(\frac{N}{f(x*)})$
Pearson's χ^2 test	$\sum_{i,j} \frac{(f_{ij} - \bar{f}_{ij})}{\bar{f}_{ij}}$
Information gain	$\sum_{u \in \{x, \bar{x}\}} \sum_{v \in \{y, \bar{y}\}} p(uv) \log \frac{p(uv)}{p(u*)p(*v)}$
Odds ratio	$\frac{ad}{bc}$
Log odds ratio	$\log \frac{ad}{bc}$
Bi-normal separation	$|F^{-1}(p(x|y)) - F^{-1}(p(x|\bar{y}))|$ (*)
Pointwise mutual information	$\log \frac{p(xy)}{p(x*)p(*y)}$
Relevance frequency	$\log_2(2 + \frac{a}{\max(b,1)})$
Relevance frequency$_{OR}$	$\log_2(2 + \frac{a}{\max(b,1)})(1 - (p(x|y) - p(x|\bar{y})))$
Relevance frequency$_{\chi^2}$	$\log_2(2 + \frac{a}{\max(b,1)})|p(x|\bar{y}) - p(x|y)|$
Jensen-Shannon divergence	$-(p(*\bar{y})p(x|y) + p(*y)p(x|\bar{y})) \log(p(*\bar{y})p(x|y) + p(*y)p(x|\bar{y}))$
	$+p(*\bar{y})p(x|y) \log(p(x|y)) + p(*y)p(x|\bar{y}) \log(p(x|\bar{y}))$
Inverse Class frequency (ICF)	$\log(\frac{M}{f_c(x)})$
One-way Klosgen	$\sqrt{p(xy)}(p(y|x) - p(*y))$
Loevinger	$1 - \frac{p(x*)p(*\bar{y})}{p(x\bar{y})}$

(*) F^{-1} is the inverse Normal cumulative distribution function.

2.2   Notations

We consider a corpus D of N documents and d a particular document of D. Let x denotes a nominal feature of d representing either, (1) t a term that occurs in d, (2) (t, n) a term that occurs at least n times in d, or (3) (t, p) a term which first position is lower or equal to p in the document d.

Each document can belong to one or many categories (labels or classes) $c_1, c_2, \ldots, c_M$. We denote by y a particular category c_i. We denote by $\bar{x}$ the fact that the feature x is not present in d and by $\bar{y}$ the fact that d does not belong to the category y. The number of documents containing the feature x and belonging to the category y is denoted by $f(xy)$ and represents the document frequency. In general, $f(uv)$ denotes the number of documents containing u and belonging to v, u being x, $\bar{x}$ or $*$ (documents containing any term) and v being y, $\bar{y}$ or $*$ (documents belonging to any category). These frequencies are represented in the contingency Table (Table 3) in which the number of documents is denoted by N, $f(xy)$ by a and f_{11}, $f(x\bar{y})$ by b and f_{12}, and so on.

Table 3. Two-way contingency table for nominal feature x (term) and category y (document label). $f(uv)$ denotes the number of documents containing u and belonging to v. $*$ represents any term or category.

	y	$\bar{y}$	$*$
x	$f(xy) = a$	$f(x\bar{y}) = b$	$f(x*)$
$\bar{x}$	$f(\bar{x}y) = c$	$f(\bar{x}\bar{y}) = d$	$f(\bar{x}*)$
$*$	$f(*y)$	$f(*\bar{y})$	$f(**) = N$

Many metrics are based on the estimation of the probability $P(uv)$ the probability that a document containing u belongs to the category v, u being x, $\bar{x}$ or $*$ and v being y, $\bar{y}$ or $*$. Under the maximum-likelihood hypothesis this probability is estimated by: $p(uv) = \frac{f(uv)}{N}$. Some metrics are based on the difference between the observed and the expected frequencies. The expected contingency frequencies under the null hypothesis of independence H_0 are given in Table 4.

Table 4. Expected contingency table for nominal feature x and category y. $\hat{f}(uv)$ denotes the expected number of documents containing u and belonging to v under the null hypothesis of independence H_0.

	y	$\bar{y}$	$*$
x	$\hat{f}(xy) = \frac{f(x*)f(*y)}{N}$	$\hat{f}(x\bar{y}) = \frac{f(x*)(N-f(*y))}{N}$	$\hat{f}(x*)$
$\bar{x}$	$\hat{f}(\bar{x}y) = \frac{(N-f(x*))f(*y)}{N}$	$\hat{f}(\bar{x}\bar{y}) = \frac{(N-f(x*))(N-f(*y))}{N}$	$\hat{f}(\bar{x}*)$
$*$	$\hat{f}(*y)$	$\hat{f}(*\bar{y})$	N

The number of categories containing a document that contains a feature x is denoted by $f_c(x)$ and corresponds to:

$$f_c(x) = |\{y|f(x,y) > 0\}| \tag{1}$$

2.3 Weighting Metrics

Giving a weight to a feature x associated to a term in a document labeled with y depends on the correlation between x and y in the training corpus. This correlation can be estimated by different metrics, all the metrics used in this paper depend only on these values:

- N: the number of training documents
- $f(x*)$: the number of documents containing the feature x (feature marginal frequency)
- $f(*y)$: the number of documents containing belonging to the category y (category marginal frequency)
- $f(xy)$: the number of documents containing the feature x and belonging to the category y (joint frequency)
- $f_c(x)$: is the number of categories containing (a document that contains) feature x.
- M: the number of categories

The first 12 metrics of Table 2 are those already been used for the problem of term weighting in the literature [7–11, 13, 14, 21, 22]. Compared to the first version of the present work [20] we have added the inverse class frequency metric used in [7, 13], and also a metric used in [14] (see Sect. 1.2). In their paper Nguyen et al. [14] tried two metics Kullback-Leibler divergence (KL) and Jensen-Shannon (JS). In their experiments (confirmed by ours) JS measure performs better than KL, so we kept only the first metric.

The last 2 metrics of Table 2, Loevinger and One-way Klosgen, are proposed by the authors of this paper. These metrics are collected from papers dealing with association rules and classification rules [23] and have never been used for supervised term weighting in the literature.

2.4 Benchmark

In order to compare experimentally the metrics, we use the Ohsumed corpus. Ohsumed is a test collection that includes 13,929 medical abstracts (6,286 for training and 7,643 for testing) from MEDLINE/PubMed indexed by 23 cardiovascular diseases MeSH categories. Ohsumed is small when compared to the entire MEDLINE/PubMed corpus that contains over 21 million references indexed by 27,149 descriptors in 2014 MeSH. However it was necessary in the first instance to use a small dataset for all the experiments we have done, namely 120 learn/prediction tasks with 12 metrics, 5 different weighting schemes and two machine learning methods (see Tables 9 and 10 next section).

Table 5. True/False positive/negative instance definitions. Categories are denoted by 0 and 1.

		Real category	
		1	0
Predicted category	1	*TP* (true positive)	*FP* (false positive)
	0	*TN* (true negative)	*FN* (false negative)

We have done a summary preprocessing on these data and did not use feature selection in order to compare the weighting metrics independently from other methods of selecting the terms. Each document was stemmed (Porter stemming [24]) and reduced to a vector of features representing 1-grams or 2-grams terms. Traditionally the performance of a classifier on a corpus is estimated by learning the classification on the training data and evaluating the accuracy of the prediction obtained on the evaluation data. The evaluation metrics used are the *precision* which is the proportion of documents placed in the category that are really in the category, *recall* which is the proportion of documents in the category that are actually placed in the category. *precision* and *recall* are computed by counting the true/false positive/negative instances (see Table 5 and Eq. 2).

$$precision = \frac{TP}{TP+FP} \quad recall = \frac{TP}{TP+FN} \tag{2}$$

The F_1-Score is defined as:

$$F_1\text{-Score} = \frac{2 \cdot precision \cdot recall}{precision + recall} \tag{3}$$

The microaveraged F_1-Score is computed globally for all the categories, while the macroaveraged F_1-Score is the average of the F_1-Scores computed for each category. This later measures the ability of a classifier to perform well when the distribution of the categories is unbalanced, while the microaveraged F_1-Score gives a global view of the document classification performance.

2.5 SVM Classification

In the following we will use support vector machines for comparison with centroid classification. In a SVM, each instance (text in our case) is viewed as a p-dimensional vector and the goal is to know whether we can separate such points with a (p-1)-dimensional hyperplane. There are many hyperplanes that might separate the data. For a SVM algorithm the best hyperplane is the one that represents the largest separation (margin) between two binary classes (belongs to a category or not). For each category, we have used SVM binary classifier which learns a linear combination of the features in order to define the decision hyperplane. We adopted the SVMLight tool [25] with a linear kernel and used the default settings. Previous studies show that SVMLight performs well for text classification [7].

3   Nearest Centroid Classification

The nearest centroid approach is a popular supervised learning techniques due to its computational efficiency [7]. Nearest centroid method computes for each category a centroid prototype vector and classifies a document in the closest categories according to the similarity between their prototype vector and the document vector. When we consider the problem of term weighting in the context of nearest centroid classification, we have to determine how each term is weighted in each document vector representation (document term weighting), and how each term is weighted in the centroid prototype vector of each category (centroid term weighting). The performance of centroid-based classifiers is very sensitive to the metric adopted for weighting terms in both document and centroid prototype vectors. Our aim is to find a metric that is efficient for biomedical articles classification using the nearest centroid approach.

3.1   Document Term Weighting

In text classification each document d belongs to one or several categories in $\{c_1, c_2, \ldots, c_M\}$. For each category y, every document d is transformed to a vector:

$$W_d = (w_y(x_1, d), w_y(x_2, d), \ldots, w_y(x_n, d))$$ (4)

where each feature x is weighted by :

$$w_y(x, d) = w_{TF}(x, d) \times w_{DF}(x, y)$$ (5)

The term frequency weight $w_{TF}(x, d)$ depends on the frequency of x in the document d (see Table 6). The document frequency weight $w_{DF}(x, y)$ is one of the metrics described in Table 2.

Each feature x can be either:

– a term feature t in the classical model,
– or a term frequency feature (t, n) and/or a term position (t, p) feature as defined in Sect. 2.1.

For the classical term representation, following [10], we have experimented three possible term frequency weights: RTF, LTF or ITF (see Table 6). For our

Table 6. Experimented term frequency weights as a function of the frequency $tf(x, d)$ of a feature x in a document d

TFW	Value	Description
BIN(x,d)	1 if $tf(x, d) > 0$, 0 otherwise	Binary weight
RTF(x,d)	$tf(x, d)$	Raw term frequency
LTF(x,d)	$\log(1 + tf(x, d))$	Term frequency logarithm
ITF(x,d)	$1 - \frac{1}{1 - tf(x,d)}$	Inverse term frequency

term representation, we use only binary term weights ($w_{TF}(x,d) = \mathrm{BIN}(x,d)$), because the frequency of the term is already considered in the extended term representation $x = (t,n)$ or $x = (t,p)$.

3.2 Centroid Term Weighting

Nearest centroid method computes for each category y a prototype vector:

$$W_y = (w_y(x_1), w_y(x_2), \ldots, w_y(x_n)) \tag{6}$$

There are different ways to define the prototype vector W_y for category y. The Table 7 gives the centroid definitions we use in this work. The centroid prototype vector is traditionally defined as an Arithmetical Average Centroid (AAC). As we observed in Sect. 1.2, the Moment Generating Function Centroid (MGFC) [14] can be seen as a generalization of the Class Feature Centroid (CFC) [7], where:

- the inner-class part $e^{\frac{f(xy)}{f(*y)}}$ is generalized to any term frequency weight $w_{TF}(x,d)$,
- and the inter-class part (ICF metric) is replaced by any document frequency weight $w_{DF}(x,y)$.

Table 7. Centroid definitions

Name	Term weight
AAC	$w_y^{AAC}(x) = \frac{\sum_{d \in y} w_y(x,d)}{f(*y)}$
CFC	$w_y^{CFC}(x) = e^{\frac{f(xy)}{f(*y)}} \log(\frac{M}{f_c(x)})$
MGFC	$w_y^{MGFC}(x) = e^{\frac{\sum_{d \in y} w_{TF}(x,d)}{f(*y)}} w_{DF}(x,y)$

In the case of extended term representation we use binary term weights: $w_{TF}(x,d) = \mathrm{BIN}(x,d)$. This means that the Eq. 5 becomes:

$$w_y(x,d) = \mathrm{BIN}(x,d) \times w_{DF}(x,y) \tag{7}$$

In this particular case the centroid definitions presented in Table 7 can be simplified as shown in Table 8. For example in the case of the arithmetical average centroid we have:

$$w_y^{AAC}(x) = \frac{\sum_{d \in y} w_y(x,d)}{f(*y)} = \frac{\sum_{d \in y} \mathrm{BIN}(x,d) \times w_{DF}(x,y)}{f(*y)} \tag{8}$$

As $\sum_{d \in y} \mathrm{BIN}(x,d)$ correponds to the number of document containing the feature x and belonging to the category y, it can be writen $f(xy)$ using our notations, and we have:

$$w_y^{AAC}(x) = \frac{f(xy)}{f(*y)} w_{DF}(x,y) \tag{9}$$

3.3   Centroid Classification

The degree of closeness between the vector representing the document d and the prototype vector for category y is defined by a similarity measure $\text{SIM}(d, y)$. Generally the cosine similarity measure is used for this purpose:

$$\text{SIM}(d, y) = \frac{W_d . W_y}{\|W_d\| \, \|W_y\|} = \frac{\sum_{x \in d} w_y(x, d) w_y(x)}{\sqrt{\sum_{x \in d} w_y^2(x, d)} \sqrt{\sum_{x \in F} w_y^2(x)}} \tag{10}$$

Generally the cosine similarity measure is used for this purpose, however other measures could be used. In particular, Guan et al. [7] use a denormalized cosine similarity for comparing a CFC prototype with a document vector. We also tried other similarity measures, but cosine provide the best results.

Table 8. Centroid definitions in case of binary term weights

Name	Term weight
AAC	$w_y^{AAC}(x) = \frac{f(xy)}{f(*y)} w_{DF}(x, y)$
CFC	$w_y^{CFC}(x) = e^{\frac{f(xy)}{f(*y)}} \log(\frac{M}{f_c(x)})$
MGFC	$w_y^{MGFC}(x) = e^{\frac{f(xy)}{f(*y)}} w_{DF}(x, y)$

After the prototype vector for each category is obtained, nearest centroid method classifies a document d by taking the closest category y according to the similarity measure $\text{SIM}(d, y)$. In the case of multi-label classification a document can belong to multiple categories, in this case if we know the number of categories k associated to the document d, we take the k closest categories. We predict the number k by learning the number of categories associated to a document following the method used in [26].

4   Results

In order to estimate the performance of both our model and the two metrics we propose, we have compared the F_1-Score of SVM and nearest centroid classification on Ohsumed documents with classical and extended term representations using different weighting schemes.

For each document frequency weight metric w_{DF} we have experimented five weighting schemes:

- raw term frequency weight (w_{TF} = RTF) for term features t
- term frequency logarithm weight (w_{TF} = LTF) for term features t
- inverse term frequency weight (w_{TF} = ITF) for term features t
- binary term frequency weight (w_{TF} = BIN) for term frequency features (t, n)

– binary term frequency weight ($w_{TF} = $ BIN) for term frequency features (t, n) and term position features (t, p)

The Table 9 reports the microaveraged and macroaveraged F_1-Score obtained with SVM classification considering different term representations and weighting

Table 9. F_1-Scores of SVM classifier with different weighting metrics and term representations: classical term features t, term frequency features (t, n) and combined term frequency and term position features $(t, n)\&(t, p)$

Term representation	t	t	t	(t,n)	(t,n)&(t,p)
Term frequency weight	RTF	LTF	ITF	BIN	BIN
Microaveraged F_1-score					
One-way Klosgen	**0.587**	**0.604**	**0.609**	**0.631**	**0.639**
Pearson's χ^2	0.593	0.598	0.600	0.618	0.629
Odds ratio	0.563	0.582	0.590	0.617	0.629
Loevinger	**0.563**	**0.579**	**0.586**	**0.614**	**0.626**
Bi-normal separation	0.553	0.586	0.593	0.614	0.623
Information gain	0.570	0.583	0.586	0.603	0.615
Relevance frequency$_{\chi^2}$	0.548	0.568	0.571	0.590	0.602
Log odds ratio	0.497	0.545	0.556	0.587	0.600
Jensen-Shannon divergence	0.579	0.559	0.575	0.577	0.595
Relevance frequency$_{OR}$	0.475	0.531	0.541	0.571	0.588
Relevance frequency	0.460	0.521	0.535	0.564	0.583
Pointwise mutual information	0.459	0.520	0.533	0.566	0.582
IDF	0.296	0.363	0.380	0.417	0.444
ICF	0.250	0.306	0.319	0.361	0.371
Macroaveraged F_1-Score					
One-way Klosgen	**0.538**	**0.569**	**0.575**	**0.595**	**0.602**
Pearson's χ^2	0.553	0.562	0.568	0.587	0.598
Odds ratio	0.520	0.545	0.553	0.576	0.594
Loevinger	**0.518**	**0.541**	**0.550**	**0.580**	**0.590**
Information gain	0.529	0.547	0.550	0.560	0.578
Relevance frequency$_{\chi^2}$	0.501	0.522	0.524	0.540	0.565
Bi-normal separation	0.468	0.513	0.521	0.552	0.564
Jensen-Shannon divergence	0.498	0.519	0.520	0.537	0.552
Log odds ratio	0.401	0.461	0.476	0.510	0.534
Relevance frequency$_{OR}$	0.384	0.450	0.462	0.504	0.523
Relevance frequency	0.365	0.435	0.453	0.486	0.515
Pointwise mutual information	0.353	0.421	0.439	0.480	0.501
IDF	0.185	0.237	0.255	0.289	0.319
ICF	0.141	0.180	0.192	0.230	0.241

Table 10. F_1-Scores of arithmetical average centroid (AAC) classifier with different weighting metrics and term representations: classical term features t, term frequency features (t, n) and combined term frequency and term position features $(t, n)\&(t, p)$

Term representation	t	t	t	(t,n)	(t,n)&(t,p)
Term frequency weight	RTF	LTF	ITF	BIN	BIN
Microaveraged F_1-Score					
ICF	0.502	0.579	0.605	0.634	0.646
Loevinger	**0.585**	**0.603**	**0.610**	**0.620**	**0.628**
IDF	0.490	0.564	0.584	0.604	0.616
Pointwise mutual information	0.506	0.548	0.563	0.585	0.593
Log odds ratio	0.516	0.547	0.559	0.577	0.590
Odds ratio	0.515	0.527	0.530	0.547	0.563
Relevance frequency	0.394	0.465	0.489	0.513	0.532
Bi-normal separation	0.461	0.491	0.500	0.518	0.531
Relevance frequency$_{OR}$	0.388	0.456	0.479	0.503	0.518
One-way Klosgen	**0.468**	**0.481**	**0.486**	**0.495**	**0.502**
Pearson's χ^2	0.444	0.452	0.456	0.463	0.468
Jensen-Shannon divergence	0.381	0.400	0.408	0.420	0.422
Information gain	0.345	0.346	0.347	0.357	0.373
Relevance frequency$_{\chi^2}$	0.319	0.334	0.335	0.325	0.330
Macroaveraged F_1-Score					
ICF	0.466	0.547	0.580	0.617	0.627
Loevinger	**0.568**	**0.584**	**0.591**	**0.604**	**0.611**
IDF	0.441	0.536	0.560	0.590	0.603
Log odds ratio	0.492	0.535	0.548	0.570	0.579
Pointwise mutual information	0.476	0.528	0.545	0.570	0.577
Odds ratio	0.517	0.524	0.526	0.543	0.564
Relevance frequency	0.384	0.463	0.491	0.522	0.542
Relevance frequency$_{OR}$	0.390	0.458	0.481	0.509	0.526
Bi-normal separation	0.439	0.476	0.484	0.505	0.513
One-way Klosgen	**0.457**	**0.468**	**0.473**	**0.485**	**0.491**
Pearson's χ^2	0.440	0.444	0.446	0.454	0.462
Jensen-Shannon divergence	0.391	0.399	0.401	0.407	0.414
Information gain	0.366	0.375	0.384	0.400	0.401
Relevance frequency$_{\chi^2}$	0.337	0.352	0.368	0.360	0.365

Table 11. F_1-Scores of moment generating function centroid (MGFC) classifier with (t,n) &(t,p) term representation and different weighting metrics

Microaveraged F_1-Score		Macroaveraged F_1-Score	
Loevinger	**0.667**	**Loevinger**	**0.647**
Odds ratio	0.665	Odds ratio	0.641
Log odds ratio	0.642	Log odds ratio	0.635
Bi-normal separation	0.610	Bi-normal separation	0.617
One-way Klosgen	**0.609**	Bi-normal separation	0.617
Pointwise mutual information	0.560	**One-way Klosgen**	**0.593**
Pearson's χ^2	0.546	Relevance frequency$_{OR}$	0.553
ICF	0.539	Relevance frequency	0.535
Gain ratio	0.518	Pearson's χ^2	0.533
Information gain	0.518	ICF	0.524
Relevance frequency$_{\chi^2}$	0.456	Gain ratio	0.500
Relevance frequency$_{OR}$	0.371	Information gain	0.500
Jensen-Shannon divergence	0.347	IDF	0.482
Relevance frequency	0.342	Relevance frequency$_{\chi^2}$	0.472
IDF	0.332	Jensen-Shannon divergence	0.409

metrics (the five table columns represent the five weighting schemes). After calculation of the F_1-Score for each classifier, the metrics are ranked in descending order of the best weighting scheme score. It can be seen that by using the One-way Klosgen metric we obtain the best classification scores on Ohsumed data. It is also clearly observed from these results that the proposed representation model (t, n) & (t, p) performs significantly better than the classical representation (the three first columns) and achieves the best performances in all experiments in terms of microaveraged F1-scores for all the metrics.

Table 10 provides the F_1-Scores obtained with nearest centroid classifier that use arithmetical average centroids (AAC). It can be seen that ICF metric obtain the best classification scores on Ohsumed data. However, as we can see in Table 11, by using moment generating function centroids (MGFC), rather than AAC, our proposed metric Loevinger performs significantly better than arithmetical average centroids with ICF and SVM with One-way Klosgen metric. considering both microaveraged and macroaveraged F1-scores. We can also notice from these experiments on Ohsumed data that with Odds ratio metric we obtain results that are close to Loevinger metric when we use MGFC. At our knowledge, Odds ratio was only used with AAC in the literature and the accuracy of this metric was not known with MGFC that was only used with ICF, Jensen-Shannon and Kullback-Leibler metrics.

5   Conclusion

In this paper, we have proposed two term weighting metrics that have not been previously used for term weighting in the literature. We showed that these metrics improve significantly the classification of Ohsumed biomedical documents. We also showed that by using moment generating function centroids, a nearest centroid classifier with Loevinger metric performs significantly better than SVM on a biomedical text classification task. In another work [18] we show that using a SVM classifier which combines the outputs of SVM classifiers that utilize different metrics improves the classification. However, this approach is time-consuming and cannot be used for huge text classification problems. At the contrary, the present work intend to assess our approach with large-scale experiments on all MEDLINE/PubMed corpus with all MeSH categories. We can achieve this goal in a future work thanks to efficiency of the nearest centroid classification method.

References

1. MEDLINE/PubMed. http://www.ncbi.nlm.nih.gov/pubmed
2. Wahle, M., Widdows, D., Herskovic, J.R., Bernstam, E.V., Cohen, T.: Deterministic binary vectors for efficient automated indexing of medline/pubmed abstracts. In: AMIA Annual Symposium Proceedings, vol. 2012, p. 940. American Medical Informatics Association (2012)
3. Huang, M., Névéol, A., Zhiyong, L.: Recommending mesh terms for annotating biomedical articles. J. Am. Med. Inf. Assoc. **18**(5), 660–667 (2011)
4. Vasuki, V., Cohen, T.: Reflective random indexing for semi-automatic indexing of the biomedical literature. J. Biomed. Inf. **43**(5), 694–700 (2010)
5. Sebastiani, F.: Machine learning in automated text categorization. ACM Comput. Surv. **34**(1), 1–47 (2002)
6. Aggarwal, C.C., Zhai, C.: A survey of text classification algorithms. In: Aggarwal, C.C., Zhai, C. (eds.) Mining Text Data, pp. 163–222. Springer, New York (2012)
7. Guan, H., Zhou, J., Guo, M.: A class-feature-centroid classifier for text categorization. In: Quemada et al. [27], pp. 201–210
8. Debole, F., Sebastiani, F.: Supervised term weighting for automated text categorization. In: Proceedings of the ACM Symposium on Applied Computing (SAC), March 9–12, Melbourne, FL, USA, pp. 784–788. ACM (2003)
9. Forman, G.: BNS feature scaling: an improved representation over tf-idf for svm text classification. In: Shanahan, J.G., Amer-Yahia, S., Manolescu, I., Zhang, Y., Evans, D.A., Kolcz, A., Choi, K.-S., Chowdhury, A., (eds.) Proceedings of the 17th ACM Conference on Information and Knowledge Management, CIKM 2008, Napa Valley, California, USA, October 26–30, pp. 263–270. ACM (2008)
10. Lan, M., Tan, C.L., Su, J., Lu, Y.: Supervised and traditional term weighting methods for automatic text categorization. IEEE Trans. Pattern Anal. Mach. Intell. **31**(4), 721–735 (2009)
11. Altınçay, H., Erenel, Z.: Using the absolute difference of term occurrence probabilities in binary text categorization. Appl. Intell. **36**(1), 148–160 (2012)
12. Han, E.-H.S., Karypis, G.: Centroid-based document classification: analysis and experimental results. In: Zighed, D.A., Komorowski, J., Żytkow, J.M. (eds.) PKDD 2000. LNCS (LNAI), vol. 1910, pp. 424–431. Springer, Heidelberg (2000)

13. Ren, F., Sohrab, M.G.: Class-indexing-based term weighting for automatic text classification. Inf. Sci. **236**, 109–125 (2013)
14. Nguyen, T.T., Chang, K., Hui, S.C.: Supervised term weighting centroid-based classifiers for text categorization. Knowl. Inf. Syst. **35**(1), 61–85 (2013)
15. Leibler, A.: On information and sufficiency. Ann. Math. Stat. **22**(1), 79–86 (1951)
16. Ali, S.M., Silvey, S.D.: A general class of coefficients of divergence of one distribution from another. J. R. Stat. Soc. Ser. B (Methodological), pp. 131–142 (1966)
17. Lin, J.: Divergence measures based on the shannon entropy. IEEE Trans. Inf. Theor. **37**(1), 145–151 (2006)
18. Haddoud, M., Mokhtari, A., Lecroq, T., Abdeddaïm, S.: Combining supervised term weighting metrics for SVM text classification with extended term representation. Knowl. Inf. Syst. 1–23 (2016)
19. Hersh, W., Buckley, C., Leone, T.J., Hickam, D.: OHSUMED: an interactive retrieval evaluation and new large test collection for research. In: Croft, B.W., van Rijsbergen, C.J. (eds.) SIGIR '94, pp. 192–201. Springer, London (1994)
20. Haddoud, M., Mokhtari, A., Lecroq, T., Abdeddaïm, S.: Supervised term weights for biomedical text classification. In: Proceedings of the 12th International Meeting on Computational Intelligence Methods for Bioinformatics and Biostatistics, CIBB, Naples, Italy, September 10–12, pp. 55–60 (2015)
21. Deng, Z.-H., Tang, S., Yang, D., Li, M.Z.L.-Y., Xie, K.-Q.: A comparative study on feature weight in text categorization. In: Yu, J.X., Lin, X., Lu, H., Zhang, Y. (eds.) APWeb 2004. LNCS, vol. 3007, pp. 588–597. Springer, Heidelberg (2004)
22. Liu, Y., Loh, H.T., Sun, A.: Imbalanced text classification: a term weighting approach. Expert Syst. Appl. **36**(1), 690–701 (2009)
23. Geng, L., Hamilton, H.J.: Interestingness measures for data mining: a survey. ACM Comput. Surv. **38**(3), 9 (2006)
24. Porter, F.M.: An algorithm for suffix stripping. Program **14**(3), 130–137 (1980)
25. Joachims, T.: Making large-scale SVM learning practical. In: Schölkopf, B., Burges, C., Smola, A. (eds.) Advances in Kernel Methods-Support Vector Learning, pp. 169–184. MIT Press, Cambridge (1999). Chap. 11
26. Tang, L., Rajan, S., Narayanan, V.K.: Large scale multi-label classification via metalabeler. In: Quemada et al. [27], pp. 211–220
27. Quemada, J., León, G., Maarek, Y.S., Nejdl, W., (eds.) Proceedings of the 18th International Conference on World Wide Web (WWW 2009), Madrid, Spain, 20-24 April 2009. ACM (2009)

Alignment Free Dissimilarities
for Nucleosome Classification

Giosué Lo Bosco[1,2(✉)]

[1] Dipartimento di Matematica e Informatica,
Università Degli Studi di Palermo, Palermo, Italy
giosue.lobosco@unipa.it
[2] Dipartimento di Scienze per L'Innovazione e le Tecnologie Abilitanti,
Istituto Euro Mediterraneo di Scienza e Tecnologia, Palermo, Italy

Abstract. Epigenetic mechanisms such as nucleosome positioning, histone modifications and DNA methylation play an important role in the regulation of cell type-specific gene activities, yet how epigenetic patterns are established and maintained remains poorly understood. Recent studies have shown a role of DNA sequences in recruitment of epigenetic regulators. For this reason, the use of more suitable similarities or dissimilarity between DNA sequences could help in the context of epigenetic studies. In particular, alignment-free dissimilarities have already been successfully applied to identify distinct sequence features that are associated with epigenetic patterns and to predict epigenomic profiles. In this work, we focalize the study on the problem of nucleosome classification, providing a benchmark study of 6 alignment free dissimilarity measures between sequences, belonging to the categories of *geometric-based*, *correlation-based*, *information-based* and *compression based*. Their comparisons have been done versus an alignment based dissimilarity, by measuring the performance of several nearest neighbour classifiers that incorporate each one the considered dissimilarities. Results computed on three dataset of nucleosome forming and inhibiting sequences, shows that among the alignment free dissimilarities, the geometric and correlation are the more suitable for the purpose of nucleosome classification, making them a more efficient alternative to the alignment-based similarity measures, which nevertheless are yet the preferred choice when dealing with sequence similarity measurements.

Keywords: k-mers · L-tuples · Alignment free DNA sequence dissimilarities · Nucleosome classification · Epigenetic · Knn classifier

1 Introduction

The primary repeating unit of chromatin is the nucleosome, which consists of 147 bp of DNA wrapped 1.67 times around an octamer of core histone proteins [1]. The N-terminal ends of the histones are unstructured and called the histone tails. Many amino acid residues on the histone tails can be covalently modified,

© Springer International Publishing Switzerland 2016
C. Angelini et al. (Eds.): CIBB 2015, LNBI 9874, pp. 114–128, 2016.
DOI: 10.1007/978-3-319-44332-4_9

and many of these modifications have distinct biological functions [2]. The field of epigenomics has been growing rapidly in the past few years, in part because of the development of genome-wide profiling technologies and new specialized computational approaches [3–5]. The underlying mechanisms for the establishment and maintenance of cell type-specific epigenetic patterns are complex and involve the dynamic interactions among multiple classes of factors [6], whose relative contribution remains poorly understood. One fundamental question is to what extent the epigenomic patterns are orchestrated by the underlying DNA sequence. Previous studies have shown that, for the case of nucleosome positioning, the DNA sequence plays an important role [7], so that methods based on sequence similarity have been developed to predict genome-wide patterns, sometimes with great accuracy. In particular, initial biological hypotheses about sequence similarity has been traditionally generated by using sequence alignment methods. Several algorithms that target specific goals such as global alignment, local alignment, with or without overlapping have been proposed [8–10]. The main issue of alignment methods is that their computational complexity escalates as a power function of the length of the related sequences. Despite the recent efforts in improving their computational efficiency [11,12], the applications of alignment methods are not unlimited. They are based on the main assumption that functional elements are related to sequence substrings whose relative order is also conserved. Unfortunately there are cases showing that this can be violated, such as the cis-regulatory element sequences where there is little evidence suggesting that the order between different elements would have any significant effect in regulating gene expression. In the case of nucleosome identification, it has been demonstrated that the most informative sequence features are those that are traditionally viewed as degenerative, such as CpG density and poly-A tract [13], posing a severe challenge for alignment-based methods. In the meantime, the development of alignment-free methods [14] has provided a promising alternative to overcoming such challenges. The interested reader can find a review about several applications of alignment free methods in the field of epigenomics in the work by Pinello et al. [15]. In this work we present a study about nucleosome classification, providing a benchmark study of 6 alignment free dissimilarity measures between sequences, when incorporated into a nearest neighbour classifier. The used dissimilarities belong to the categories of *geometric-based*, *correlation-based*, *information-based* and *compression based*. The performance of the resulting nearest neighbour classifiers are compared with a classifier that incorporate an alignment based dissimilarity. The paper is organized as follows: in the next section such dissimilarities will be formally described, in Sect. 3 the experimental design and related results will be presented and discussed. Concluding remarks will be given in Sect. 4.

2 Dissimilarity Functions

Let $\mathbb{X}$ be a set. A function $\delta : \mathbb{X} \times \mathbb{X} \to \mathbb{R}$ is a *dissimilarity* on $\mathbb{X}$ if, $\forall\, \mathbf{x}, \mathbf{y} \in \mathbb{X}$, it satisfies the following three conditions:

1. $\delta(\mathbf{x}, \mathbf{y}) \geq 0$ (*non-negativity*);
2. $\delta(\mathbf{x}, \mathbf{y}) = \delta(\mathbf{y}, \mathbf{x})$ (*symmetry*);
3. $\delta(\mathbf{x}, \mathbf{x}) = 0$;

A generic DNA sequence $s \in \Sigma^*$ is a string of finite length whose symbols are taken in the nucleotide alphabet $\Sigma = \{A, T, C, G\}$. An alignment based dissimilarity between two DNA sequences s and t is a dissimilarity that make use of an alignment algorithm, conversely an alignment free dissimilarity is established without making us of any alignment process. In general, we can think to an alignment free dissimilarity, as a dissimilarity computed on a particular co-domain $X = \phi(\Sigma^*)$ mapped by a particular mapping function ϕ.

2.1 Dissimilarities Computed by Alignment

The *alignment based* class of dissimilarities are defined by making use of an *alignment algorithm* [16]. Alignment algorithms are mainly categorized into global and local methods. The first category is related to the solution of a global optimization problem that imposes the alignment algorithm to span the entire length of the two sequences, and is most useful when the sequences are similar and quite of equal size. The local alignment algorithms identify instead regions of similarity within long sequences that are often widely divergent overall. They are more difficult to compute because of the additional challenge of identifying the similarity regions. In both cases, after the alignment process, every symbol at each position of the two strings can be the same (match case), can be different (mismatch case) or one of the letters aligns to a *gap* in the other sequence (insertion or deletion case). Assuming that it is possible to assign a predefined score to each kind of mismatch, insertion or deletion errors, alignment algorithms search for the alignment of highest total score. A common way of defining an alignment based dissimilarity (ABD for short) is by using the score related to the alignment in the following way:

$$ABD(s, t) = \left(1 - \frac{score(s, t)}{score(s, s)}\right) \times \left(1 - \frac{score(s, t)}{score(t, t)}\right) \qquad (1)$$

Note that in the rare case where the score between sequences is greater than the score when aligning a sequence with itself, one assumes the dissimilarity equal to zero.

2.2 Alignment Free Dissimilarities Computed on L-tuple Codomain

In the case of L-tuple codomain representation, the mapping function ϕ project s into a vector $\mathbf{x}_s$ that enumerates the frequency of occurrence of a finite set of fixed length words $W = \{w_i, .., w_m\}$ in the string s. W is the set of *L-tuples* (or *k-mers*), i.e. a set containing any string of fixed length L whose symbols are taken in the nucleotide alphabet $\Sigma = \{A, T, C, G\}$. As result, each sequence s will be mapped to a vector $\mathbf{x}_s \in \mathbb{R}^m$ with $m = 4^L$, such that the component $(\mathbf{x}_s)_i$

counts the occurrence of the $i-th$ L-tuple into the string s. The counting process uses a window of length L that is run by steps of one through the sequence, from string position 1 to $M - L + 1$, being M the length of the string s.

The L tuple representation has shown to be very effective for classification of several kind of DNA sequences [17–20] and for the specific case of nucleosomic related sequences it represents an effective representation for the best performing nucleosome classification methodologies [5,21]. Anyway, it is also important to remark that this representation involves an exponential space complexity with respect to the length L of the representation. To this purpose, several solutions for the selection of relevant k-mers have been proposed [22,23] also for the specific case of nucleosome identification [27]. Anyway, L represents a parameter of the representation. In the euclidean space $\mathbb{R}^m$, it is possible to categorize dissimilarity functions according to three broad classes: *geometric-based*, *correlation-based* and *information-based*. Functions in the first class capture the concept of *physical* distance between two objects. They are strongly influenced by the magnitude of changes in the measured components of vectors $\mathbf{x}$ and $\mathbf{y}$, making them sensitive to noise and outliers. Functions in the second class capture dependencies between the coordinates of two vectors. In particular, they usually have the benefit of capturing positive, negative and linear relationships between two vectors. Functions in the third class are defined via well known quantities in information theory such as entropy and mutual information. They have the advantage of capturing statistical dependencies between two data points, even if they are not linear. We now formally define the functions of interest for this work, starting with the geometric ones.

The *Euclidean* or *2-norm* dissimilarity is defined as follows:

$$d_e(\mathbf{x}, \mathbf{y}) = \sqrt[2]{\sum_{i=1}^{m}(x_i - y_i)^2} \tag{2}$$

where $\mathbf{x} = (x_1, \ldots, x_m)$, $\mathbf{y} = (y_1, \ldots, y_m)$, where m represents the number of components.

Among the correlation-based dissimilarities, the most known is the *Pearson* disssimilarity d_r:

$$d_r(\mathbf{x}, \mathbf{y}) = 1 - r = 1 - \frac{\sum_{i=1}^{m}(x_i - \bar{x})(y_i - \bar{y})}{\sum_{i=1}^{m}(x_j - \bar{x})^2 \sum_{j=1}^{m}(y_j - \bar{y})^2} \tag{3}$$

where $\bar{x} = \frac{1}{m}\sum_i x_i$, $\bar{y} = \frac{1}{m}\sum_i y_i$. The *Cosine Distance*, is another example of correlation-based dissimilarity, and can be defined as:

$$d_{cos}(\mathbf{x}, \mathbf{y}) = 1 - \left(\frac{\mathbf{x} \cdot \mathbf{y}}{\sqrt{\mathbf{x} \cdot \mathbf{x}}\sqrt{\mathbf{y} \cdot \mathbf{y}}}\right) \tag{4}$$

that corresponds to 1 minus the cosine of the angle between the two vectors $\mathbf{x}$ and $\mathbf{y}$.

Finally, the symmetrical Kullback-Leibler dissimilarity between two vectors **x** and **y** belongs to the class of information-based dissimilarities, and is so defined:

$$d_{kl}(\mathbf{x}, \mathbf{y}) = \frac{\sum_i p_{x_i} log_2 \frac{p_{x_i}}{p_{y_i}} + \sum_j p_{y_j} log_2 \frac{p_{y_j}}{p_{x_j}}}{2} \qquad (5)$$

where $p_{x_i} = \frac{x_i}{M-L+1}$. This dissimilarity is able to measure the difference between the probability distributions of the L-tuple representation of two sequences.

2.3   Alignment Free Dissimilarities Computed by Compression Measures

The *compression based* class of dissimilarities are defined by making use of a *string compressor* to be considered as an auto-mapping C of Σ^*. They are based on the concept of Kolmogorov complexity of a string, which intuitively can be viewed as a measure of the computational resources needed to generate such string. Formally, the conditional Kolmogorov complexity $KC(s|t)$ between two generic strings s and t is the length of the shortest binary program written in a generic programming language that computes s giving t as input. The Kolmogorov complexity $KC(s)$ of a string s is defined as $KC(s|\lambda)$ where λ stands for the empty string. Li et al. [24] have defines the so called *Universal similarity metric* (USM for short) for strings based on the Kolmogorov complexity:

$$USM(s, t) = \frac{max\{KC(s|t^*), KC(t|s^*)\}}{max\{KC(s), KC(t)\}} \qquad (6)$$

where w^* denotes the shortest program that produces the generic sequence w on an empty input. Unfortunately, USM is not computable; therefore, it is commonly approximated by the *Universal Compression Dissimilarity* whose definition is in the following :

$$UCD(s, t) = \frac{max\{|C(st)| - |C(s)|, |C(ts)| - |C(t)|\}}{max\{|C(s)|, |C(t)|\}} \qquad (7)$$

where st and ts denote the concatenations of the sequences s and t, C is the mapping associated to a compression algorithm and $C(s)$ its output on a string s, $|.|$ denotes the length of a string. Details about UCD and other approximation of USM can be found in the work of Ferragina et al. [25]. The main advantage on using *compression based* dissimilarities, is that they are able to measure the combinatorial properties of a string. Note that in the case of nucleosome classification, combinatorial properties such as periodicities of particular dinucleotides [21] have been observed in several species. This class of dissimilarities have been used for general sequence classification problem [26] and also to distinguish nucleosome-enriched and depleted regions [27].

3   Experiments and Results

3.1   Materials and Methods

For the experiments we have considered three datasets of DNA sequences under-lying nucleosomes belonging to the following three species: (i) *Homo sapiens (HM)*; (ii) *Caenorhabditis elegans (CE)* and (iii) *Drosophila melanogaster (DM)*. Details about all the steps of data extraction and filtering of the three datasets can be found in the work by Guo et al. [5] and in the references therein. Each of the three datasets is composed by two classes of samples: the nucleosome-forming sequence samples (positive data) and the linkers or nucleosome-inhibiting sequence samples (negative data). The HM dataset con-tains $2,273$ positives and $2,300$ negatives, the CE $2,567$ positives and $2,608$ neg-atives and the DM $2,900$ positives and $2,850$ negatives. The length of a generic sequence is 147 bp. The three dataset can be downloaded from the url that the authors have provided [28]. All the dissimilarities have been compared in terms of classification performances of a related K *nearest neighbor* (Knn) classifier. This kind of classifier represents the more suitable for dissimilarity comparisons because, apart form the dissimilarity, it only makes use of the notion of neigh-bourhood. For completeness, in the following we formally describe how it works. Let R be the number of classes, T_i the training set of elements for a class i and δ a distance between elements. Let $Y(\mathbf{x})$ be the set of K elements closest to an unlabelled element $\mathbf{x}$ with respect to δ, then the Knn classifier assign to $\mathbf{x}$ the class j using the following assignment rule:

$$j = \arg\max_{1 \leq i \leq R}(|Y(\mathbf{x}) \cap T_i|) \tag{8}$$

This rule means that the unlabelled element x is classified by assigning the label which is most frequent among the K training samples nearest to that point. Note that K is a parameter of the method, and its best choice depends upon the data. Generally, larger values of K are preferred since the effect of noise on the classification can be reduced.

In our experiment, we have computed a total of 3 performance measure for the Knn classifier related to a particular dissimilarity: *Accuracy (A), Sensitivity (Se) and Specificity (Sp)*. In the following, we recall their definitions:

$$A = \frac{TP+TN}{TP+FN+FP+TN}, \quad Se = \frac{TP}{TP+FN}, \quad Sp = \frac{TN}{FP+TN} \tag{9}$$

where the prefix T (true) indicates the number of correctly classified sequences, F (false) the incorrect ones, P the positives class and N the neg-atives class.

3.2   Results

Starting from a dataset S of n sequences, the training and test of the Knn classifier have been selected using a 10 fold cross validation schema. For all the considered dissimilarities, except for the alignment and compression based ones, we have used as numerical dataset for the classifier a matrix D_S of size $n \times 4^L$, such that $(D_S)_i^j$ stores the counting c_i^j of the $j - th$ L-tuple w_j into a sequence s_i of the dataset. In this case, we have computed the experiments for different L ranging from 2 to 6. For the alignment and compression based dissimilarities we have used the set S directly. Regarding the K value of the Knn classifier, we have decided to compute experiments for each odd value (the number of classes is $R = 2$) ranging in the integer interval $\{1, ., 21\}$.

We have computed the accuracy results of 7 Knn classifiers, referred as *Knn-ABD-nw, Knn-euclidean, Knn-correlation, Knn-cosine, Knn-Kullback, Knn-UCD-deflate* and *Knn-UCD-gzip*. The first classifier is referred to a knn classifier that incorporates the ABD distance (see Subsect. 2.1), with Needlmenan-Wunsch as global alignment algorithm [8]. The choice of this global alignment algorithm is due to the equal size of all the sequences, i.e. 147 bp as mentioned before. The other classifiers are related to the alignment free distances defined in Subsects. 2.2 and 2.3. The last two differentiate themselves in the use of the compression algorithm, i.e. the deflate compression algorithm (*UCD-defalte*) and the gzip compression algorithm (*UCD-gzip*).

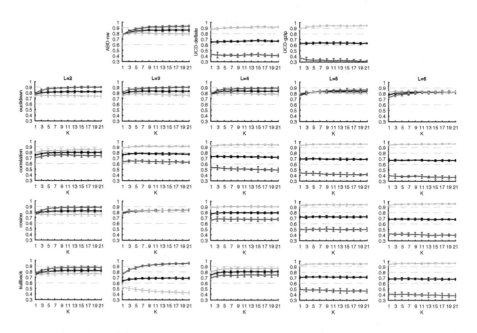

Fig. 1. Accuracy (black), Sensitivity (gray) and Specificity (light gray) plots of the Knn classifiers, for different neighbors K in the range $\{1, ., 21\}$, and for different L-tuple lengths L in the range $\{2, ., 6\}$ in the case of Caenorhabditis elegans (CE) dataset. Error bars for each K are also reported. The first three plots in the first line are referred to ABD-nw, UCD-deflate and UCD gzip and does not depend on the L-tuple length L.

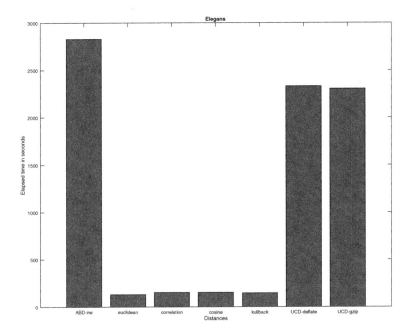

Fig. 2. Elapsed times in seconds of each classifier for the classification of one fold in the case of the Caenorhabditis elegans (CE) dataset

Figures 1, 3 and 5 show the mean accuracy, sensitivity, specificity values computed among all the considered neighbours K, for each values of $L = \{2, ., 6\}$. Note that ABD and UCD dissimilarities are independent from the L parameter.

Table 1 show mean (μ) and standard deviation (σ) of the 3 performance measure (in percentage) reached by each one of the classifiers, eventually for different L (L-tuple length). In the case of ABD and UCD in consideration of their independence from L, the results are replicated. Accuracies, Sensitivities and Specificities of the two best performers are depicted in bold. We have also verified that the difference between the performance metrics of the best classifiers versus the others are statistically significant by computing a two sample paired t-test with significance level of 5 %. It means that a bold value in the table show that the corresponding classifier has reached the best value of the performance metric, that is also significantly different from all the others with a p-value lower than 0.05 (maximum p-value $= 2,7 \times 10^{-4}$).

The elapsed times in seconds of each knn-classifier for the classification of one fold are shown in Figs. 2, 4 and 6.

Results show that the alignment based distance ABD-nw is the best in terms of accuracy and sensitivity for CE and DM datasets, but not for the case of HM dataset. Anyway, such superiority in terms of accuracy versus the alignment free dissimilarity (at most 2 % with the euclidean), seems to be not justified by their ratio of elapsed computation times. In fact knn-euclidean is more than 20 times

Table 1. In column, for each L in the range $\{2,..,6\}$ the mean (μ) and standard deviation (σ) best values of the K-nn classifier accuracy (A), sensitivity (Se) and specificity (Sp) computed on 10 folds in the cases of the Caenorhabditis elegans (CE), Drosophila melanogaster (DM) and Homo sapiens (HM) dataset, for each one of the considered classifiers. In bold, the values with the best values for each dataset.

	L=2 A μ	σ	Se μ	σ	Sp μ	σ	L=3 A μ	σ	Se μ	σ	Sp μ	σ	L=4 A μ	σ	Se μ	σ	Sp μ	σ	L=5 A μ	σ	Se μ	σ	Sp μ	σ	L=6 A μ	σ	Se μ	σ	Sp μ	σ
CE-Knn-ABD-nw	**86**	2	**92**	2	80	2	**86**	2	**92**	2	80	2	**86**	2	**92**	2	80	2	**86**	2	**92**	2	80	2	**86**	2	**92**	2	80	2
CE-Knn-euclidean	82	1	90	1	74	3	83	1	91	1	74	3	84	3	89	2	78	3	83	1	86	2	81	2	83	2	82	3	83	1
CE-Knn-correlation	80	1	74	2	86	2	78	1	65	2	92	2	73	1	52	3	92	1	69	1	43	3	**95**	1	67	1	38	1	**95**	1
CE-Knn-cosine	82	1	89	2	75	2	83	1	84	2	83	2	79	2	68	2	90	1	72	1	50	3	94	1	68	1	41	3	**95**	1
CE-Knn-Kullback	82	1	87	2	76	2	68	1	90	1	47	3	81	3	75	2	87	2	72	2	47	2	92	1	68	1	39	3	**96**	1
CE-Knn-UCD-deflate	67	1	41	2	92	1	67	1	41	2	92	1	67	2	41	2	92	2	67	1	41	2	92	1	67	1	41	1	92	1
CE-Knn-UCD-gzip	63	2	32	3	94	2	63	2	32	3	94	2	63	2	32	3	94	2	63	2	32	3	94	2	63	2	32	3	94	2
DM-Knn-ABD-nw	**78**	1	**87**	2	69	2	**78**	1	**87**	2	69	2	**78**	2	**87**	2	69	2	**78**	1	**87**	2	69	2	**78**	1	**87**	2	69	2
DM-Knn-euclidean	76	1	78	1	74	1	77	1	80	1	74	1	76	3	79	3	74	2	75	1	80	1	71	2	75	1	80	1	70	2
DM-Knn-correlation	72	1	73	2	69	2	70	2	75	2	69	2	69	1	71	2	66	2	68	1	62	2	74	2	68	2	52	2	**84**	2
DM-Knn-cosine	77	2	79	2	74	2	77	2	82	2	74	2	73	3	79	3	67	3	70	1	68	1	73	2	69	2	55	2	**83**	2
DM-Knn-Kullback	76	2	78	3	75	3	69	2	79	4	61	3	72	2	**85**	3	57	3	65	2	75	3	54	2	70	2	68	3	73	3
DM-Knn-UCD-deflate	58	1	69	4	47	3	58	1	69	4	47	3	58	1	69	4	47	3	58	1	69	4	47	3	58	1	69	4	47	3
DM-Knn-UCD-gzip	**77**	1	79	2	74	1	**77**	1	79	2	74	1	**77**	1	79	2	74	1	**77**	1	79	2	74	1	**77**	1	79	2	74	1
HM-Knn-ABD-nw	81	2	88	2	74	2	81	2	88	2	74	2	81	2	88	2	74	2	81	2	88	2	74	2	81	2	88	2	74	2
HM-Knn-euclidean	80	1	86	1	75	1	84	1	86	1	78	1	**85**	1	**91**	1	80	3	**85**	3	**91**	1	87	1	**84**	1	**90**	1	79	1
HM-Knn-correlation	79	1	82	2	76	2	83	1	82	2	76	2	83	3	79	3	84	3	81	2	74	3	**87**	2	79	1	70	3	**88**	2
HM-Knn-cosine	80	1	85	3	76	2	85	2	88	1	81	2	85	1	85	2	81	3	83	1	80	2	86	1	81	1	73	2	**87**	3
HM-Knn-Kullback	80	2	85	3	75	2	84	1	85	3	75	2	85	4	86	2	84	2	83	1	79	1	**87**	2	80	1	72	2	**88**	2
HM-Knn-UCD-deflate	73	1	68	3	77	2	73	1	68	3	77	2	73	3	68	3	77	2	73	1	68	3	77	2	73	1	68	3	77	2
HM-Knn-UCD-gzip	71	3	62	4	80	3	71	3	62	4	80	3	71	3	62	4	80	3	71	3	62	4	80	3	71	3	62	4	80	3

faster than knn-ABD-nw in computing a one fold classification (see Figs. 2, 4 and 6). This is not surprising since it is noteworthy that the complexity of the Needlmenan-Wunsch global alignment algorithm is $O(M * N)$ being M and N the lengths of the two sequences.

Among the alignment free dissimilarities based classifiers, Knn-euclidean reaches 91 % of sensitivity on CE and HM dataset, and accuracies greater than 83 %, 76 %, 84 % on CE, DM, HM datasets respectively. This makes the euclidean distance the best among the alignment free measures. Note that this performance are comparable, and in the case of sensitivity superior, to the results presented in [5]. Another interesting property is that the Knn-euclidean classifier is also invariant, in terms of accuracy, to the used L (see Figs. 1, 3 and 5). Knn-cosine is also a good performer in terms of accuracy (> 83 %, $L = 3$ CE, > 76 %, $L = 3$, DM). Knn-kullback and Knn-UCD-gzip seem only suitable for the more difficult dataset to classify, i.e. the DM dataset (best accuracy 77 %). Knn-correlation is one of the worst performer, showing only good specificity values for $L = 6$. Finally, it is also observable that the best choice for L of the *geometric-based*, *correlation-based* and *information-based* dissimilarity is 3 and 4, and that all the related classifiers' sensitivities and accuracies tend to decrease while L increases. In the case of compression based dissimilarities, Figs. 1, 3 and 5 show clearly an

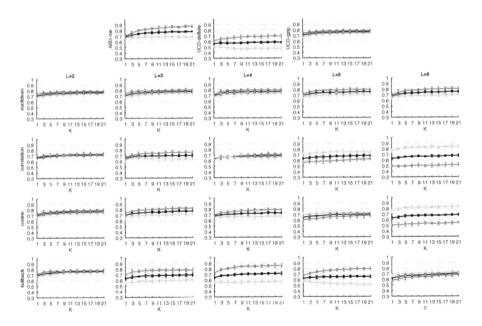

Fig. 3. Accuracy (black), Sensitivity (gray) and Specificity (light gray) plots of the Knn classifiers, for different neighbors K in the range $\{1,..,21\}$, and for different L-tuple lengths L in the range $\{2,..,6\}$ in the case of Drosophila melanogaster (DM) dataset. Error bars for each K are also reported. The first three plots in the first line are referred to ABD-nw, UCD-deflate and UCD gzip and does not depend on the L-tuple length L.

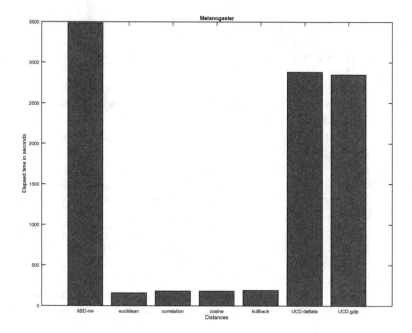

Fig. 4. Elapsed times in seconds of each classifier for the classification of one fold in the case of the Drosophila melanogaster (DM) dataset

increasing trend of all the performance indices while K increases. This could lead to the suggestion of increasing K, but, taking into consideration that the computation cost required by the compression based dissimilarity is significantly greater than the cost required by other ones, this will surely affect the computing time required by the Knn classifier.

3.3 Implementation Notes

The algorithms that implement the knn classifiers have been developed in Matlab^{TM} release 2014b, and are available upon request to the author.

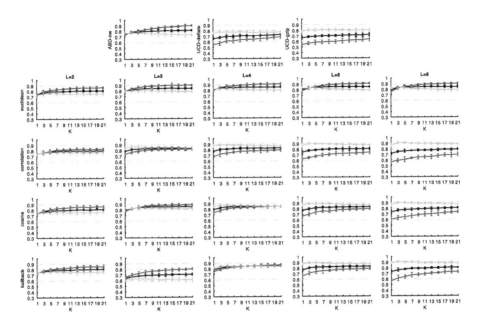

Fig. 5. Accuracy (black), Sensitivity (gray) and Specificity (light gray) plots of the Knn classifiers, for different neighbors K in the range $\{1,..,21\}$, and for different L-tuple lengths L in the range $\{2,..,6\}$ in the case of Homo sapiens (HM) dataset. Error bars for each K are also reported. The first three plots in the first line are referred to ABD-nw, UCD-deflate and UCD gzip and does not depend on the L-tuple length L.

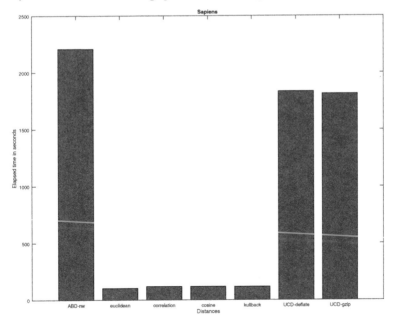

Fig. 6. Elapsed times in seconds of each classifier for the classification of one fold in the case of the Homo sapiens (HM) dataset

4   Conclusion

In this paper it is presented a benchmark study of 6 alignment free dissimilarities between sequences, belonging to the classes of *geometric-based, correlation-based information-based* and *compression-based,* for the purpose of nucleosome classification. Such dissimilarities are computed on a co-domain of the sequence set after a particular mapping process, and are alignment free in the sense that does not make any use of string alignment algorithms. The comparisons between the dissimilarities has been carried out in terms of performance comparisons among the related K nearest neighbour classifiers and versus a K nearest neighbour classifier that incorporates an alignment based dissimilarity. Experiments have been carried out on three public datasets of nucleosome sequences, showing that the geometric based dissimilarities computed on the $L-tuple$ representation of sequences lead to reach very good classification results, also in terms of computational efficiency. Several nucleosome classification methodologies have been recently proposed, sometimes implementing very complex algorithms, also from the computational point of view. This work shows that nucleosomes can be successfully and efficiently predicted by using only sequence information, by a very simple classification paradigm such as the Knn classifier, adopting common alignment free dissimilarity measures.

Acknowledgments. Part of this work was carried out using instruments provided by the Euro-Mediterranean Institute of Science and Technology, and funded with the Italian National Operational Programme for Research and Competitiveness 2007–2013 grant awarded to the project titled "CyberBrain-Polo di innovazione" (Project code: PONa3_00210, European Regional Development Fund).

References

1. Kornberg, R.D., Lorch, Y.: Twenty-five years of the nucleosome, fundamental particle of the eukaryote chromosome. Cell **98**, 285–294 (1999)
2. Jenuwein, T., Allis, C.: Translating the histone code. Science **293**(5532), 1074–1080 (2001)
3. Yuan, G.C., Liu, Y.J., Dion, M.F., Slack, M.D., Wu, L.F., Altschuler, S.J., Rando, O.J.: Genome-scale identification of nucleosome positions in S. cerevisiae. Science **309**(5734), 626–630 (2005)
4. Di Gesù, V., Lo Bosco, G., Pinello, L., Yuan, G.C., Corona, D.F.V.: A multi-layer method to study genome-scale positions of nucleosomes. Genomics **93**(2), 140–145 (2009)
5. Guo, S.-H., Deng, E.-Z., Xu, L.-Q., Ding, H., Lin, H., Chen, W., Chou, K.-C.: iNuc-PseKNC: a sequence-based predictor for predicting nucleosome positioning in genomes with pseudo k-tuple nucleotide composition. Bioinformatics **30**(11), 1522–1529 (2014)
6. Kouzarides, T.: Chromatin modifications and their function. Cell **128**(4), 693–705 (2007)

7. Struhl, K., Segal, E.: Determinants of nucleosome positioning. Nat. Struct. Mol. Biol. **20**(3), 267–273 (2013)
8. Needleman, S.B., Wunsch, C.D.: A general method applicable to the search for similarities in the amino acid sequence of two proteins. J. Mol. Biol. **48**, 443–453 (1970)
9. Smith, T.F., Waterman, M.S.: Identification of common molecular subsequences. J. Mol. Biol. **147**, 195–197 (1981)
10. Gotoh, O.: An improved algorithm for matching biological sequences. J. Mol. Biol. **162**, 705–708 (1982)
11. Altschul, S., Gish, W., Miller, W., et al.: Basic local alignment search tool. J. Mol. Biol. **25**(3), 403–410 (1990)
12. Lipman, D., Pearson, W.: Rapid and sensitive protein similarity searches. Science **227**(4693), 1435–1441 (1985)
13. Yuan, G.C.: Linking genome to epigenome. Wiley Interdisc. Rev. Syst. Biol. Med. **4**(3), 297–309 (2012)
14. Vinga, S., Almeida, J.: Alignment-free sequence comparisona review. Bioinformatics **19**(4), 513–523 (2003)
15. Pinello, L., Lo Bosco, G., Yuan, G.-C.: Applications of alignment-free methods in epigenomics. Briefings Bioinf. **15**(3), 419–430 (2013)
16. Durbin, R., Eddy, S.R., Krogh, A., Mitchison, G.: Biological Sequence Analysis. Cambridge University Press, Cambridge (1998)
17. La Rosa, M., Fiannaca, A., Rizzo, R., Urso, A.: Genomic sequence classification using probabilistic topic modeling. In: Formenti, E., Tagliaferri, R., Wit, E. (eds.) CIBB 2013. LNCS, vol. 8452, pp. 49–61. Springer, Heidelberg (2014)
18. La Rosa, M., Fiannaca, A., Rizzo, R., Urso, A.: Probabilistic topic modeling for the analysis and classification of genomic sequences. BMC Bioinformatics **16**(S6) (2015)
19. Rosa, M., Fiannaca, A., Rizzo, R., Urso, A.: A k-mer-based barcode DNA classification methodology based on spectral representation and a neural gas network. Artif. Intell. Med. **64**(3), 173–184 (2015)
20. Rizzo, R., Fiannaca, A., Rosa, M., Urso, A.: The general regression neural network to classify barcode and mini-barcode DNA. CIBB 2014. LNCS, vol. 8623, pp. 142–155. Springer, Heidelberg (2015)
21. Yuan, G.C., Liu, J.S.: Genomic sequence is highly predictive of local nucleosome depletion. PLoS Comput. Biol. **4**(1), e13 (2008)
22. Giancarlo, R., Rombo, S.E., Utro, F.: Epigenomic k-mer dictionaries: shedding light on how sequence composition influences in vivo nucleosome positioning. Bioinformatics **31**(18), 2939–2946 (2015)
23. Lo Bosco, G., Pinello, L.: A new feature selection methodology for k-mers representation of DNA sequences. In: Di Serio, C., Liò, P., Nonis, A., Tagliaferri, R. (eds.) CIBB 2014. LNCS, vol. 8623, pp. 99–108. Springer, Heidelberg (2015)
24. Li, M., Chen, X., Li, X., Ma, B., Vitanyi, P.M.B.: The similarity metric. IEEE Trans. Inf. Theor. **50**(12), 3250–3264 (2004)
25. Ferragina, P., Giancarlo, R., Greco, V., et al.: Compression based classification of biological sequences and structures. BMC Bioinf. **8**(252) (2007)
26. La Rosa, M., Fiannaca, A., Rizzo, R., Urso, A.: A study of compression–based methods for the analysis of barcode sequences. In: Peterson, L.E., Masulli, F., Russo, G. (eds.) CIBB 2012. LNCS, vol. 7845, pp. 105–116. Springer, Heidelberg (2013)

27. Utro, F., Di Benedetto, V., Corona, D.F.V., Giancarlo, R.: The intrinsic combinatorial organization and information theoretic content of a sequence are correlated to the DNA encoded nucleosome organization of eukaryotic genomes. Bioinformatics **32**(6), 835–842 (2016)
28. http://lin.uestc.edu.cn/server/iNucPseKNC/dataset

A Deep Learning Approach to DNA Sequence Classification

Riccardo Rizzo, Antonino Fiannaca, Massimo La Rosa$^{(\boxtimes)}$, and Alfonso Urso

ICAR-CNR, National Research Council of Italy,
Via Ugo La Malfa 153, 90146 Palermo, Italy
{ricrizzo,fiannaca,larosa,urso}@pa.icar.cnr.it

Abstract. Deep learning neural networks are capable to extract significant features from raw data, and to use these features for classification tasks. In this work we present a deep learning neural network for DNA sequence classification based on spectral sequence representation. The framework is tested on a dataset of 16S genes and its performances, in terms of accuracy and F1 score, are compared to the General Regression Neural Network, already tested on a similar problem, as well as naive Bayes, random forest and support vector machine classifiers. The obtained results demonstrate that the deep learning approach outperformed all the other classifiers when considering classification of small sequence fragment 500 bp long.

Keywords: Deep learning · Convolutional neural network · Barcode classification

1 Introduction

In the last years there has been a great interest in the neural networks, due to the so called deep architecture or deep learning networks. The term "deep" refers intuitively to the number of layers that are used in these networks, and, more precisely, it is related to the path from an input node to the output node in the network (considering the network as a directed graph) [1]. Deep learning architectures are able to extract the features used for classification tasks from input patterns [2]. Among the deep learning architecture, it is usually comprised the LeNet-5 network, or Convolutional Neural Network (CNN), a neural network that is inspired by the visual system's structure [3]. This network was used for character recognition in the original paper, and for image processing [4] and speech detection [5].

The main drawback of the deep learning methods is that it is still impossible to reuse the knowledge acquired by the network; deep networks are still black boxes and it is complicated to correct wrong answers or to understand the reasons of a good one. As proved in [6], it is possible to build artificial images with no recognizable objects in it, but classified with high confidence in specific categories, as "chair" or "lion", by a deep neural network.

© Springer International Publishing Switzerland 2016
C. Angelini et al. (Eds.): CIBB 2015, LNBI 9874, pp. 129–140, 2016.
DOI: 10.1007/978-3-319-44332-4_10

The application of these techniques to gene classification requires a fixed dimension representation of the sequences like the spectral representation based on k–mers occurrences. This representation was used for sequence classification in many works. In particular in our work [7] it is noticed that some k–mers are much more important than the other for sequence representation, this means that in the representing vectors there are details that should be taken into account. This observation resembles the problem of feature extraction from image and this idea is at the core of the present work.

In this work we want to understand if the convolutional network is capable to identify and to use these features for sequence classification, outperforming the classifiers proposed in the past. To do that, we considered a dataset of 16S rRNA sequences that, for bacteria, represent the genes used for taxa assignment thanks to their strong conservation among similar species [8].

2   Materials and Methods

2.1   Spectral Representation

Each biological species, as demonstrated by [9], has a proper modal spectrum, that can distinguish it from the other ones. Therefore, spectral representation has been successfully used in many bioinformatics works, implementing several algorithms: support vector machine (SVM) and K-nearest neighbour (K-NN)[10], general regression neural network (GRNN) [7,11], neural gas (NG) [12,13], topic models [14,15], logic formulas [16]. Given a fixed value k, a spectral representation is a vector of size 4^k. Its components are computed by counting the occurrences of small DNA snippets of length k, called k-mers, which are extracted from the genomic sequences by means of a sliding window, with step $= 1$ and length $= k$. In case of k-mers containing one or more undefined nucleotides, for example the "N" character, they are discarded. Since the CNN is able to discover and exploit those k-mers representing distinctive features, we adopt the so called "bag-of-words" model, which does not take into account the position of k-mers in the original sequence. This procedure is summarized in Fig. 1. The main computational advantages of using this representation are: (1) to obtain a fixed-size vector representation of genomic sequences and (2) to take into account only distinctive k-mers.

2.2   Convolutional Neural Network

The convolutional neural networks are made of a very large number of connections and layers. The one used in this work is a modified version of the LeNet-5 network introduced by LeCun et al. in [3] and it is implemented using the python Theano package for deep learning [17,18].

The modified LeNet-5 proposed network is made of two lower layers of convolutional and max-pooling processing elements, followed by two "traditional" fully connected Multi Layer Perceptron (MLP) processing layers, so that there are 6 processing layers.

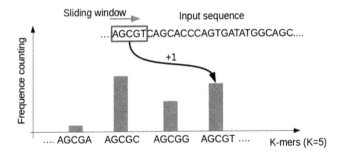

Fig. 1. k–mers spectral representation. The sliding window on the sequence selects the k–mer counted; the bar graph represent the frequency count obtained

The convolutional layers calculate L 1-D convolutions between the kernel vectors w^l, of size $2n+1$, and the input signal x:

$$q^l(i) = \sum_{u=-n}^{n} w^l(u)x(i-u) \qquad l = 1, 2 \ldots, L \tag{1}$$

The dimension (dim) of q is defined as:

$$dim(q) = \frac{dim(x) - (dim(w) - 1)}{\text{size of max-pooling}} \tag{2}$$

In Eq. 1 $q^l(i)$ is the component i of the l-th output vector and $w^l(u)$ is the component u of the l-th kernel vector. Then a bias term b^l is added and a non–linear function is applied:

$$h^l(i) = \tanh(q^l(i) + b^l) \qquad l = 1, 2 \ldots, L \tag{3}$$

The vector h^l is the output of the convolutional layer. The max-pooling is a non-linear down-sampling layer. In these processing layers the input vector is partitioned into a set of non-overlapping regions (of 2 elements in this implementation) and, for each subregion, the maximum value is considered as output. This processing layer reduces the complexity for the higher layers and operates a sort of translational invariance.

Figure 2 shows the network architecture: the overall network is depicted in the lower part, and a detail of the two convolutional layers is reported in the upper part of the figure. The first layer of convolutional kernels, named $kernel\_0$, is made of $L = 10$ kernels of dimension 5 (so that $n = 2$). From a spectral representation vector made of 1024 components this layer produces 10 vectors of 1024 dimensions that the pooling layer reduces to the 10 feature maps of 510 dimensions. These vectors are the input for the second convolutional layer. The second layer of kernel ($kernel\_1$) is made of $L = 20$ kernels of dimension 5. In both cases the max-pooling layer has dimension 2. Convolution and max-pooling are usually considered together and they are represented in the lower part of Fig. 2 as two highly connected blocks.

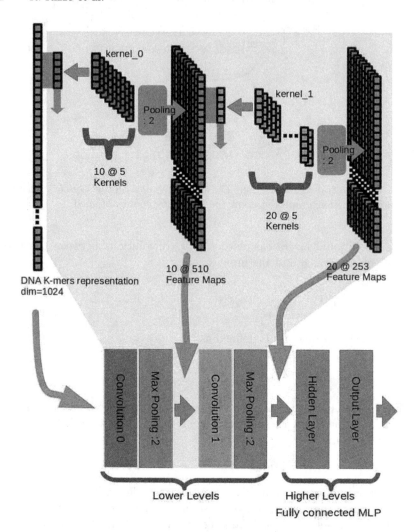

Fig. 2. The architecture of the used network: in the lower part of the figure an overall schema of the CNN network, in the upper part a detail of the lower levels of the network with the kernels and the output signal of the two blocks of convolution and max pooling.

The two upper level layers corresponds to a traditional fully-connected MLP: the first layer of the MLP operates on the total number of output from the lower level (the output is flattened to a 1-D vector) and the total number of units in the Hidden Layer is 500. The output layer has one unit for each class.

2.3 Dataset of 16S Sequences

The 16S rRNA sequences have been downloaded from the RDP Ribosomal Database Project II repository [19], release 10.27. We randomly selected 1000 sequences from each of the three most common bacteria phyla, Actinobacteria, Firmicutes, Proteobacteria, collecting in total 3000 sequences. All the sequences have length greater than 1200 bp, they are classified as type strain, i.e. they are the best representative of their own species, and they are certified as "of good quality" by the RDP database. Table 1 reports the number of taxonomic categories, or taxa, from phylum to genus. It clearly shows that, even if the dataset is balanced at phylum level, it becomes more and more unbalanced for the other taxonomic categories, reaching 393 different groups of 16S sequences at genus level. Figure 3 shows a boxplot representing, for each taxa, the number of samples contained in the categories of each taxa (vertical axis). This figure, together with the Table 1, gives an idea of the classification task, for example the distribution of samples in each Class varies from 1 to nearly 1000 (there are 5 classes of 998, 994, 5, 2 and 1 sample).

Table 1. 16S bacteria dataset composition.

Three main bacteria phyla	Number of categories for each taxa				
	Phylum	Class	Order	Family	Genus
Actinobacteria	1	1	3	12	79
Firmicutes	1	2	3	19	110
Proteobacteria	1	2	13	34	204

3 Results

Experimental tests have been carried out using the algorithm and the dataset described in Sect. 2. Two kinds of experiments have been made. In the first case, using a ten fold cross validation scheme, the prediction performances of the CNN have been tested at each taxonomic rank (from phylum to genus) and considering full length sequences. In the second case, the ten-fold cross validation scheme was repeated considering as test set the sequence fragments of shorter size, 500 bp long, obtained randomly extracting 500 consecutive nucleotides from the original full length sequences. This way, we wanted to asses if the network is able to correctly predicts the taxonomic rank of the test sequences even if they just contain only a small part (500 bp) of the original information content, usually composed of about 1400 bp. This situation is often dealt with in metagenomic studies [20] where, considering environmental species, only a small part of the 16S gene is actually sequenced. In our experiments, we set the k-mers size to k = 5, as done in other works adopting the spectral representation, e.g. in [7, 10]. The CNN has been run considering two different kernels sizes: $kernel\_0 = kernel\_1 = 5$ in the first run; $kernel\_0 = 25, kernel\_1 = 15$ in the second run. From here on, the first kernels configuration will be named *kern_1*, whereas the second one will be named *kern_2*. In both configurations the training phase has been run for 200 epochs.

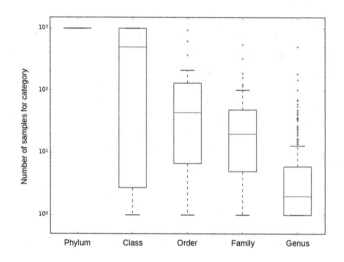

Fig. 3. 16S bacteria dataset distribution.

Classification scores have been computed in terms of accuracy and F1 score:

$$Accuracy = \frac{TP + FN}{TP + FP + TN + FN} \tag{4}$$

$$F1 = \frac{2TP}{2TP + FP + FN} = 2 * \frac{precision * recall}{precision + recall} \tag{5}$$

where TP are true positives, FP are false positives, TN are true negatives, FN are false negatives.

We compared our approach with another classifier based on the GRNN algorithm [21], presented in our previous work [7] for the classification of barcode and mini-barcode sequences. Moreover we compared the obtained classification scores with regard to three other state-of-the-art classifiers, namely naive Bayes (NB) [22], random forest (RF) [23] and SVM [24].

The GRNN is a one-pass training neural network, usually adopted for regression purposes, that we adapted for the classification of barcode sequences of animal species, taking into account the COI gene. Moreover we developed three different versions of the GRNN, each one implementing a different distance model: euclidean distance, city-block (Manhattan) distance, Jaccard distance.

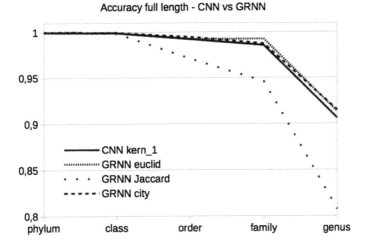

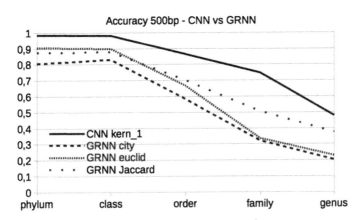

Fig. 4. Accuracy scores for full length sequences (upper chart) and 500 bp sequences (lower chart). Comparison between CNN and GRNN with different distance models.

Experiments with NB, RF and SVM classifiers have been done using the WEKA experimenter platform [25]. NB and RF were used with default parameters. As for SVM, Weka uses the LibSVM library [26] that implements the "one-against-one" approach [27] for multiclass classification. We adopted a Gaussian radial basis kernel and the parameters have been optimised by means of a grid search over a set of parameters values.

In our experiments, the CNN network with *kern_1* configuration always provided better results with respect to the CNN with *kern_2* configuration. For this reason, in the following we will only discuss the results obtained with *kern_1* configuration.

As for the comparison with the GRNN, all the classification results have been summarized in the charts of Figs. 4 and 5, showing respectively accuracy

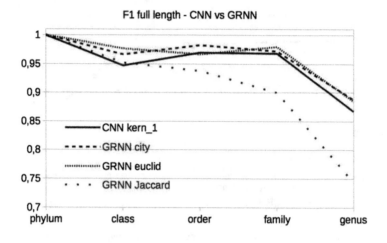

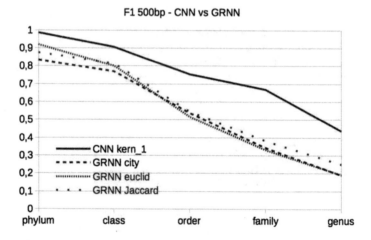

Fig. 5. F1 scores for full length sequences (upper chart) and 500 bp sequences (lower chart). Comparison between CNN and GRNN with different distance models.

and F1 score. Considering the full length sequences, it is evident that our approach based on the CNN network, with *kern_1* configuration, reaches almost identical scores, with variance lesser than 1 %, with regards with the GRNN classifiers based on the euclidean and the city block distance models. Otherwise the GRNN with Jaccard distance model produced lower results. The CNN performances are obtained with a grater execution time with regard to the GRNN network, because the CNN neural network has a training phase that is necessary to obtain the minimization of the classification error (that is a function of the kernel parameters), while the GRNN algorithm has just one parameter, and does not need a training phase. The GRNN processing time lasted about 10 min on average, whereas the CNN proposed architecture has a training time of about 22 min on average and it gives better performances.

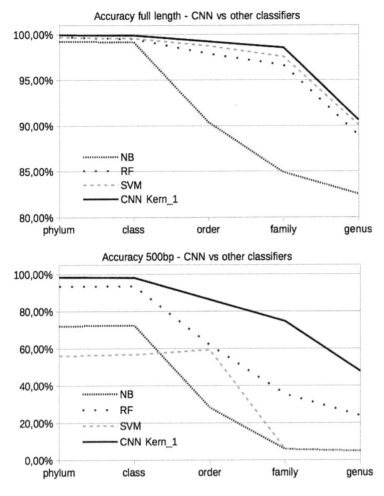

Fig. 6. Accuracy scores for full length sequences (upper chart) and 500 bp sequences (lower chart). Comparison among CNN and NB, RF and SVM classifiers.

Classification scores considering 500 bp sequences showed very interesting results. Our CNN approach, with *kern_1* configuration, clearly outperforms all the other classifiers in terms of accuracy at all taxonomic levels. Only at genus level, accuracy score does not reach the 50 %: this behaviour can be explained considering the great number of different genera (393) of the dataset. With regards to the F1 score chart (Fig. 5), the CNN with *kern_1* configuration performs similar to the other classifiers for all taxa. There once again the CNN with *kern_1* configuration always reaches the highest scores, demonstrating that our approach has a better true positive rate, that is the percentage of retrieving correctly classified samples.

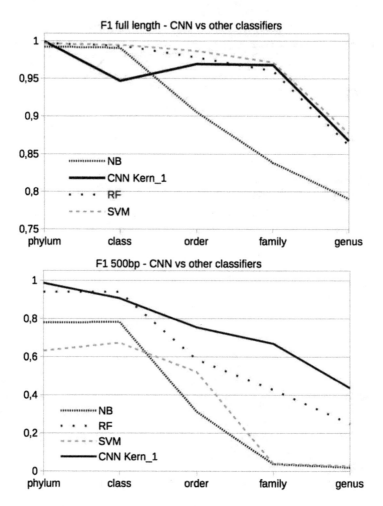

Fig. 7. F1 scores for full length sequences (upper chart) and 500 bp sequences (lower chart). Comparison among CNN and NB, RF and SVM classifiers.

In Figs. 6 and 7, accuracy and F1 score of CNN versus the NB, RF and VM classifiers are showed. Considering the full length sequences, once again the CNN network obtained very similar results in terms of accuracy with regards to the other classifiers. At family and genus level, CNN reached better F1 score, while at genus level it is below of about ten percentage points with respect to the other classifiers, and SVM in particular.

On the other hand, considering 500 bp sequences, CNN outperforms all the other classifiers considering all the three performances score, demonstrating that the CNN network is the only algorithm that build a model that is able to generalize in order to correctly classify genomic sequences even in only a short fragment is provided.

4   Conclusion

In this work we presented a classifier of 16S bacterial genomic sequences based on spectral representation and convolutional neural networks. The spectral representation is obtained as k-mers frequencies along the sequences; the CNN belongs to the so called "deep learning" algorithms. Experiments were carried out with the aim of classify both full length sequences and sequence fragments. In metagenomic studies, for example, only a small part of DNA sequences is often available. Our CNN classifier has been compared, in terms of accuracy and F1 score, with three state-of-the-art classifiers, NB, RF, SVM, and finally with one of our previously developed classifier based on GRNN. If considering full length sequences all the algorithms provided very similar good results (between 95–99 %) at each taxonomic level. In the near future we are going to further test this deep learning approach considering more complex sequence representation methods, like for instance those provided by the Pse-in-One web server [28].

References

1. Bengio, Y.: Learning deep architectures for AI. Found. Trends Mach. Learn. **2**(1), 1–127 (2009)
2. Coates, A., Ng, A.Y., Lee, H.: An analysis of single-layer networks in unsupervised feature learning. In: International Conference on Artificial Intelligence and Statistics, pp. 215–223 (2011)
3. LeCun, Y., Bottou, L., Bengio, Y., Haffner, P.: Gradient-based learning applied to document recognition. Proc. IEEE **86**(11), 2278–2324 (1998)
4. Farabet, C., Couprie, C., Najman, L., LeCun, Y.: Learning hierarchical features for scene labeling. IEEE Trans. Pattern Anal. Mach. Intell. **35**(8), 1915–1929 (2013)
5. Sukittanon, S., Surendran, A., Platt, J., Burges, C.: Convolutional Networks for Speech Detection. In: Interspeech, International Speech Communication Association (2004)
6. Nguyen, A., Yosinski, J., Clune, J.: Deep neural networks are easily fooled: high confidence predictions for unrecognizable images. In: 2015 IEEE Conference on Computer Vision and Pattern Recognition (CVPR), pp. 427–436. IEEE (2015)
7. Rizzo, R., Fiannaca, A., La Rosa, M., Urso, A.: The general regression neural network to classify barcode and mini-barcode DNA. In: di Serio, C., Liò, P., Nonis, A., Tagliaferri, R. (eds.) CIBB 2014. LNCS, vol. 8623, pp. 142–155. Springer, Heidelberg (2015)
8. Drancourt, M., Bollet, C., Carlioz, A., Martelin, R., Gayral, J.P., Raoult, D.: 16S Ribosomal DNA sequence analysis of a large collection of environmental and clinical unidentifiable bacterial isolates. J. Clin. Microbiol. **38**(10), 3623–3630 (2000)
9. Chor, B., Horn, D., Goldman, N., Levy, Y., Massingham, T.: Genomic DNA k-mer spectra: models and modalities. Genome Biol. **10**(10), R108 (2009)
10. Kuksa, P., Pavlovic, V.: Efficient alignment-free DNA barcode analytics. BMC Bioinform. **10**(Suppl. 14), S9 (2009)
11. La Rosa, M., Fiannaca, A., Rizzo, R., Urso, A.: DNA Barcode classification using general regression neural network with different distance models. In: Zazzu, V., Ferraro, M.B., Guarracino, M.R. (eds.) Mathematical Models in Biology, pp. 119–132. Springer, Heidelberg (2015)

12. Fiannaca, A., La Rosa, M., Rizzo, R., Urso, A.: Analysis of DNA barcode sequences using neural gas and spectral representation. In: Iliadis, L., Papadopoulos, H., Jayne, C. (eds.) EANN 2013, Part II. CCIS, vol. 384, pp. 212–221. Springer, Heidelberg (2013)

13. Fiannaca, A., La Rosa, M., Rizzo, R., Urso, A.: A k-mer-based barcode DNA classification methodology based on spectral representation and a neural gas network. Artif. Intell. Med. **64**(3), 173–184 (2015)

14. La Rosa, M., Fiannaca, A., Rizzo, R., Urso, A.: Genomic sequence classification using probabilistic topic modeling. In: Formenti, E., Tagliaferri, R., Wit, E. (eds.) CIBB 2013. LNCS, vol. 8452, pp. 49–61. Springer, Heidelberg (2014)

15. La Rosa, M., Fiannaca, A., Rizzo, R., Urso, A.: Probabilistic topic modeling for the analysis and classification of genomic sequences. BMC Bioinform. **16**(Suppl 6), S2 (2015)

16. Weitschek, E., Cunial, F., Felici, G.: Classifying bacterial genomes with compact logic formulas on k-Mer frequencies. In: 2014 25th International Workshop on Database and Expert Systems Applications, pp. 69–73. IEEE (2014)

17. Bastien, F., Lamblin, P., Pascanu, R., Bergstra, J., Godfellow, I., Bergeron, A., Bouchard, N., Warde-Farley, D., Bengio, Y.: Theano: new features and speed improvements. In: NIPS 2012 Deep Learning Workshop (2012)

18. Bergstra, J., Breuleux, O., Bastien, F., Lamblin, P., Pascanu, R., Desjardins, G., Turian, J., Warde-Farley, D., Bengio, Y.: Theano: a CPU and GPU math compiler in Python. In: 9th Python in Science Conference, pp. 1–7 (2010)

19. Cole, J.R., Wang, Q., Cardenas, E., Fish, J., Chai, B., Farris, R.J., Kulam-Syed-Mohideen, A.S., McGarrell, D.M., Marsh, T., Garrity, G.M., Tiedje, J.M.: The Ribosomal Database Project: improved alignments and new tools for rRNA analysis. Nucleic Acids Res. **37**(Database issue), D141–D145 (2009)

20. Wooley, J.C., Godzik, A., Friedberg, I.: A primer on metagenomics. PLoS Comput. Biol. **6**(2), e1000667 (2010)

21. Specht, D.F.: A general regression neural network. IEEE Trans. Neural Netw. **2**(6), 568–576 (1991)

22. John, G.H.G., Langley, P.: Estimating continuous distributions in bayesian classifiers. In: Besnard, P., Hanks, S. (eds.) Proceedings of the Eleventh Conference on Uncertainty in Artificial Intelligence, Montreal, Quebec, Canada, vol. 1, pp. 338–345. Morgan Kaufmann, San Franisco, CA (1995)

23. Breiman, L.: Random forests. Mach. Learn. **45**, 5–32 (2001)

24. Scholkopf, B., Smola, A.: Learning with Kernels. MIT Press, Cambridge (2002)

25. Hall, M., Frank, E., Holmes, G., Pfahringer, B., Reutemann, P., Witten, I.H.: The WEKA data mining software. ACM SIGKDD Explor. Newsl. **11**(1), 10–18 (2009)

26. Chang, C.C., Lin, C.J.: LIBSVM: a library for support vector machines. ACM Trans. Intell. Syst. Technol. **2**(3), 1–27 (2011)

27. Knerr, S., Personnaz, L., Dreyfus, G.: Single-layer learning revisited: a stepwise procedure for building and training a neural network. In: Soulié, F.F., Hérault, J. (eds.) Neurocomputing, pp. 41–50. Springer, Heidelberg (1990)

28. Liu, B., Liu, F., Wang, X., Chen, J., Fang, L., Chou, K.C.: Pse-in-One: a web server for generating various modes of pseudo components of DNA, RNA, and protein sequences. Nucleic Acids Res. **43**(W1), W65–W71 (2015)

Clustering Protein Structures with Hadoop

Giacomo Paschina[1], Luca Roverelli[1], Daniele D'Agostino[1],
Federica Chiappori[2], and Ivan Merelli[2(✉)]

[1] Institute of Applied Mathematics and Information Technologies "E. Magenes",
National Research Council of Italy, Genoa, Italy
{paschina,roverelli,dagostino}@ge.imati.cnr.it
[2] Institute of Biomedical Technologies,
National Research Council of Italy, Segrate, MI, Italy
{federica.chiappori,ivan.merelli}@itb.cnr.it

Abstract. Machine learning is a widely used technique in structural biology, since the analysis of large conformational ensembles originated from single protein structures (e.g. derived from NMR experiments or molecular dynamics simulations) can be approached by partitioning the original dataset into sensible subsets, revealing important structural and dynamics behaviours. Clustering is a good unsupervised approach for dealing with these ensembles of structures, in order to identify stable conformations and driving characteristics shared by the different structures. A common problem of the applications that implement protein clustering is the scalability of the performance, in particular concerning the data load into memory. In this work we show how it is possible to improve the parallel performance of the GROMOS clustering algorithm by using Hadoop. The preliminary results show the validity of this approach, providing a hint for future development in this field.

Keywords: Hadoop · Clustering protein structures · Molecular dynamics · Data parallel

1 Introduction

A common task in unsupervised machine learning and data analysis is clustering. This means a method to partition a discrete metric space into sensible subsets. The exact setup and procedures may vary, but the general idea is to group data points with similar features together. This might reveal some structure in the data set, or it can help in simplifying the data set by dealing with entire clusters instead of many individual data points. In this context, hierarchical clustering is very important since it does not produce a prescribed number of clusters, but a dendrogram that allows the user to decide which is a reasonable number of clusters, in order to create a partition of the data accordingly.

These concepts apply very well to structural biology, in which protein-structure prediction experiments and protein-folding simulation software generate large ensembles of candidate structures using different starting conditions.

© Springer International Publishing Switzerland 2016
C. Angelini et al. (Eds.): CIBB 2015, LNBI 9874, pp. 141–153, 2016.
DOI: 10.1007/978-3-319-44332-4_11

In particular, molecular dynamics simulations produce trajectories of atom positions, velocities, and energies as a function of time, providing a sampling of the different conformations achieved by a given macromolecule. As simulations on the 100 ns–1 μs time scale are becoming routinely, with sampled configurations stored on the picosecond time scale, the resulting trajectories contain large amounts of data.

Data-mining techniques, like hierarchical clustering, provide a valuable tool to make sense of the information available in these trajectories [2]. Since protein conformations that are more frequently assumed during a trajectory represent at best the real structure of the macromolecule, clustering is a suitable approach to improve both consistency and accuracy of the most probable protein conformations, which can be very important for achieving good results in docking screenings [3]. From the computational point of view this process is very time consuming, since there are dozens or even hundreds of thousand structures to compare. Many algorithms have been implemented in parallel in order to overcome this problem, such as MAX_CLUST [1] and FAST_PROTEIN_CLUSTER [4]. The latter is based on an efficient GPU-accelerated solution, although not specifically designed to analyse molecular dynamics trajectories.

Nonetheless, analysing the scalability of these software a clear bottleneck emerges, which is the large number of I/O operations necessary for data acquisition and possible use of virtual memory when processing these large structures. This is particularly true for bunches of PDB structures, which are text files, while trajectories are usually stored in binary files. It is therefore clear that a large amount of computational time is spent in reading the structures for the clustering operation. In this paper we propose a proof-of-concept Apache Hadoop based solutions in order to parallelize the clustering on partitions of the original dataset.

The Apache Hadoop framework is an ecosystem of related projects and tools. Mostly they are based on the Hadoop Distributed File System (HDFS), which stores data on commodity hardware, providing very high aggregate bandwidth across the cluster. We will exploit in particular MapReduce, a programming model for large scale data processing based on HDFS that is well suited for data-intensive problems. Both are based on two principles: jobs must be moved near data for their analysis and the high availability of Hadoop-based systems, because hardware failures are very frequent.

To implement this solution we started from the GROMOS [5] algorithm for clustering protein structures, which is provided as part of the GROMACS [6] package. Although GROMOS is not parallel in its released implementation, it is widely used to cluster conformation from molecular dynamics simulations. The algorithm is fast, but it suffers, as introduced, of a slow loading of the structures in memory. In the following we present a preliminary implementation of a solution based on the Hadoop-MapReduce architecture and the preliminary results.

The remainder of this work is structured as follows. In Sect. 2 the clustering algorithm is presented and Apache Hadoop is introduced. In Sect. 3 a real-life test case of protein conformations clustering is presented. In Sect. 4 the implementation is sketched. Section 5 presents the results achieved in terms of performance, while in Sect. 6 some conclusions are drawn about the presented work.

2   Materials and Methods

In this section we present the details of the GROMOS algorithm and the Apache Hadoop platform that has been used for our tests.

GROMOS Algorithm. Clustering involves partitioning models into sets of similar structures. The input of these algorithms is a distance matrix in which are reported, for each sampled structural conformation of the molecular dynamics trajectory, the RMSD distances, time frame after time frame, of the C_α atoms that compose the backbone of the protein. The GROMOS algorithm relies on the nearest neighbour algorithm, which is a non-parametric method used for classification and regression [7]. In particular, the GROMOS implementation of this algorithm is an iterative process:

- the neighbours of each data point are defined according to a cut-off distance c;
- the point with largest neighbourhood defines the "best" cluster, corresponding to a stable conformation of the protein;
- all the points belonging to the cluster are removed;
- the algorithm is iterated until all data have been assigned to a cluster and removed.

In Fig. 1 the different steps of the algorithm are shown on a simple matrix representing the conformation distances. For each column below threshold distances are evidenced and the best cluster is identified accordingly. Then, all the conformations belonging to this cluster are eliminated from the matrix and the algorithm is iterated.

Apache Hadoop and Map Reduce. Apache Hadoop [8] is an open source framework for distributed storage and processing of large sets of data on commodity hardware. Hadoop enables to quickly mine massive amounts of structured and unstructured data. Numerous Apache Software Foundation projects make up the services required by an enterprise to deploy, integrate and work with Hadoop (MapReduce, Spark, Storm, Hive, HBase and many others). Each project has been developed to deliver an explicit function and each has its own community of developers and individual release cycles. In the analysis of Big Data, Apache Hadoop is probably the most popular approach available to researchers, and for this reason is receiving a great attention from the Bioinformatics research field [9].

The following modules compose the Apache Hadoop framework:

- Hadoop Common: it contains libraries and utilities needed by other Hadoop modules.

	t1	t2	t3	t4	t5	t6	t7	t8	t9	t10	t11	t12
t1	0	0.91	0.44	0.83	0.98	0.65	0.73	0.67	0.54	0.40	0.56	0.60
t2	0.91	0	0.55	0.23	0.56	0.59	0.52	0.60	0.19	0.88	0.76	0.88
t3	0.44	0.55	0	0.22	0.85	0.86	0.94	0.59	0.78	0.77	0.44	0.61
t4	0.83	0.23	0.22	0	0.63	0.45	0.40	0.55	0.90	0.60	0.16	0.55
t5	0.98	0.56	0.85	0.63	0	0.12	0.32	0.77	0.55	0.99	0.90	0.66
t6	0.65	0.59	0.86	0.45	0.12	0	0.51	0.12	0.67	0.56	0.54	0.59
t7	0.73	0.52	0.94	0.40	0.32	0.51	0	0.56	0.89	0.76	0.77	0.90
t8	0.67	0.60	0.59	0.55	0.77	0.12	0.56	0	0.51	0.80	0.51	0.89
t9	0.54	0.19	078	0.90	0.55	0.67	0.89	0.51	0	0.23	0.77	0.54
t10	0.40	0.88	0.77	0.60	0.99	0.56	0.76	0.80	0.23	0	0.91	0.69
t11	0.56	0.76	0.44	0.16	0.90	0.54	0.77	0.51	0.77	0.91	0	0.45
t12	0.60	0.88	0.61	0.55	0.66	0.59	0.90	0.89	0.54	0.69	0.45	0
#	3	3	4	6	3	4	3	2	3	3	4	2

	t1	t5	t8	t9	t10	t12
t1	0	0.98	0.67	0.54	0.40	0.60
t5	0.98	0	0.77	0.55	0.99	0.66
t8	0.67	0.77	0	0.51	0.80	0.89
t9	0.54	0.55	0.51	0	0.23	0.54
t10	0.40	0.99	0.80	0.23	0	0.69
t12	0.60	0.66	0.89	0.54	0.69	0

Fig. 1. An example of the GROMOS algorithm on a table describing the distances between different conformations of the same protein. Underlined distances are below the 0.50 cut-off threshold. The orange column is the one with the largest neighbourhood. Blue columns are removed from the matrix during the first iteration as they belong to the first cluster. (Color figure online)

- Hadoop Distributed File System (HDFS): a distributed file system that stores data on commodity machines, providing very high aggregate bandwidth across the cluster.
- Hadoop YARN: a resource-management platform that is responsible for allocating computational resources in clusters for scheduling users' applications.

Hadoop Distributed File System (HDFS) is the core technology for the efficient scale out storage layer, and is designed to run across low-cost commodity hardware. HDFS is a distributed, scalable, and portable file-system written in Java for the Hadoop framework. A Hadoop cluster has nominally a single NameNode plus a cluster of DataNodes, although redundancy options are available for the NameNode due to its criticality. Each DataNode serves up blocks of data over the network using a block protocol specific to HDFS. The file system uses TCP/IP sockets for communication. Clients use remote procedure call (RPC) to communicate between each other.

HDFS stores large files (typically in the range of gigabytes to terabytes) across multiple machines. It achieves reliability by replicating the data across multiple hosts, and hence theoretically does not require RAID storage on hosts (but to increase I/O performance some RAID configurations are still useful). With the default replication value of 3, data is stored on three nodes. DataNodes can talk to each other to rebalance data, to move file copies around, and to keep high the replication of data. HDFS is not fully POSIX-compliant, because the requirements for a POSIX file-system differ from the target goals for a Hadoop application. The trade off of not having a fully POSIX-compliant file-system is increased performance for data throughput and support for non-POSIX operations such as Append.

Apache Hadoop YARN is the pre-requisite for Enterprise Hadoop as it provides the resource management and pluggable architecture for enabling a wide variety of data access methods to operate on data stored in Hadoop with predictable performance and service levels. YARN is the architectural core of Hadoop that allows multiple data processing engines such as interactive SQL, real-time streaming, data science and batch processing to handle data stored in a single platform, unlocking an entirely new approach to analytics.

YARN is composed of a global ResourceManager and a per-node NodeManager, which form the generic system for managing applications in a distributed manner. The ResourceManager is the ultimate authority that arbitrates resources among all applications in the system. Each application has a dedicated ApplicationMaster, which is a framework-specific entity that negotiates resources from the ResourceManager and works with the NodeManager(s) to execute and monitor the component tasks. Moreover, each application runs into a Container instantiated by the NodeManager.

MapReduce [10] is the natural companion of Hadoop, since it consists of a framework for writing applications that process large amounts of structured and unstructured data in parallel across a cluster of thousands of machines, in a reliable and fault-tolerant manner. The MapReduce engine consists of one Job-Tracker, to which client applications submit MapReduce jobs. The JobTracker schedules work to available TaskTracker nodes in the cluster, trying to keep the work as close to the data as possible.

With a server-aware file system, the JobTracker knows which node contains the data, and which other machines are nearby. If the work cannot be hosted on the actual node where the data resides, priority is given to nodes in the same rack. This reduces network traffic on the main backbone network. If a TaskTracker fails or times out, that part of the job is rescheduled. The TaskTracker on each node spawns off a separate Java Virtual Machine process to prevent the TaskTracker itself from failing if the running job crashes the JVM. A heartbeat is sent from the TaskTracker to the JobTracker every few minutes to check its status. The JobTracker and TaskTracker status and information is exposed by Jetty and can be viewed from a web browser.

In early versions of Hadoop, when a JobTracker failed all the on going work was lost, but actually there are checkpoints along with the process and the

JobTracker records what it is up to in the file system. When a JobTracker starts up, it looks for any such data, so that it can restart work from where it left off. The allocation of work to TaskTrackers is very simple. Every TaskTracker has a number of available slots and every active map or reduce task takes up one slot. The JobTracker allocates work to the tracker nearest to the data with an available slot. There is no consideration of the current system load of the allocated machine, and hence its actual availability. If one TaskTracker is very slow, it can delay the entire MapReduce job, especially towards the end of a job, where everything can end up waiting for the slowest task. With speculative execution enabled, however, a single task can be executed on multiple slave nodes.

An advantage of using HDFS is data awareness between the JobTracker and the TaskTracker. The JobTracker schedules map or reduce jobs to TaskTrackers with an awareness of the data location. For example: if node A contains data (x,y,z) and node B contains data (a,b,c), the JobTracker schedules node B to perform map or reduce tasks on (a,b,c) and node A would be scheduled to perform map or reduce tasks on (x,y,z). This reduces the amount of traffic that goes over the network and prevents unnecessary data transfer. When Hadoop is used with other file systems this advantage is not always available. This can have a significant impact on job-completion times, which has been demonstrated when running data-intensive jobs. A few applications can be solved with a single MapReduce job, while the others are based on several MapReduce steps which run in series to accomplish specific tasks, i.e. $Map_1 \rightarrow Reduce_1 \rightarrow Map_2 \rightarrow Reduce_2 \rightarrow Map_3 \ldots$

3   Case Study: Hsp70

In the context of this work, the input dataset for testing the clustering algorithm consisted of a molecular dynamics (MD) trajectory of the human protein Hsp70 in complex with ADP ligand. Hsp70 is a molecular chaperone, which prevents the incorrect folding of other proteins, or it is involved in protein translocation though the mitochondrial membrane [11]. This protein belongs to a large protein family, which is present in different organisms, from bacteria to human. Hsp70 displays two active conformations the closed bound to ADP and the open bound to ATP. Due to the absence of crystallized human structure, simulated protein was build with homology modelling based on bacterial DnaK in ATP conformation [12]. The analysis consists in evaluating the effect of the nucleotide exchange (ATP to ADP) on protein conformation (open to closed). To identify the more stable conformation, that is the conformation assumed more frequently during the trajectory, cluster analysis is a useful tool.

In detail, the complex has about 600 atoms. To evaluate the protein conformation two independent trajectories of 25 ns and 100 ns of MD simulation of the solvated complex were obtained with Gromacs 4.0 [13]. Both trajectories were skipped every 50 frames, and we obtained two trajectories of 500 and 2000 conformations [14]. The two files have a size of, respectively, 200 and 800 MBs,

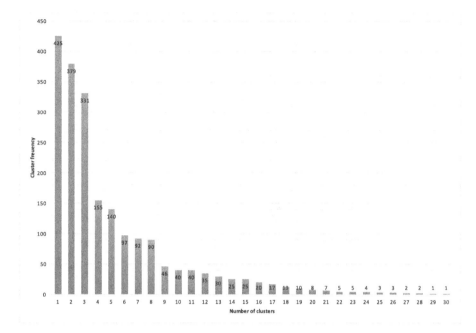

Fig. 2. Histogram of the cluster population distribution in the Hsp70 example with 2000 conformations. The GROMOS algorithm creates 30 clusters of decreasing dimension. The second and third clusters represent, respectively, the open and closed conformation of the HPS70 protein, while the first cluster represents an intermediate conformation between the two.

while the original one has a size of about 40 GB [15]. Cut-off was set at 0.5 nm, in agreement with the number of atoms to be clustered and to the protein conformation type. This is also the default value for GROMOS: higher values result in less clusters, composed by a larger number of points and vice versa for smaller values.

We performed the clustering analysis using the *g_cluster* tool provided by the GROMACS suite. Considering the largest test, at the end of clustering analysis we obtain 30 clusters distributed as shown in Fig. 2. The first three clusters can be considered representative for this simulation. As shown in Fig. 3 the central structure of the first cluster obtained is an intermediate conformation between the open and the closed one, showing that molecular dynamics simulation is a useful tool to evaluate a conformational change and clustering allow to identify a representative conformation. The second test presents a subsampling of this distribution, with a very similar profile for what concerns the first three clusters.

4   Implementation

In this section we discuss our strategy to achieve a proof-of-concept implementation of GROMOS relying on Apache Hadoop.

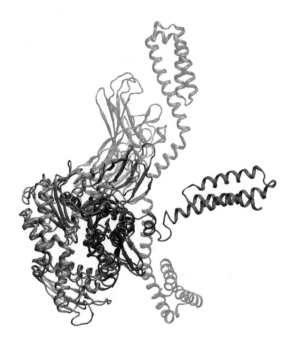

Fig. 3. Central structure of the first cluster. Hsp70 sub-domains are shown in different colours for the open conformations, in solid colours the central structure of the first cluster, in transparent colours the starting conformation, and in transparent grey the closed conformations. (Color figure online)

Data Structure and Pre-processing. Figure 4 shows the data structure contained in the trajectories file. Each row in the file contains the atom id, the atom type, the residue name, the chain to which it belongs, the residue number, the x, y, z coordinates, the occupancy, and the B-factor. Before the execution of the analysis the trajectory file is stored on HDFS and therefore split into several parts.

In this step we extract, in parallel, only the trajectories related to the atoms involved in the computation. This is implemented with several map tasks (typically one for each file split) that retrieve data relying on the type of atom of interest (step A). In a preliminary implementation we re-created a data file with only these data with a Reducer task. Considering that these data are used only in the next step we disregarded the use of the reducer task. In conclusion the input pairs correspond to the row of the trajectory file, having the line number as key. The output is composed by a subset of these pairs, those corresponding to C_α, where the time instant is added. The schema of these (key; value1+value2+value3+) pairs is:

```
(linenumber; Atom_id| CA |...| x | y | z |...| time_instant)
```

ATOM	33	MG	MG	A	2	46.670	61.020	48.290	1.00	0.00
ATOM	34	N	LYS	B	3	64.740	68.010	60.640	1.00	0.00
ATOM	35	H1	LYS	B	3	65.620	67.730	61.030	1.00	0.00
ATOM	36	H2	LYS	B	3	64.650	68.990	60.800	1.00	0.00
ATOM	37	H3	LYS	B	3	64.030	67.520	61.150	1.00	0.00
ATOM	38	CA	LYS	B	3	64.660	67.690	59.210	1.00	0.00
ATOM	39	CB	LYS	B	3	64.890	66.200	58.960	1.00	0.00
ATOM	40	CG	LYS	B	3	66.330	65.800	58.620	1.00	0.00
ATOM	41	CD	LYS	B	3	67.320	66.190	59.720	1.00	0.00
ATOM	42	CE	LYS	B	3	68.760	65.730	59.520	1.00	0.00
ATOM	43	NZ	LYS	B	3	68.940	64.300	59.820	1.00	0.00
ATOM	44	HZ1	LYS	B	3	69.910	64.060	59.810	1.00	0.00
ATOM	45	HZ2	LYS	B	3	68.590	64.100	60.740	1.00	0.00
ATOM	46	HZ3	LYS	B	3	68.480	63.720	59.150	1.00	0.00
ATOM	47	C	LYS	B	3	63.320	68.090	58.610	1.00	0.00
ATOM	48	O	LYS	B	3	62.340	68.210	59.350	1.00	0.00
ATOM	49	N	ILE	B	4	63.250	68.320	57.300	1.00	0.00
ATOM	50	H	ILE	B	4	64.060	68.320	56.710	1.00	0.00
ATOM	51	CA	ILE	B	4	62.000	68.690	56.610	1.00	0.00
ATOM	52	CB	ILE	B	4	62.260	69.480	55.330	1.00	0.00

Fig. 4. The data structure within trajectories file.

Retrieval of the Coordinates from Trajectory Files. Subsequently, every Map task reads portions of the new trajectories file and associates to each atom the x, y, z coordinates assumed at a given time instant (step B). The schema of these (key,values) pairs is:

$$(\text{Atom}_{id}; \quad \text{xyz}_{t0} \quad | \quad \text{xyz}_{t40} \quad | \quad \text{xyz}_{t122})$$

At the end of the Map tasks, a Reducer collects all data and creates a new single file containing the coordinates of all atoms for each frame. The data structure used by Map Task to temporary store the coordinates consists of an id that identifies the atom and a list in which each node is formed by an object that wraps the coordinates taken by the atom at a given time instant. In this way we get a list of nodes in which each node represents a time frame with inside the coordinates taken from the atom at that instant. The schema of these (key,values) pairs is:

$$(\text{Atom}_{id}; \quad \text{xyz}_{t0} \quad | \quad \text{xyz}_{t1} \quad | \quad \text{xyz}_{t2} \quad | \ldots | \quad \text{xyz}_{tstop})$$

Creation of Distances Matrix. Once we have created the file containing all the coordinates of all atoms for each frame as described in step B, new Map tasks are performed to calculate the Euclidean distance of the positions taken by each atom at each time instant (step C). The schema of these (key,values) pairs is:

$$(\text{Atom}_{id}; \quad \text{dist}(t0, t1) \quad | \quad \text{dist}(t0, t2) \quad | \quad \ldots)$$

After completing these tasks, a Reducer collects data by summing the various distances of all atoms at each time frame to obtain the variation of the centre of gravity of the entire molecule with respect to all the others frames. The schema of these (key,values) pairs is:

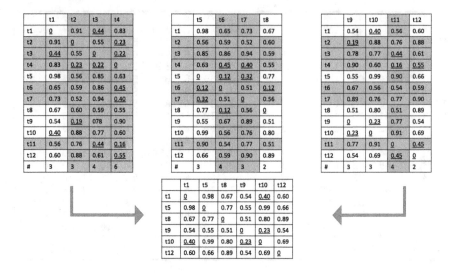

Fig. 5. Apache Hadoop implementation of the GROMOS algorithm. (Color figure online)

$$(\text{Conformation}_i; \ \text{rmsd}(\text{conformation}_i, \text{conformation}_j) \ | \ \dots \ )$$

Clustering. The distances matrix resulting from step C is processed by several Map tasks: the matrix is divided in different parts and each part is elaborated by a single Map task that produces partial frame clusters on the basis of a cut-off established a priori (in our case it is set to 0.5 nm, as said before). The schema of these (key,values) pairs is:

$$(\text{Conformation}_i; \ \text{isneighbour}_0 \ | \ \text{isneighbour}_{14} \ | \ \text{isneighbour}_{122})$$

A Reducer task manages the data and defines the final frame clusters, selecting the biggest one, and removing the frame associated with the largest cluster from the other ones (step D). The schema of these (key,values) pairs is:

$$(\text{Conformation}_i; \ \text{numneighbours})$$

The elaboration is repeated until clusters consume themselves. In this case, the processing is the union of a parallel part (the elaboration of parts of matrix by Map tasks) and a serial part (the reorganization of clusters and the cycle until the end of the clusters by the Reducer task). The schema of the Apache Hadoop computation of the same matrix presented in Fig. 1 is presented here in Fig. 5.

5   Results

The computer cluster named *imatihpc* has been used to experiment our Hadoop based solution and it is composed by three computational nodes and a front-end node, each of those presents the following hardware configuration: two 6-core Intel Xeon E5645 CPUs, 32 GB of RAM, 2 TB of SATA hard disk. Nodes are linked together with a Gigabit network connection. The Hadoop cluster follows the physical configuration of the *imatihpc* cluster on which it is implemented: the front-end node acts as the master while the 3 nodes used for the computation act as a slave, with an HDFS file system of 3 TB.

The exact reproducibility of the achieved clusters, and therefore the consistency of the biological results, is clearly an essential prerequisite for any parallel implementation of this algorithm. Considering our Hadoop implementation, results are exactly the same of the ones achieved with the sequential algorithm. Therefore, we have 30 clusters for the large example, representing all the protein conformations achieved during the Molecular Dynamics simulations, and a sub-sampling of these for the small example in which, however, the distribution of the first three clusters is clearly conserved.

Performances are shown in Table 1. They are based on the average on 10 executions. The Hadoop framework automatically determines the number of parallel processes. Typically, a MapTask is created for each map split, and there is a map split for each input file or, for files larger than the data block size of HDFS, with 64 MB as default value, there is a MapTask for each data block. This means that in the first test (500 conformations) 2 MapTask are launched, and 13 in the second test (2000 conformations). The number of ReduceTasks is normally set to 1, in order to create the result at a global level and to produce results also when one or more nodes of the cluster fail.

The first step of the algorithm (A - Data Structure and pre-processing) achieves good results because a good overlapping between the map and reduce tasks. Similar considerations hold true for the second step (B - Retrieval of the coordinates from trajectories file). In both cases the speedup in the first case is around 2, in the second around 10. The last two steps (C - Creation of distances matrix, D - clustering) instead suffer of the framework overheads and of the bottleneck due to the reduce task, presenting therefore a speedup of 1.3 and 7 respectively.

Table 1. Elapsed times (Sec) of the sequential and MapReduce versions of the Gromos algorithm subdivided in the steps described in Sect. 4.

	500			2000		
	A	B	C+D	A	B	C+D
Sequential	3.40	1.2	54.75	19.38	5.06	2184.10
MapReduce	1.70	0.6	41.88	1.8	0.49	312.50

6   Conclusions

Structure clustering is very important in the analysis of Molecular Dynamics trajectories, since it allow the identification of stable conformations. The number of structures to analyse in this kind of analysis is very high because the increasing computational power allows the generation of very large simulations, which poses important scalability problems to clustering algorithm. Here we present a Hadoop based solution to accelerate the loading of data into memory, which is the real bottleneck of this algorithm. The presented results show interesting performance figures, which support the use of Hadoop when analysing very large data files.

Acknowledgments. This paper has been supported by the Italian Ministry of Education and Research (MIUR) through the Flagship (PB05) InterOmics, HIRMA (RBAP11YS7K), and the European MIMOMICS projects.

References

1. MaxCluster - A tool for Protein Structure Comparison and Clustering. http://www.sbg.bio.ic.ac.uk/maxcluster
2. Chiappori, F., Merelli, I., Milanesi, L., Marabotti, A.: Static and dynamic interactions between GALK enzyme and known inhibitors: guidelines to design new drugs for galactosemic patients. Eur. J. Med. Chem. **63**, 423–434 (2013)
3. D'Ursi, P., Chiappori, F., Merelli, I., Cozzi, P., Rovida, E., Milanesi, L.: Virtual screening pipeline and ligand modelling for H5N1 neuraminidase. Biochem. Biophys. Res. Commun. **383**(4), 445–449 (2009)
4. Hung, L.H., Samudrala, R.: fast_protein_cluster: parallel and optimized clustering of large-scale protein modeling data. Bioinformatics **30**(12), 1774–1776 (2014)
5. Daura, X., Gademann, K., Jaun, B., Seebach, D., van Gunsteren, W.F., Mark, A.E.: Peptide folding: when simulation meets experiment. Angew. Chem. Int. Ed. **38**(1–2), 236–240 (1999)
6. Berendsen, H.J.C., van der Spoel, D., van Drunen, R.: GROMACS: a message-passing parallel molecular dynamics implementation. Comput. Phys. Commun. **91**, 43–56 (1995)
7. Altman, N.S.: An introduction to Kernel and nearest-neighbor nonparametric regression. Am. Stat. **46**(3), 175–185 (1995)
8. White, T.: Hadoop: The Definitive Guide. O'Reilly Media Inc., Sebastopol (2009)
9. Merelli, I., Prez-Snchez, H., Gesing, S., D'Agostino, D.: Managing, analysing, and integrating big data in medical bioinformatics: open problems and future perspectives. BioMed Res. Int. (2014). Article ID: 134023
10. Dean, J., Ghemawat, S.: MapReduce: simplified data processing on large clusters. Commun. ACM **51**(1), 107–113 (2008)
11. Mayer, M.P., Bukau, B.: Hsp70 chaperones: cellular functions and molecular mechanism. Cell. Mol. Life Sci. **62**(6), 670–684 (2005)
12. Kityk, R., Kopp, J., Sinning, I., Mayer, M.P.: Structure and dynamics of the ATP-bound open conformation of Hsp70 chaperones. Mol. Cell. **48**(6), 863–874 (2012)
13. van der Spoel, D., Lindahl, E., Hess, B., Groenhof, G., Mark, A.E., Berendsen, H.J.C.: GROMACS: fast, flexible, and free. J. Comput. Chem. **26**, 1701–1718 (2005)

14. Chiappori, F., Merelli, I., Colombo, G., Milanesi, L., Morra, G.: Molecular mechanism of allosteric communication in Hsp70 revealed by molecular dynamics simulations. PLoS Comput. Biol. **8**(12), e1002844 (2012)
15. Chiappori, F., Milanesi, L., Merelli, I.: HPC analysis of multiple binding sites communication and allosteric modulations in drug design: the HSP case study. Curr. Drug Targets (2015)
16. Eadline, D.: Is Hadoop the New HPC? http://www.admin-magazine.com/HPC/Articles/Is-Hadoop-the-New-HPC

Comparative Analysis of MALDI-TOF Mass Spectrometric Data in Proteomics: A Case Study

Eugenio Del Prete[1,2(✉)], Diego d'Esposito[1], Maria Fiorella Mazzeo[1],
Rosa Anna Siciliano[1], and Angelo Facchiano[1]

[1] Istituto di Scienze dell'Alimentazione, CNR,
Via Roma 64, 83100 Avellino, Italy
{eugenio.delprete,fmazzeo,rsiciliano,
angelo.facchiano}@isa.cnr.it, diegodesposito@alice.it
[2] Dipartimento di Scienze, Università della Basilicata,
Viale dell'Ateneo Lucano 10, 85100 Potenza, Italy

Abstract. Mass spectrometry is a well-known technology used for the analysis of pure compounds as well as mixtures, widely applied in large-scale studies such proteomic studies. The result of mass spectrometric analyses is a mass spectrum, a profile of mass/charge values and corresponding intensity values originated from the analyzed compounds. In the case of large-scale analyses, raw mass spectra comparisons are difficult due to different drawback typologies: data defects, unusual distributions, underlying disturbs and noise, bad data calibration. A bunch of data elaborations is essential, from data processing to feature extraction, in order to obtain a list of peaks from different mass spectra. In this work, a workflow has been developed to process raw mass spectra and compare the new tidy ones with the aim of defining a robust procedure, suitable for real applications and reusable for different kind of studies. A similarity measure has been used for comparison purposes, in order to verify similarity among replicates and differences among analyzed samples, and a clustering method has been performed on fish species, in order to discover how they cluster statistically. A case study is shown with the application of the processing method to data obtained from the analysis of different fish species.

Keywords: Proteomics · Mass spectrometry · MALDI-TOF · Similarity measures · R environment

1 Scientific Background

Proteomics is a scientific discipline focused on the identification and structural analysis of all the proteins expressed by a biological system (such as an organism, a tissue etc.) in a determined state and of proteins that change their expression level when the biological system is subjected to a perturbation, for instance, a pathological condition. Proteomic studies produce large-scale datasets and together with other "omics" disciplines, have a fundamental role in biomedical research. Mass spectrometry is the analytical technology used in proteomics and metabolomics, which allows analyzing proteins, peptides and different metabolites. Matrix-Assisted Laser Desorption Ionization-Time of Flight mass

© Springer International Publishing Switzerland 2016
C. Angelini et al. (Eds.): CIBB 2015, LNBI 9874, pp. 154–164, 2016.
DOI: 10.1007/978-3-319-44332-4_12

spectrometry (MALDI-TOF-MS) is one of the most used technique in proteomics. In MALDI-TOF-MS, samples are mixed with an appropriate matrix (an organic compound with a strong optical absorption in the UV range), and, upon a laser irradiation, sample molecules are ionized and desorbed as gaseous ions. These ions are accelerated and transferred to the TOF analyzer. The ions, accelerated by the electric field, acquired velocities that are inversely proportional to their mass/charge, i.e. heavier particles reach lower speeds values, therefore ions are separated in the TOF and, from the TOF measures, the ion mass/charge value can be calculated. This technology gives the benefit of getting large amounts of data in a short time and with a high-resolution, accuracy and sensitivity of molecular mass measurements [1, 2]. The most common data representation is a mass spectrum, an indented profile where mass/charge values are reported in abscissa and the corresponding intensities are reported in ordinate. The intensity is usually shown in percentage, in relationship with the tallest (base) peak. The comparison of mass spectra data is a common task in proteomics, in order to detect signals that can represent a signature of each group. The comparison can be performed among replicates from the same sample or, more interestingly, among different sample groups.

2   Materials and Methods

2.1   Data and Tools

Mass spectra used to set up the procedure have been obtained from the analysis of protein extracts of fish muscle by MALDI-TOF-MS, as previously described in [3]. In particular, mass spectra of sixteen different fish species, four belonging to the same *genus* (*Diplodus*) and the others belonging to different *genera*, have been chosen for this study, as shown in Table 1. Related common names have been retrieval from FishBase database, available online [4].

Table 1. Fishes species, with common English name

Scientific name	Common name
Auxis thazard	Frigate tuna
Coryphoena hippurus	Common dolphinfish
Diplodus annularis	Annular seabream
Diplodus puntazzo	Sharpsnout seabream
Diplodus sargus	White seabream
Diplodus vulgaris	Common two-banded seabream
Engraulis encrasicolus	European anchovy
Lophius piscatorius	Angler
Pagellus erythrinus	Common pandora
Perca fluviatilis	European perch
Sarda sarda	Atlantic bonito
Sardina pilchardus	European pilchard
Scomber scombrus	Atlantic mackerel
Solea solea	Common sole
Thunnus albacares	Yellowfin tuna
Trachurus trachurus	Atlantic horse mackerel

Each fish has been represented by six mass spectra, three replicates for two samples. The work has been performed in *R* environment, mainly using the following packages: *MALDIquant, MALDIquantForeign* [5], *OrgMassSpecR* [6] and *pvclust* [7].

2.2 Data Processing

The workflow of our mass spectrometry data analysis is shown in Fig. 1. The software related to the bio-spectrometry workstation supplies a data matrix, where the first column represents the m/z values (or m/z ratios, or masses) and the second one represents their intensity; a header is also present, with the information about the id number and the base (tallest) peak for each experiment. Metadata from header have been stored by means of regular expressions, mass spectrum data have been extracted from a .txt file and transformed in a special object, for a simpler access. This object is constituted by class type, number and range of m/z values, range of intensity values, memory usage and mass spectrum name (if available). After raw data storage, two prearranged controls have been verified: if the mass spectrum is empty and if the distances between two consecutive mass points are equal or monotonically increasing.

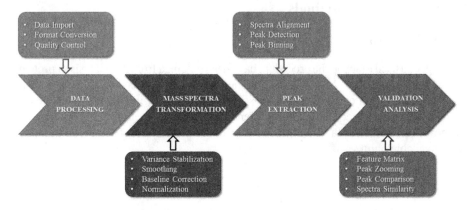

Fig. 1. Mass spectrum data analysis workflow.

2.3 Mass Spectra Transformation

Due to specific features of MALDI-TOF-MS [8], peak intensity values could vary between different measurements on the same substance or sample. Hence, a strengthened normalization is compulsory for comparing several mass spectra from the same sample and different experiments. Global and local transformations have been performed on mass spectra: one for a normalization on each entire mass spectrum, one for a normalization on different parts of the mass spectrum. In particular:

- variance stabilization, in order to shift the data for a better graphical result and to avoid a dependency between variance and mean, for example, using a square root transformation [9];

- smoothing, in order to reduce noise coming from artefacts or other underlying disturbs and, consequently, to improve the signal-to-noise ratio, for example, using the Savitsky-Golay filter [10];
- baseline correction, in order to control the amplification of chemical noise in the low mass range, for example, using the Statistics-sensitive Non-linear Iterative Peak-clipping (SNIP) algorithm [11];
- normalization, in order to preserve the proportionality between the intensity of different peaks of the mass spectrum, for example, using Total Ion Current (TIC) method, the most common normalization technique for MALDI-TOF data [12].

Some details are available. Before the variance stabilization, the m/z range has been trimmed in 4000–14000, in order to eliminate upper-lower bounds noise for the following elaborations. The window dimension used for the smoothing step has been chosen in a heuristic way, taking in account that its dimension should be smaller than the full width at half maximum (FWHM) of the peaks: the zone considered is the one near the highest peak for each fish species (one random sample for species). The resulting average value is 11 points. The number of iterations about the baseline correction is 25 (among the attempts 25, 50, 75, 100), because it is the value which is more adaptive with the spectra profiles, as shown in Fig. 2.

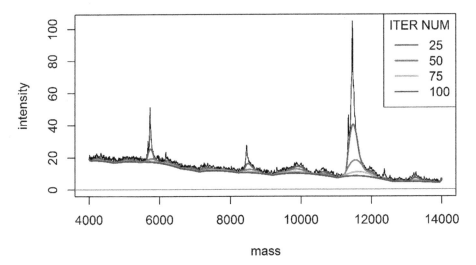

Fig. 2. Different number of iterations about the baseline correction algorithm. The most adaptive profile is the one related to 25.

2.4 Peak Extraction

The set of the most important peaks from a mass spectrum represents a simpler fingerprint for the sample. At first, an alignment procedure is required to obtain a preliminary correspondence between the highest peaks from mass spectrum replicates and to preserve an average mass spectrum, useful for following comparison purposes. Local Weighted Scatterplot Smoothing (LOWESS) technique has been chosen as warping algorithm for the alignment, because phase errors need a correction due to their non-additive nature: it is different from other regression method (linear, polynomial), because it is applied on data subsets, with better performances in extracting local variability of data. After having estimated an overall noise on the mass spectra, peaks have been locally detected with the Median Absolute Deviation (MAD) method [13]: *a priori* knowledge of mass spectra can help in selecting a suitable window size (as for the smoothing) for data subsets and signal-to-noise ratio. In particular, the signal-to-noise ratio has been selected with a value of 3, chosen after some proofs from 1 to 5 values, as a good compromise between number of remaining peaks and further noise eliminated (Fig. 3).

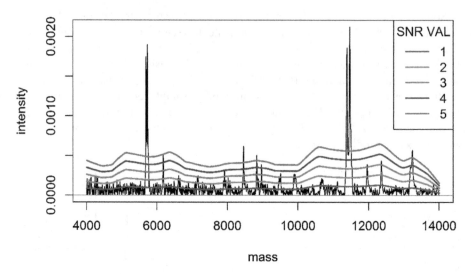

Fig. 3. Different signal-to-noise ratio (SNR) values for peaks detection algorithm. A good suboptimal solution can be 3, in order to eliminate the mostly part of noise, without reducing the peaks number too much.

Peak extraction shows that mass values are very similar, but not the same. A binning step has been executed, thus only one abscissa value represents the sets of aligned peaks from each mass spectrum.

2.5   Validation Analysis

Feature matrix can be useful for further statistical analysis: missing values have been interpolated from all the mass spectra and only the peaks over a selected frequency threshold have been kept. It is possible that only one interval of m/z values can be characteristic for a sample, thus a zoom on the mass spectrum or on the peaks can be more descriptive than the entire dataset. Moreover, numerical and graphical analyses of different mass spectra from the same sample have been performed. A good similarity measure is the cosine correlation, also called dot product, which considers two lists of peaks as vectors, and the cosine of the angle establishes how much they are similar [14, 15]. Cosine correlation has been calculated with *OrgMassSpecR* package, and a head-to-tail plot between a reference and a target mass spectrum has been shown for a visual comparison about peaks and a consequent mass spectrum discrimination. Because of the strict relationship between Pearson's correlation and cosine correlation [16], it is possible:

- to extract a dissimilarity measurement (similar to Pearson's distance) and to create a distance matrix among the mass spectra, for studying their clusterization;
- to provide a statistical validation on cosine correlation, using a t-Student's test with a significance level of 0.95. All the p-values have been improved by means of the classical Bonferroni-Holm and Benjamini-Hochberg corrections, concerning a multiple comparisons problem.

3   Results

Starting from raw or partially processed mass spectra, it is possible to obtain a refined one with a smoothed trend, a low level of noise and evident peaks, which characterize the chosen dataset. In Fig. 4, five replicate mass spectra from *D. puntazzo* have been assembled together, in order to get a clear profile for this sample. The peaks have been marked with a cross: 150 intensity values have been recognized as peaks after the processing, approximately 1.5 % of the total mass spectrum length, and ten of them are the highest, over a threshold of 25 % about the presence in all the samples for one species. Thus, three m/z regions in this average mass spectrum include the most intense peaks: the first region

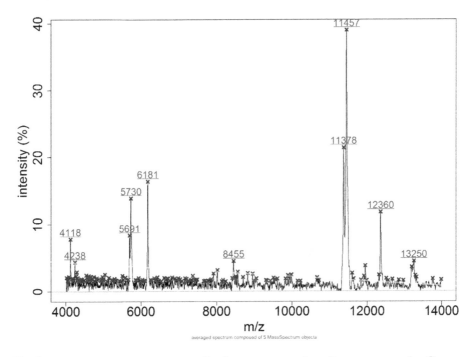

Fig. 4. Average mass spectrum extracted by five mass spectra from *D. puntazzo* samples. Crosses show all the detected peaks, numbers highlight the most important ten of them (more than 25 % about the presence in all the samples for one species).

around 4000, the second around 6000, the third around 12000, and they can be assumed as identifiers (the isolated peaks in 8455 and in 13250 require a more accurate study).

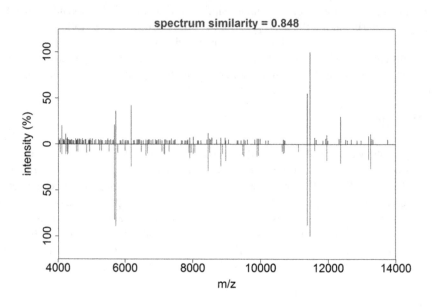

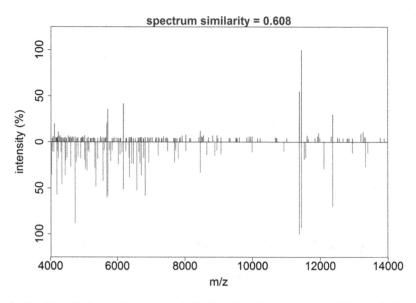

Fig. 5. Head-to-tail plots with spectrum similarity values (trimmed version). Top: plot between *D. puntazzo* average mass spectrum peaks (upper) and *D. puntazzo* sample mass spectrum peaks (lower). Bottom: plot between *D. puntazzo* average mass spectrum peaks (upper) and *A. thazard* sample mass spectrum peaks (lower).

Two comparisons have been performed over the average mass spectrum, as shown in Fig. 5. After randomly choosing a mass spectrum from *D. puntazzo* to make a comparison and using the others to construct the average mass spectrum, head-to-tail graph shows that sample mass spectrum is very similar (taking in account the peaks) to the average one. Moreover, a quantitative measurement assures a similarity of 83.8 %, thus, as expected, it is possible to say that sample mass spectrum is coherent with the average one. At the same time, the average mass spectrum can be used for performing a control over a mass spectrum extracted from a different dataset. The external control has been made with an *A. thazard* mass spectrum: both graphical effect - different peak profile - and cosine correlation calculation - a similarity of 61.3 % - reveal that the new mass spectrum is not coherent with the average one. The elimination of the noise in the range 4000–14000 does not show relevant modification in the evaluation.

For further information, the first similarity percentage becomes 84.8 %, while the second one becomes 60.3 %: thus, peripheral mass spectrum noise, in the m/z regions 2000–4000 and 14000–20000, weights only for the 1 %. The Leave-One-Out methodology, performed iteratively for the construction of the average mass spectra, has showed as *Diplodus Puntazzo* has mean 0.862 and standard deviation 0.021, in term of multiple evaluations of the spectrum similarity. The same methodology has been applied for each fish family, underlying that, in our database, *Diplodus Puntazzo* is the 'worst' case, as shown in Table 2. Thus, a value of 0.840 can be considered as (the lowest) threshold, used to say if a mass spectrum comes from the same species or not. The result with (*) has been considered as an outlier.

Table 2. Mean similarity score, calculated with Leave-One-Out technique

Scientific name	Mean ± standard deviation
Auxis thazard	0.982 ± 0.010
Coryphoena hippurus	0.984 ± 0.006
Diplodus annularis	0.986 ± 0.012
Diplodus puntazzo	0.862 ± 0.021
Diplodus sargus	0.959 ± 0.011
Diplodus vulgaris	0.972 ± 0.026
Engraulis encrasicolus	0.980 ± 0.008
Lophius piscatorius	0.985 ± 0.010
Pagellus erythrinus	0.985 ± 0.010
Perca fluviatilis	0.865 ± 0.116 (*)
Sarda sarda	0.981 ± 0.006
Sardina pilchardus	0.978 ± 0.005
Scomber scombrus	0.979 ± 0.007
Solea solea	0.935 ± 0.028
Thunnus albacares	0.981 ± 0.010
Trachurus trachurus	0.973 ± 0.004

In Fig. 6, dataset dendrogram, built from the distance matrix, has been shown. Each fish *genus* has a well-defined cluster, under a cut-off of 0.3. The most valuable results are:

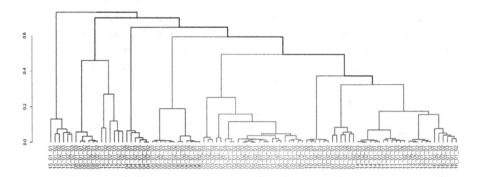

Fig. 6. Dataset dendrogram, in which the code represents fish_sample_replicate. Legend for fish: **01** *D. vulgaris*, **02** *D. sargus*, **03** *D. puntazzo*, **04** *D. annularis*, **05** *P. erythrinus*, **06** *L. piscatorius*, **07** *T. trachurus*, **08** *A. thazard*, **09** *E. encrasicolus*, **10** *C. hippurus*, **11** *S. sarda*, **12** *P. fluviatilis*, **13** *S. pilchardus*, **14** *S. scombrus*, **15** *S. solea*, **16** *T. albacares*

1. the light blue cluster - with fish species 01, 02, 03, 04 - which represents the *Diplodus genus*;
2. the last purple cluster - with fish species 11, 13, 14, 16 - which represents three fishes from the same family (*Sarda sarda*, *Scomber scombrus* and *Thunnus albacares*) and one from a singular species (*Sardina pilchardus*).

The correction about multiple comparisons problem has been made especially to validate the intra-family cosine correlations. The difference in the use of Bonferroni-Holm, or Benjamini-Hochberg, is not so sensible in the intra-family cosine correlations (whereas the number of validated extra-family cosine correlations increases considerably, around five times, with the second method).

In Fig. 7, another dendrogram has been built from the feature matrix, more restrictive in terms of peaks, and with a different R package (*pvclust*) which concerns a hierarchical clustering with bootstrapping: the two previous results are unchanged.

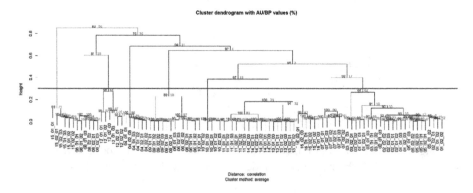

Fig. 7. Hierarchical clustering with bootstrapping for the same dataset. The legend is the same as in Fig. 4. The cut-off is at 0.3.

4   Conclusion

The proposed workflow starts from importing raw mass spectra data and it ends with analyzing and comparing tidy mass spectra data: intermediate steps concern mass spectra and peaks processing. Peak extraction allows implementing a comparison between an average mass spectrum and a sample one, in order to perform a comparison and to determine if the sample mass spectrum is graphically coherent with the average one, as shown in the first example, or not, as shown in the second one. Similarity measure (cosine correlation) and Pearson's distance quantify the difference among mass spectra. At last, fish species has been clustered, revealing that the clusters, tied to mass spectrum peaks, have a connection in term of phylogenetic classification.

Possible future applications of this work include the creation of a tool for sample identification, on the basis of comparison to reference datasets, as already performed in the case of bacteria identification through the analysis of intact cells by MALDI-TOF-MS [17], or for sample classification, especially useful in biomedical application, for instance to classify pathological or non-pathological samples [18]. From a statistical point of view, it is possible to perform a discriminant analysis (linear or diagonal) to find the peaks that are typical for a specific species [19] and to apply over them a variable selection by means of cross validation [20], even if it should pay attention about the hard reduction of peaks number during the feature matrix creation.

Acknowledgments. This work is partially supported by the Flagship InterOmics Project (PB.P05), funded and supported by the Italian Ministry of Education, University and Research and Italian National Research Council organizations. This work is also partially supported by a dedicated grant from the Italian Ministry of Economy and Finance to CNR and ENEA for the Project "Innovazione e Sviluppo del Mezzogiorno e Conoscenze Integrate per Sostenibilità ed Innovazione del Made in Italy Agroalimentare (CISIA)" Legge n.191/2009.

References

1. Karas, M., Hillenkamp, F.: Laser desorption ionization of proteins with molecular masses exceeding 10,000 daltons. Anal. Chem. **60**(20), 2299–2301 (1988)
2. Cotter, R.J.: Time-of-Flight Mass Spectrometry: Instrumentation and Applications in Biological Research. American Chemical Society, Washington D.C. (1997)
3. Mazzeo, M.F., De Giulio, B., Guerriero, G., Ciarcia, G., Malorni, A., Russo, G.L., Siciliano, R.A.: Fish authentication by MALDI-TOF mass spectrometry. J. Agric. Food Chem. **56**(23), 11071–11076 (2008)
4. Froese, R., Pauly, D.: FishBase. World Wide Web Electronic Publication (2015). http://www.fishbase.org/
5. Gibb, S., Strimmer, K.: MALDIquant: a versatile R package for the analysis of mass spectrometry data. Bioinformatics **28**(17), 2270–2271 (2012). http://strimmerlab.org/software/maldiquant/
6. Stein, S.E., Scott, D.R.: Optimization and testing of mass spectral library search algorithms for compound identification. J. Am. Soc. Mass Spectrom. **5**, 859–866 (1994)
7. Suzuki, R., Shimodaira, H.: pvclust: an R package for assessing the uncertainty in hierarchical clustering. Bioinformatics **22**(12), 1540–1542 (2006)

8. Duncan, M.W., Roder, H., Hunsucker, S.W.: Quantitative matrix-assisted laser desorption/ ionization mass spectrometry. Briefings Funct. Genomics Proteomics 7(5), 355–370 (2008)
9. Quinn, G.P., Keough, M.J.: Experimental Design and Data Analysis for Biologists, pp. 64–67. Cambridge University Press, Cambridge (2002)
10. Bromba, M.U.A., Ziegler, H.: Application hints for Savitzky-Golay digital smoothing filters. Anal. Chem. 53, 1583–1586 (1981)
11. Morhac, M.: An algorithm for determination of peak regions and baseline elimination in spectroscopic data. Nucl. Instr. Meth. Phys. Res. A 600(2), 478–487 (2009)
12. Dieterle, F., Ross, A., Schlotterbeck, G., Senn, H.: Probabilistic quotient normalization as robust method to account for dilution of complex biological mixtures. Application in 1H NMR metabonomics. Anal. Chem. 78(13), 4281–4290 (2006)
13. Khalil, H.H., Rahmat, R.O.K., Mahmoud, W.A.: Estimation of noise in gray-scale and colored images using Median Absolute Deviation (MAD). In: 3rd International Conference on Geometric Modeling and Imaging (Modern Techniques and Applications), pp. 92–97 (2008)
14. Seongho, K., Xiang, Z.: Comparative analysis of mass spectral similarity measures on peak alignment for comprehensive two-dimensional gas chromatography mass spectrometry. In: Computational and Mathematical Methods in Medicine. Hindawi Publishing Corp. (2013)
15. Seongho, K., Aiqin, F., Bing, W., Jaesik, J., Xiang, Z.: An optimal peak alignment for comprehensive two-dimensional gas chromatography mass spectrometry using mixture similarity measure. Bioinformatics 27(12), 1660–1666 (2011)
16. Gniazdowski, Z.: Geometric interpretation of a correlation. Zesz. Nauk. Warszawskiej Wyższej Szkoły Informatyki 9(7), 27–35 (2013)
17. Mazzeo, M.F., Sorrentino, A., Gaita, M., Cacace, G., Di Stasio, M., Facchiano, A., Comi, G., Malorni, A., Siciliano, R.A.: MALDI-TOF mass spectrometry for the discrimination of foodborne microorganisms. Appl. Environ. Microbiol. 72(2), 1180–1189 (2006)
18. Siciliano, R.A., Mazzeo, M.F., Spada, V., Facchiano, A., d'Acierno, A., Stocchero, M., De Franciscis, P., Colacurci, N., Sannolo, N., Miraglia, N.: Rapid peptidomic profiling of peritoneal fluid by MALDI-TOF mass spectrometry for the identification of biomarkers of endometriosis. Gynecol. Endocrinol. 30, 872–876 (2014)
19. Ahdesmaki, M., Strimmer, K.: Feature selection in omics prediction problems using cat scores and false nondiscovery rate control. Ann. Appl. Statist. 4(1), 503–519 (2010)
20. Kuhn, M.: Building predictive models in R using the caret package. J. Stat. Softw. 28(5), 1–26 (2008). http://topepo.github.io/caret/index.html

Binary Particle Swarm Optimization Versus Hybrid Genetic Algorithm for Inferring Well Supported Phylogenetic Trees

Bassam Alkindy[1,2(✉)], Bashar Al-Nuaimi[1], Christophe Guyeux[1],
Jean-François Couchot[1], Michel Salomon[1], Reem Alsrraj[1],
and Laurent Philippe[1]

[1] FEMTO-ST Institute, UMR 6174 CNRS, DISC Computer Science Department,
University of Bourgogne Franche-Comté, Besançon, France
{bassam.al-kindly,christophe.guyeux}@univ-fcomte.fr
[2] Department of Computer Science, University of Mustansiriyah, Baghdad, Iraq

Abstract. The amount of completely sequenced chloroplast genomes increases rapidly every day, leading to the possibility to build large-scale phylogenetic trees of plant species. Considering a subset of close plant species defined according to their chloroplasts, the phylogenetic tree that can be inferred by their core genes is not necessarily well supported, due to the possible occurrence of "problematic" genes (i.e., homoplasy, incomplete lineage sorting, horizontal gene transfers, etc.) which may blur the phylogenetic signal. However, a trustworthy phylogenetic tree can still be obtained provided such a number of blurring genes is reduced. The problem is thus to determine the largest subset of core genes that produces the best-supported tree. To discard problematic genes and due to the overwhelming number of possible combinations, this article focuses on how to extract the largest subset of sequences in order to obtain the most supported species tree. Due to computational complexity, a distributed Binary Particle Swarm Optimization (BPSO) is proposed in sequential and distributed fashions. Obtained results from both versions of the BPSO are compared with those computed using an hybrid approach embedding both genetic algorithms and statistical tests. The proposal has been applied to different cases of plant families, leading to encouraging results for these families.

Keywords: Chloroplasts · Phylogeny · Genetic algorithms · Lasso test · Binary Particle Swarm Optimization

1 Introduction

The multiplication of completely sequenced chloroplast genomes should normally lead to the ability to infer reliable phylogenetic trees for plant species. This is due to the existence of trustworthy coding sequence prediction and annotation software specific to chloroplasts (like DOGMA [1]) and of accurate sequence

© Springer International Publishing Switzerland 2016
C. Angelini et al. (Eds.): CIBB 2015, LNBI 9874, pp. 165–179, 2016.
DOI: 10.1007/978-3-319-44332-4_13

alignment tools. Additionally, given a set of biomolecular sequences or characters, various well-established approaches have been developed in recent years to deduce their phylogenetic relationship, encompassing methods based on Bayesian inference or maximum likelihood [2]. Robustness aspects of the produced trees can be evaluated too, for instance through bootstrap analyses. In other words, given a set of close plant species, their core genome (the set of genes in common) is as large and accurately detected as possible, to hope to be able to finally obtain a well-supported phylogenetic tree. However, all genes of the core genome are not necessarily constrained in a similar way, some genes having a larger ability to evolve than other ones due to their lower importance: such minority genes tell their own story instead of the species one, blurring so the phylogenetic information. The link between the robustness and accuracy of the phylogenetic tree, and the amount of data used for this reconstruction, is not yet completely understood. More precisely, if we consider a set of species reduced to lists of gene sequences, we have an obvious dependence between the chosen subset of sequences and the obtained tree (topology, branch length, and/or robustness). This dependence is usually regarded by the mean of gene trees merged in a phylogenetic network. This article investigates the converse approach: it starts by the union of whole core genes and tries to remove the ones that blear the phylogenetic signals. More precisely, the objective is to find the largest part of the genomes that produces a phylogenetic tree as supported as possible, reflecting by doing so the relationship of the largest part of the sequences under consideration.

Due to an overwhelming number of combinations to investigate, a brute force approach is a nonsense, which explains why heuristics are considered.

A previous work [3] has proposed the use of an ad hoc Genetic Algorithm (GA) to solve the problem of finding the largest subset of core genes producing a phylogenetic tree as supported as possible. However, in some situations, this algorithm fails to solve the optimization problem due to a low convergence rate. The proposal of this research work is thus to investigate the application of the Binary Particle Swarm Optimization (BPSO) to face our optimization challenge, and to compare it to the GA one. A new algorithm has been proposed and applied, in a distributed manner using supercomputing facilities, to investigate the phylogeny of various families of plant species.

This article is indeed an extended and improved version of the work published in the CIBB proceedings book [4]. New contributions encompass a second version of the BPSO for phylogenetic studies together with its distributed algorithm. The two BPSO versions are evaluated on a large number of new group of species. New experimental results have been thus obtained with this BPSO based approaches and with the genetic algorithm and further compared.

The remainder of this article is organized as follows. Section 2 gives a general presentation of the problem, further recalls how to extract the restricted set of core genes, and next presents various tools for constructing the phylogenetic tree from the hybrid approach. It ends with a brief description of the BPSO metaheuristic. Section 3 describes the way the metaheuristic approach is applied to solve problematic supports in biomolecular based phylogenies, considering the particular case of *Rosales* order. The distributed version of BPSO algorithm using MPI is also discussed. Obtained results and comparisons with GA approaches are detailed in Sect. 4. Finally, this paper ends with a conclusion section, in which the article is summarized and intended future work is outlined.

2   Presentation of the Problem

2.1   General Presentation

Let us consider a set of chloroplast genomes that have been annotated using DOGMA [1]. Following [5,6], we have then access to the restricted core genome [5] (genes present everywhere) of these species, whose size is about one hundred genes when the species are close enough. Sequences are further aligned using MUSCLE [7] and the RAxML [2] tool infers the corresponding phylogenetic tree. If the resulting tree is well-supported (*i.e.*, if all bootstrap values are larger than 95) we can indeed reasonably consider that the phylogeny of these species is resolved.

In a case where some branches are not well supported, we can wonder whether a few genes can be incriminated in this lack of support. If so, we face an optimization problem: *find the most supported tree using the largest subset of core genes*. Obviously, a brute force approach investigating all possible combinations of genes is intractable, as it leads to 2^n phylogenetic tree inferences for a core genome of size n. To solve this optimization problem, we have formerly proposed in [3] a general pipeline detailed in Fig. 1. In this pipeline, the stage of phylogenetic tree analysis mixes both genetic algorithm with LASSO tests in order to discover problematic genes. However, deeper experimental investigations summarized in Table 1 have shown that such a pipeline does not succeed to predict the phylogeny of some particular plant orders: in 14 groups of species the pipeline produces a score of bootstrap lower than 95 (the b column). It is important to understand what the bootstrap value represents before we can get a good response for what is "good" or "poor" support.

Bootstrapping is a resembling analysis that involves taking columns of characters out of the analysis, rebuilding the tree, and testing if the same nodes are recovered. This is done through many (100 or 1000, quite often) iterations. If, for example, you recover the same node through 95 of 100 iterations of taking out one character and resampling your tree, then you have a good idea that the node is well supported. If we get low support, this suggests that only few characters support that node, as removing characters at random from your matrix leads to a different reconstruction of that node.

We thus wonder whether a binary particle swarm optimization approach can outperform the GA when finding the largest subset of core genes producing the most supported phylogenetic tree (GA replaced by the BPSO in the "Phylogenetic tree analysis" box of Fig. 1).

Let us now give the general idea behind particle swarm optimization.

Table 1. Results of genetic algorithm approach on various families.

Group	occ	c	# taxa	b	Terminus	Likelihood	Outgroup
Gossypium_group_0	85	84	12	26	1	−84187.03	*Theo_cacao*
Ericales	674	84	9	67	3	−86819.86	*Dauc_carota*
Eucalyptus_group_1	83	82	12	48	1	−62898.18	*Cory_gummifera*
Caryophyllales	75	74	10	52	1	−145296.95	*Goss_capitis-viridis*
Brassicaceae_group_0	78	77	13	64	1	−101056.76	*Cari_papaya*
Orobanchaceae	26	25	7	69	1	−19365.69	*Olea_maroccana*
Eucalyptus_group_2	87	86	11	71	1	−72840.23	*Stoc_quadrifida*
Malpighiales	422	78	10	96	3	−91014.86	*Mill_pinnata*
Pinaceae_group_0	76	75	6	80	1	−76813.22	*Juni_virginiana*
Pinus	80	79	11	80	1	−69688.94	*Pice_sitchensis*
Bambusoideae	83	81	11	80	3	−60431.89	*Oryz_nivara*
Chlorophyta_group_0	231	24	8	81	3	−22983.83	*Olea_europaea*
Marchantiophyta	65	64	5	82	1	−117881.12	*Pice_abies*
Lamiales_group_0	78	77	8	83	1	−109528.47	*Caps_annuum*
Rosales	81	80	10	88	1	−108449.4	*Glyc_soja*
Eucalyptus_group_0	2254	85	11	90	3	−57607.06	*Allo_ternata*
Prasinophyceae	39	43	4	97	1	−66458.26	*Oltm_viridis*
Asparagales	32	73	11	98	1	−88067.37	*Acor_americanus*
Magnoliidae_group_0	326	79	4	98	3	−85319.31	*Sacc_SP80-3280*
Gossypium_group_1	66	83	11	98	1	−81027.85	*Theo_cacao*
Triticeae	40	80	10	98	1	−72822.71	*Loli_perenne*
Corymbia	90	85	5	98	2	−65712.51	*Euca_salmonophloia*
Moniliformopses	60	59	13	100	1	−187044.23	*Prax_clematidea*
Magnoliophyta_group_0	31	81	7	100	1	−136306.99	*Taxu_mairei*
Liliopsida_group_0	31	73	7	100	1	−119953.04	*Drim_granadensis*
basal_Magnoliophyta	31	83	5	100	1	−117094.87	*Ascl_nivea*
Araucariales	31	89	5	100	1	−112285.58	*Taxu_mairei*
Araceae	31	75	6	100	1	−110245.74	*Arun_gigantea*
Embryophyta_group_0	31	77	4	100	1	−106803.89	*Stau_punctulatum*
Cupressales	87	78	11	100	2	−101871.03	*Podo_totara*
Ranunculales	31	71	5	100	1	−100882.34	*Cruc_wallichii*
Saxifragales	31	84	4	100	1	−100376.12	*Aral_undulata*
Spermatophyta_group_0	31	79	4	100	1	−94718.95	*Mars_crenata*
Proteales	31	85	4	100	1	−92357.77	*Trig_doichangensis*
Poaceae_group_0	31	74	5	100	1	−89665.65	*Typh_latifolia*
Oleaceae	36	82	6	100	1	−84357.82	*Boea_hygrometrica*
Arecaceae	31	79	4	100	1	−81649.52	*Aegi_geniculata*
PACMAD_clade	31	79	9	100	1	−80549.79	*Bamb_emeiensis*
eudicotyledons_group_0	31	73	4	100	1	−80237.7	*Eryc_pusilla*
Poeae	31	80	4	100	1	−78164.34	*Trit_aestivum*
Trebouxiophyceae	31	41	7	100	1	−77826.4	*Ostr_tauri*
Myrtaceae_group_0	31	80	5	100	1	−76080.59	*Oeno_glazioviana*
Onagraceae	31	81	5	100	1	−75131.08	*Euca_cloeziana*
Geraniales	31	33	6	100	1	−73472.77	*Ango_floribunda*
Ehrhartoideae	31	81	5	100	1	−72192.88	*Phyl_henonis*
Picea	31	85	4	100	1	−68947.4	*Pinu_massoniana*
Streptophyta_group_0	31	35	7	100	1	−68373.57	*Oedo_cardiacum*
Gnetidae	31	53	5	100	1	−61403.83	*Cusc_exaltata*
Euglenozoa	29	26	4	100	3	−8889.56	*Lath_sativus*

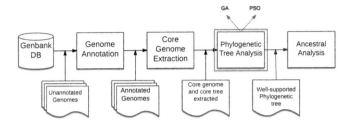

Fig. 1. Overview of the proposed pipeline

2.2 Binary Particle Swarm Optimization

Particle Swarm Optimization (PSO) is a stochastic optimization technique developed by Eberhart and Kennedy in 1995 [8]. PSOs have been successfully applied on various optimization problems like function optimization, artificial neural network training, and fuzzy system control. In this metaheuristic, particles follow a very simple behavior that is to learn from the success of neighboring individuals. An emergent behavior enables individual swarm members, particles, to take benefit from the discoveries, or from previous experiences, of the other particles that have obtained more accurate solutions. In the case of the standard binary PSO model [9], the particle position is a vector of N parameters that can be set as "yes" or "no", "true" or "false", "include" or "not include", *etc.* (binary values). A function associates to such kind of vector a score (real number) according to the optimization problem. The objective is then to define a way to move the particles in the N dimensional binary search space so that they produce the optimal binary vector w.r.t. the scoring function.

In more details, each particle i is represented by a binary vector X_i (its position). Its length N corresponds to the dimension of the search space, that is, the number of binary parameters to investigate. A 1 in coordinate j of this vector means that the associated j-th parameter is selected. A swarm of n particles is then a list of n vectors of positions $(X_1, X_2, \ldots, X_n)$ together with their associated velocities $V = (V_1, V_2, \ldots, V_n)$, which are N-dimensional vectors of real numbers between 0 and 1. These latter are initially set randomly. At each iteration, a new velocity vector is computed as follows:

$$V_i(t+1) = wV_i(t) + \phi_1 \left(P_i^{best} - X_i \right) + \phi_2 \left(P_g^{best} - X_i \right) \tag{1}$$

where w, ϕ_1, and ϕ_2 are weighted parameters setting the level of each three trends for the particle, which are respectively to continue in its adventurous direction, to move in the direction of its own best position P_i^{best}, or to follow the gregarious instinct to the global best known solution P_g^{best}. Both P_i^{best} and P_g^{best} are computed according to the scoring function.

The new position of the particle is then obtained using the equation below:

$$X_{ij}(t+1) = \begin{cases} 1 & \text{if } r_{ij} \leq Sig(V_{ij}(t+1)), \\ 0 & \text{otherwise,} \end{cases} \tag{2}$$

where r_{ij} is a threshold that depends on both the particle i and the parameter j, while the *Sig* function is the sigmoid one [9], that is:

$$Sig(V_{ij}(t+1)) = \frac{1}{1 + e^{-V_{ij}(t+1)}} \tag{3}$$

Let us now recall how to use a BPSO approach to solve our optimization problem related to phylogeny [4].

3   Particle Swarm for Phylogenetic Investigations

3.1   BPSO Applied to Phylogeny

In order to illustrate how to use the BPSO approach, we have considered the *Rosales* order, which has already been analyzed in [3] using a hybrid genetic algorithm and Lasso test approach. The *Rosales* order is constituted by 9 ingroup species and 1 outgroup (*Mollissima*), as described in Table 2. They have been annotated using DOGMA and their core genome has been computed according to the method described in [5, 6]. Its size is equal to 82 genes. Unfortunately, the phylogeny cannot be resolved directly neither by considering all these core genes nor by considering any of the 82 combinations of 81 core genes.

Table 2. Genomes information of *Rosales* species under consideration

Species	Accession	Seq.length	Family	Genus
Chiloensis	NC_019601	155603 bp	*Rosaceae*	*Fragaria*
Bracteata	NC_018766	129788 bp	*Rosaceae*	*Fragaria*
Vesca	NC_015206	155691 bp	*Rosaceae*	*Fragaria*
Virginiana	NC_019602	155621 bp	*Rosaceae*	*Fragaria*
Kansuensis	NC_023956	157736 bp	*Rosaceae*	*Prunus*
Persica	NC_014697	157790 bp	*Rosaceae*	*Prunus*
Pyrifolia	NC_015996	159922 bp	*Rosaceae*	*Pyrus*
Rupicola	NC_016921	156612 bp	*Rosaceae*	*Pentactina*
Indica	NC_008359	158484 bp	*Moraceae*	*Morus*
Mollissima	NC_014674	160799 bp	*Fagaceae*	*Castanea*

As some branches are not well supported, we can wonder whether a few genes can be incriminated in this lack of support, for a large variety of reasons encompassing homoplasy, stochastic errors, undetected paralogy, incomplete lineage sorting, horizontal gene transfers, or even hybridization. If so, we face the optimization problem presented previously: *find the most supported tree using the largest subset of core genes.*

Genes of the core genome are now supposed to be lexicographically ordered. Each subset S of the core genome is thus associated with a unique binary word W of length n: for each i, $1 \leq i \leq n$, W_i is 1 if the i-th core gene is in S and 0 otherwise. Any n-length binary word W can be associated with its percentage p of 1's and the lowest bootstrap b of the phylogenetic tree we obtain when considering the subset of genes associated to W. Each word W is thus associated with a fitness score value $\mathcal{F} = \frac{b+p}{2}$.

In the BPSO context the search space is then $\{0,1\}^N$, where $N = 82$ in *Rosales*. Each node of this N-cube is associated with the set of following data: its subset of core genes, the deduced phylogenetic tree, its lowest bootstrap b and the percentage p of considered core genes, and, finally, the score $\frac{b+p}{2}$. Notice that two close nodes of the N-cube have two close percentages of core genes. We thus have to construct two phylogenies based on close sequences, leading with a high probability to the same topology with close bootstraps. In other words, the score remains essentially unchanged when moving from a node to one of its neighbors. It allows to find optimal solutions using approaches like BPSO.

During swarm initialization, the L particles (set to 10 in our experiments) of a swarm are randomly distributed among all the vertices (binary words) of the N-cube that have a large percentage of 1's. The objective is then to move these particles in the cube so that they will converge to an optimal node.

At each iteration, the particle velocity is updated by taking into account its own best position and the best one considering the whole particle swarm (both identified according to the fitness value). It is influenced by constant weight factors as expressed in Eq. (1). In this one, we have set $\phi_1 = c_1 \cdot r_1$ and $\phi_2 = c_2 \cdot r_2$ where $c_1 = 1$ and $c_2 = 1$, while r_1, r_2 are random numbers belonging to $[0.1, 0.5]$, and w is the inertia weight that is computed based on the following formula:

$$w = w_{max} - \frac{w_{max} - w_{min}}{I_{max}} \times I'_{cur} \qquad (4)$$

where I_{max} represents the maximum number of iterations (or time step) and I'_{cur} is the current iteration. This equation determines the contribution rate of a particle's previous velocity and is determined as in [10].

To increase the number of included components in a particle, we reduce the interval of Eq. (1) to $[0.1, 0.5]$. For instance, if the velocity V_{ij} of an element is equal to 0.51 and $r_{ij} = 0.83$, then $Sig(0.51) = 0.62$. So $r_{ij} > Sig(V_{ij})$ and this leads to 0 in the vector element j of the particle i. By minimizing the interval, we increase the probability of having $r_{ij} < Sig(V_{ij})$ and consequently the number of 1s, which means more included elements in the particle (a larger number of core genes).

Note that a large inertia weight facilitates a global search, while a small inertia weight tends more to a local investigation [11]. In other words, a larger value of w facilitates a complete exploration, whereas small values promote exploitation of areas. This is why Eberhart and Shi [12] suggested to decrease w over time, typically from 0.9 to 0.4, thereby gradually changing from exploration to exploitation. Finally, each particle position is updated according to Eq. (2).

3.2   Distributed BPSO with MPI

Traditional PSO algorithms are time consuming in sequential mode. The distributed version shown in Fig. 2 has thus been proposed to minimize the execution time as much as possible. The general idea of the proposed algorithm is simple: a processor core is employed for each particle in order to compute its fitness value, while a last core called the master centralizes the obtained results. In other words, if we have a swarm of ten particles, we use ten cores as workers and one core as master (or supervisor).

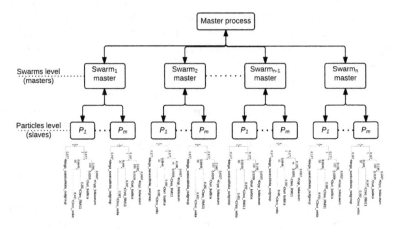

Fig. 2. The distributed structure of BPSO algorithm.

More precisely, the master initializes the particles of the swarm and distributes them to the workers. When one worker finishes its job, it sends a "terminate" signal with the fitness value to the master. This latter waits until all the workers have finished their jobs. Then, it determines the position of the particle that has the best fitness value as the global best position and sends this information to the workers that update their respective particle velocity and position. This mechanism is repeated until a particle achieves a fitness value larger than or equal to 95 with a large set of included genes. In the following, two distributed versions of the BPSO described previously are considered: in version I the equation used to update the velocity is slightly changed as shown below, and in version II we use the equations of Sect. 2.2.

Distributed BPSO Algorithm: Version I. In this version Eq. (1), which is used to update the velocity vector, is replaced by:

$$V_i(t+1) = x \cdot [V_i(t) + C_1(P_i^{best} - X_i) + C_2(P_g^{best} - X_i)] \tag{5}$$

where x, C_1, and C_2 are weighted parameters setting the level of each three trends for the particle. The default values of these parameters are $C_1 = c_1 \cdot r_1 = 2.05$,

$C_2 = c_2 \cdot r_2 = 2.05$, while x which represents the constriction coefficient is computed according to formula [13,14]:

$$x = \frac{2 \times k}{|2 - C - (\sqrt{C \times (C-4)})|}, \tag{6}$$

where k is a random value between $[0,1]$ and $C = C_1 + C_2$, where $C \geq 4$. According to Clerc [14], using a constriction coefficient results in particle convergence over time.

Distributed BPSO Algorithm: Version II. This version is a distributed approach of the sequential PSO algorithm presented previously in Sect. 2.2.

4 Phylogenetic Prediction

4.1 Genetic Algorithm Evaluation on a Large Group of Plant Species

The proposed pipeline has been tested with the genetic algorithm on various sets of close plant species. 50 subgroups, including on average from 12 to 15 chloroplasts species, encompassing 356 plant species, and already presented in this document (*c.f.* Table 1) have been used with our formerly published genetic algorithm. Obtained results with details are contained too in Table 1. Column *Occ* represents the amount of generated phylogenetic trees from the corresponding search space for each group. The column *c* represents the number of core genes included within each group. The *# taxa* column is the amount of species corresponding to the considered group. *b* is the lowest value from bootstrap analysis. The *Terminus* column contains the termination stage for each subgroup, namely: the systematic (1), random (2), or optimization (3) stage using genetic algorithm and/or Lasso test. These stages, which have been proposed in [3], correspond to the systematic deletion of 0 or 1 gene ($N + 1$ computations for N core genes), random suppression of core genes (ranging from 2 to 5 genes), and the so-called genetic algorithm on binary word populations improved by the use of a statistical test. Finally, the *Likelihood* column stores the likelihood value of the best phylogenetic tree (*i.e.*, according to the lowest bootstrap value b). A large occurrence value in this table means that the associated p-value and/or subgroup has its computation terminated in either penultimate or last pipeline stage. An occurrence of 31 is frequent due to the fact that 32 MPI threads (one master plus 31 slaves) have been launched on our supercomputing facility.

Notice that the groups in Table 1 can be divided in four parts:

– Groups of species stopped in systematic stage with weak bootstrap values. This is due to the fact that an upper time limit has been set for each group and/or subgroups, while each computed tree in these remarkable groups needed a lot of times for computations.
– Subgroups terminated during systematic stage with desired bootstrap value.

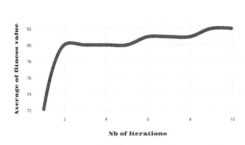

Fig. 3. Average fitness of *Rosales* order

Table 3. Best tree in each swarm

Swarm	Removed genes	$\mathcal{F}$	b
1	4	75.5	73
2	6	75.5	76
3	20	75	88
4	52	59.5	89
5	3	75.5	72
6	19	77.5	92
7	47	63.5	92
8	9	73.5	74
9	10	72.5	73
10	13	76.5	84

– Groups or subgroups terminated in random stage with desired bootstrap value.
– Finally, groups or subgroups terminated with optimization stages.

A majority of subgroups has its phylogeny satisfactorily resolved, as can be seen on all obtained trees which can be downloadable at http://meso.univ-fcomte.fr/peg/phylo. However, some problematic subgroups still remain to be investigated, which explains why the distributed BPSO is considered in the next section.

4.2 First Experiments on *Rosales* Order

In a first collection of experiments, we have implemented the proposed BPSO algorithm on a supercomputing facility. Investigated species are the ones listed in Table 2. 10 swarms having a variable number of particles have been launched 10 times, with $c_1 = 1, c_2 = 1$, and w linearly decreasing from 0.9 to 0.4. Obtained results are summarized in Table 3 that contains, for each 10 runs of each 10 swarms: the number of removed genes and the minimum bootstrap of the best tree. Remark that some bootstraps are not so far from the intended ones (larger than 95), whereas the number of removed genes are in average larger than what is desired.

Seven topologies have been obtained after either convergence or *maxIter* iterations. Only 3 of them have occurred a representative number of times, namely the Topologies 0, 2, and 4, which are depicted in Fig. 4 (see details in Table 4).

These three topologies are almost well supported, except in a few branches. We can notice that the differences in these topologies are based on the sister relationship of two species named *Fragaria vesca* and *Fragaria bracteata*, and of the relation between *Pentactina rupicola* and *Pyrus pyrifolia*. Due to its larger score and number of occurrences, we tend to select Topology 0 as the best representative of the *Rosale* phylogeny.

Table 4. Best topologies obtained from the generated trees, b is the lowest bootstrap of the best tree having this topology, $= p$ is the number of considered genes to obtain this tree.

Topology	Swarms	b	p	$\mathcal{F}$	Occurrences
0	1, 2, 3, 4, 5, 6, 7, 8, 9, 10	92	63	77.5	568
1	1, 2, 3, 4, 5, 6, 10	63	45	54	11
2	1, 2, 3, 4, 5, 6, 7, 8, 9, 10	76	67	71.5	55
3	8, 1, 2, 3, 4	56	41	48.5	5
4	1, 2, 3, 4, 5, 6, 7, 8, 9, 10	89	30	59.5	65
5	1, 3, 4, 5, 6, 9	71	33	52	9
6	5, 6	25	45	35	2

To further validate this choice, CONSEL [15] software has been used on per site likelihoods of each best tree obtained using the RAxML [2]. The CONSEL computes the p-values of various well-known statistical tests, like the so-called approximately unbiased (au), Kishino-Hasegawa (kh), Shimodaira-Hasegawa (sh), and Weighted Shimodaira-Hasegawa (wsh) tests. Obtained results are provided in Table 5, they confirm the selection of Topology 0 as the tree reflecting the best the *Rosales* phylogeny.

Table 5. The CONSEL results regarding best trees

Rank	item	obs	au	np	bp	pp	kh	sh	wkh	wsh
1	0	−1.4	0.774	0.436	0.433	0.768	0.728	0.89	0.672	0.907
2	4	1.4	0.267	0.255	0.249	0.194	0.272	0.525	0.272	0.439
3	2	3	0.364	0.312	0.317	0.037	0.328	0.389	0.328	0.383

After having verified that BPSO can be used to resolve phylogenetic issues thanks to the *Rosales* order, we now intend to deeply compare the genetic algorithm versus the swarm particle optimization. In order to do so, a large collection of group of plant species have been selected, on which we have successively launched the genetic algorithm and the BPSO one in distributed mode.

4.3 Comparison Between Distributed Version of GA and the Two Distributed Versions of BPSO

12 groups of plant genomes have been extracted from the 49 ones used in the GA evaluation. More precisely, seven "difficult" groups have been selected from those that have reached the third stage in genetic algorithm method (no resolution of phylogeny during systematic and random modes). Conversely, five "easy" groups

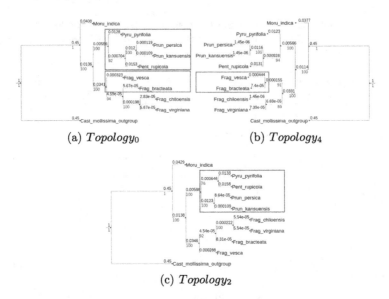

(a) $Topology_0$ (b) $Topology_4$

(c) $Topology_2$

Fig. 4. The best obtained topologies for *Rosales* order

have been added in the pool of experiments, for the sake of comparison: in these groups, the phylogeny has been resolved during the systematic mode. They have been applied on our two swarm versions, and results have been compared to the genetic algorithm ones. We have successively tested 10 and 15 particles (with each of the two algorithms), on the supercomputer facilities.

Comparisons are provided in Tables 6 and 7. In these tables, *Topo.* column stands for the number of topologies, *NbTrees* is the total number of obtained trees using 10 swarms, b is the minimum bootstrap value of selected w, $100 - p$ is the number of missing genes in w and *Occ.* is the number of occurrences of

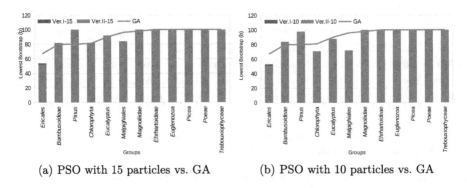

(a) PSO with 15 particles vs. GA   (b) PSO with 10 particles vs. GA

Fig. 5. PSO with 10 and 15 particles vs. GA.

Table 6. Groups from BPSO Version 1.

Group	Topo.	NbTrees	b	$\lvert c\rvert$	$100-p'$	Occ.	Swarms	Particles
Pinus	3	508	98	79	32	462	1,2,3,4,5,6,7,8,9,10	10
Pinus	3	530	94	79	11	129	1,2,3,4,5,6,7,8,9,10	15
Picea	1	100	100	85	42	100	1,2,3,4,5,6,7,8,9,10	10
Picea	1	428	100	85	13	428	1,2,3,4,5,6,7,8,9,10	15
Magnoliidae	3	750	100	79	20	613	1,2,3,4,5,6,7,8,9,10	10
Magnoliidae	3	845	100	79	19	707	1,2,3,4,5,6,7,8,9,10	15
Ericales	30	344	53	84	26	185	1,2,3,4,5,6,7,8,9,10	10
Ericales	34	555	54	84	5	363	1,2,3,4,5,6,7,8,9,10	15
Bambusoideae	8	496	72	94	37	456	1,2,3,4,5,6,7,8,9,10	10
Bambusoideae	11	694	69	94	18	621	1,2,3,4,5,6,7,8,9,10	15
Eucalyptus	16	828	86	83	7	632	1,2,3,4,5,6,7,8,9,10	10
Eucalyptus	20	1073	86	80	4	845	1,2,3,4,5,6,7,8,9,10	15
Malpighiales	34	327	65	78	35	233	1,2,3,4,5,6,7,8,9,10	10
Malpighiales	38	483	69	78	40	326	1,2,3,4,5,6,7,8,9,10	15
Chlorophyta	25	191	70	24	11	109	1,2,3,4,5,6,7,8,9,10	10
Chlorophyta	29	94	68	24	11	1	1,2,3,4,5,6,7,8,9,10	15
Euglenozoa	3	450	100	26	7	292	1,2,3,4,5,6,7,8,9,10	10
Euglenozoa	3	520	100	26	4	491	1,2,3,4,5,6,7,8,9,10	15
Ehrhartoideae	2	23	100	81	0	23	1,2,3,4,5,6,7,8,9,10	10
Ehrhartoideae	3	455	100	81	0	451	1,2,3,4,5,6,7,8,9,10	15
Trebouxiophyceae	3	409	100	41	2	405	1,2,3,4,5,6,7,8,9,10	10
Trebouxiophyceae	3	415	100	41	8	354	1,2,3,4,5,6,7,8,9,10	15
Poeae	1	971	100	80	9	971	1,2,3,4,5,6,7,8,9,10	10
Poeae	1	1399	100	80	20	1399	1,2,3,4,5,6,7,8,9,10	15

Table 7. Groups from BPSO Version 2.

Group	Topo.	NbTrees	b	$\lvert c\rvert$	$100-p'$	Occ.	Swarms	Particles
Pinus	3	615	98	79	14	275	1,2,3,4,5,6,7,8,9,10	10
Pinus	3	628	100	79	12	558	1,2,3,4,5,6,7,8,9,10	15
Picea	1	635	100	85	14	635	1,2,3,4,5,6,7,8,9,10	10
Picea	1	821	100	85	15	821	1,2,3,4,5,6,7,8,9,10	15
Magnoliidae	3	494	100	79	16	73	1,2,3,4,5,6,7,8,9,10	10
Magnoliidae	3	535	100	79	42	384	1,2,3,4,5,6,7,8,9,10	15
Bambusoideae	6	952	84	81	23	94	1,2,3,4,5,6,7,8,9,10	10
Bambusoideae	9	1450	82	81	18	113	1,2,3,4,5,6,7,8,9,10	15
Eucalyptus	17	972	88	80	18	618	1,2,3,4,5,6,7,8,9,10	10
Eucalyptus	23	1439	92	80	10	843	1,2,3,4,5,6,7,8,9,10	15
Chlorophyta	25	529	71	24	6	397	1,2,3,4,5,6,7,8,9,10	10
Chlorophyta	46	1500	82	24	11	397	1,2,3,4,5,6,7,8,9,10	15
Ericales	30	97	51	84	11	56	1,2,3,4,5,6,7,8,9,10	10
Ericales	34	1257	52	84	7	800	1,2,3,4,5,6,7,8,9,10	15
Malpighiales	35	725	72	79	25	445	1,2,3,4,5,6,7,8,9,10	10
Malpighiales	86	1464	84	79	45	359	1,2,3,4,5,6,7,8,9,10	15
Euglenozoa	3	197	100	26	1	165	1,2,3,4,5,6,7,8,9,10	10
Euglenozoa	3	450	100	26	10	393	1,2,3,4,5,6,7,8,9,10	15
Ehrhartoideae	1	24	100	81	10	24	1,2,3,4,5,6,7,8,9,10	10
Ehrhartoideae	1	20	100	81	9	20	1,2,3,4,5,6,7,8,9,10	15
Trebouxiophyceae	3	319	100	41	1	313	1,2,3,4,5,6,7,8,9,10	10
Trebouxiophyceae	3	818	100	41	2	81	1,2,3,4,5,6,7,8,9,10	15
Poeae	1	991	100	80	22	991	1,2,3,4,5,6,7,8,9,10	15
Poeae	1	1490	100	80	26	1490	1,2,3,4,5,6,7,8,9,10	15

Table 8. PSO vs GA.

Group	PSO Ver. I		PSO Ver. II		GA
	10	15	10	15	
Ericales	53	54	51	52	67
Bambusoideae	72	69	84	82	80
Pinus	98	94	98	100	80
Chlorophyta	70	68	71	82	81
Eucalyptus	86	86	88	92	90
Malpighiales	65	69	72	84	96
Magnoliidae	100	100	100	100	98
Ehrhartoideae	100	100	100	100	100
Euglenozoa	100	100	100	100	100
Picea	94	100	100	100	100
Poeae	80	80	100	100	100
Trebouxiophyceae	100	100	100	100	100

the best obtained topology from 10 swarms. As can be seen in these tables, the two versions of BPSO did not provide the same kind of results:

– In the case of *Chlorophyta*, *Pinus*, and *Bambusoideae*, the second version of the BPSO has outperformed the first one, as the minimum bootstrap b of the best tree is finally larger for at least one swarm.
– In the *Ericales* case, the first version has produced the best result.

We can also remark that *Malpighiales* has better b in GA than the two versions of PSO. For easy to solve subgroups, *Pinus* data set has got maximum bootstrap b larger than what has been obtained using the genetic algorithm, while *Picea* and *Trebouxiophyceae* have got the same values of b than in genetic algorithm. More comparison results between GA and both versions of PSOs are provided in Fig. 5.

According to this figure, we can conclude that the two approaches lead to quite equivalent bootstrap values in most data sets, while on particular subgroups obtained results are complementary. In particular, PSO often produces better bootstraps that GA (see *Magnoliidae* or on *Bambusoideae*), but with a larger number of removed genes. Finally, using 15 particles instead of 10 does not improve so much the obtained results (see Fig. 5 and Table 8).

5   Conclusion

This article has presented an original method to produce a well supported and large-scale phylogenetic tree of chloroplast species where various optimization algorithms are applied to highlight the relationships among given gene sequences.

More precisely, this method first discovers and removes blurring genes in the set of core genes by applying a bootstrap analysis for each tree produced from a subset of core genes. It then continues with integrating a discrete PSO method to provide the largest subset of sequences. Two distributed versions of this PSO-based optimization step have been developed in order to reduce the computation time and memory used. Finally, a per site analysis by the CONSEL is applied: a dedicated topological process analyses all the output trees and might use a per site analysis in order to extract the most relevant ones. Our proposed pipeline has been applied to various families of plant species. More than 65 % of phylogenetic trees produced by this pipeline have presented bootstrap values larger than 95.

Acknowledgements. All computations have been performed on the *Mésocentre de calculs* supercomputer facilities of the University of Franche-Comté.

References

1. Wyman, S.K., Jansen, R.K., Boore, J.L.: Automatic annotation of organellar genomes with dogma. Bioinformatics **20**(172004), 3252–3255 (2004). Oxford Press
2. Stamatakis, A.: Raxml version 8: a tool for phylogenetic analysis and post-analysis of large phylogenies. Bioinformatics **30**, 1312–1313 (2014)
3. AlKindy, B., Guyeux, C., Couchot, J.-F., Salomon, M., Parisod, C., Bahi, J.M.: Hybrid genetic algorithm and lasso test approach for inferring well supported phylogenetic trees based on subsets of chloroplastic core genes. In: Dediu, A.-H., Hernández-Quiroz, F., Martín-Vide, C., Rosenblueth, D.A. (eds.) AlCoB 2015. LNCS, vol. 9199, pp. 83–96. Springer, Heidelberg (2015)
4. Alsrraj, R., Alkindy, B., Guyeux, C., Philippe, L., Couchot, J.-F.: Well-supported phylogenies using largest subsets of core-genes by discrete particle swarm optimization. In: CIBB 2015, 12th International Meeting on Computational Intelligence Methods for Bioinformatics and Biostatistics, Naples, Italy, September 2015
5. Alkindy, B., Couchot, J.-F., Guyeux, C., Mouly, A., Salomon, M., Bahi, J.M.: Finding the core-genes of chloroplasts. J. Biosci. Biochem. Bioinform. **4**(5), 357–364 (2014)

6. Alkindy, B., Guyeux, C., Couchot, J.-F., Salomon, M., Bahi, J.M.: Gene similarity-based approaches for determining core-genes of chloroplasts. In: IEEE International Conference on Bioinformatics and Biomedicine (BIBM) (to Present) (2014)

7. Edgar, R.C.: Muscle: multiple sequence alignment with high accuracy and high throughput. Nucleic Acids Res. **32**(5), 1792–1797 (2004)

8. Kennedy, J., Eberhart, R.C.: Particle swarm optimization. In: Proceedings of IEEE International Conference on Neural Networks, vol. 4, pp. 1942–1948 (1995)

9. Khanesar, M.A., Tavakoli, H., Teshnehlab, M., Shoorehdeli, M.A.: Novel binary particle swarm optimization, 11 (2009). www.intechopen.com, ISBN: 978-953-7619-48-0

10. Premalatha, K., Natarajan, A.M.: Hybrid pso and ga for global maximization. Int. J. Open Probl. Compt. Math **2**(4), 597–608 (2009)

11. Poli, R., Kennedy, J., Blackwell, T.: Particle swarm optimization. Swarm Intell. **1**, 33–57 (2007). doi:10.1007/s11721-007-0002-0. (Springer Science + Business Media)

12. Eberhart, R.C., Shi, Y.: Particle swarm optimization: developments, applications and resources. In: Proceedings of the 2001 Congress on Evolutionary Computation, vol. 1, pp. 81–86. IEEE (2001)

13. Sedighizadeh, D., Masehian, E.: Particle swarm optimization methods, taxonomy and applications. Int. J. Comput. Theory Eng. **1**(5), 486–502 (2009)

14. Clerc, M.: The swarm and the queen: towards a deterministic and adaptive particle swarm optimization. In: Proceedings of the 1999 Congress on Evolutionary Computation, CEC 1999, vol. 3. IEEE (1999)

15. Shimodaira, H., Hasegawa, M.: Consel: for assessing the confidence of phylogenetic tree selection. Bioinformatics **17**(12), 1246–1247 (2001)

Computing Discrete Fine-Grained Representations of Protein Surfaces

Sebastian Daberdaku[✉] and Carlo Ferrari

Department of Information Engineering, University of Padova,
Via Gradenigo 6/B, 35131 Padova (PD), Italy
sebastian.daberdaku@dei.unipd.it, carlo.ferrari@unipd.it

Abstract. We present a voxel-based methodology for the computation of discrete representations of macromolecular surfaces at high resolutions. The procedure can calculate the three main molecular surfaces, namely van der Waals, Solvent-Accessible and Solvent-Excluded, by employing compact data structures and implementing a spatial slicing protocol. Fast Solvent-Excluded surface generation is achieved by adapting an approximate Euclidean Distance Transform algorithm. The algorithm exploits the geometrical relationship between the Solvent-Excluded and the Solvent-Accessible surfaces and limits the calculation of the distance map values to a small subset of the overall voxels representing the macromolecule. A parallelization scheme for the slicing procedure is also proposed and discussed.

Keywords: Macromolecular surface · High-resolution voxel surface · EDT

1 Introduction

Proteins are large molecules that play a vast range of biological functions by intervening in virtually all cellular activities. Each protein consists of one or more sequences of linear polymers (chains) of amino acids linked to each other by peptide covalent bonds. The physicochemical properties of its components, along with the surrounding environment (solvent), determine the proteins' specific three-dimensional (3D) shape. Proteins express their biological roles by binding selectively and with high affinity to other biomolecules, which depends on the formation of a set of weak, non-covalent bonds (hydrogen bonds, ionic bonds, van der Waals interactions) plus favourable hydrophobic interactions. Because the weakness of each individual bond, effective binding interactions require the simultaneous formation of multiple weak bonds, which is only possible if the surface contours of the interacting molecules are geometrically complementary. For this reason, many *in silico* methods for the prediction of protein functions, properties and interactions require proper representations of the molecular surface and its associated physicochemical properties.

To capture diverse aspects of the 3D geometry of proteins and macromolecules, different representations of the molecular surface have been introduced.

© Springer International Publishing Switzerland 2016
C. Angelini et al. (Eds.): CIBB 2015, LNBI 9874, pp. 180–195, 2016.
DOI: 10.1007/978-3-319-44332-4_14

Currently, the most used representations are: the van der Waals surface (vdW) [1], the Solvent-Accessible surface or Lee-Richards surface (SAS) [2] and the Solvent-Excluded surface (SES) or Connolly surface [3,4]. Surface calculations of large biomolecules, such as proteins or nucleic acids, based on their experimentally determined 3D structures (typically obtained by X-ray crystallography, NMR spectroscopy, or, increasingly, cryo-electron microscopy) have been extensively used in modern molecular biology studies [5–9].

Voxel-based molecular surface representations have received a lot of interest in many bioinformatics and computational biology applications. A voxel is the tiniest distinguishable element of a 3D object and can be thought as the equivalent of a pixel in the 3D domain. It is a discrete volume element that represents a single cell on a regularly spaced 3D grid. Voxels can be labelled with multiple values, describing various properties of a certain portion of space, and have been extensively used for visualization and analysis of scientific and medical data.

Voxelized protein surfaces are currently being employed in descriptor-based protein docking, pairwise alignment of molecules, protein shape comparison, pocket identification and FFT-based fast computations. Kihara et al. propose protein docking, shape comparison and interface identification methods based on 3D Zernike descriptors (3DZD) [10–12], which are calculated over circular surface patches of voxelized macromolecular surfaces. The voxelized representation of a molecular surface can describe the molecule's flexibility [13,14] and physicochemical property values, such as electrostatic potentials or hydrophobicity [15]. In [16] a ligand-binding pocket identification algorithm is introduced which uses a voxelized representation of the Connolly surface. In [17,18] the Fast Fourier Transform is used to efficiently match shape and electrostatic properties on surface grid points for protein docking. Protein surface atoms extraction based on a voxelized representation, which yields full atoms listings useful for studying binding regions on protein surfaces, was introduced in [19]. In [20], a voxelized protein representation is used for the identification and modelling of ligand binding areas. Grid representations of protein surfaces have also been used in cavity detection, binding-pockets identification and evaluation techniques [21–24].

Although macromolecular structural data repositories such as the Protein Data Bank (PDB) [25] have long been available, only a limited number of surface representations is provided, primarily aimed for visualization purposes. Surface calculation is an application-dependent task, resulting in multiple parametrisations based on the users' requirements. Protein surfaces are usually produced at runtime, adding high computational cost to the overall calculation. Moreover, many of the techniques and algorithms employing voxel-based molecular surfaces derive the latter from other explicit representations such as triangle mesh surfaces. Surface meshes are placed inside 3D grids, and the voxels whose centres are intersected by or within a certain distance from the mesh faces are marked as occupied (typically with $1, 0$ otherwise), resulting in tremendous accuracy loss for the final representation [26–28]. These naïve voxelization methods cannot guarantee two important requirements that voxelized surfaces must exhibit: separability and minimality [29]. The separability requirement ensures that the resulting voxelized surface is connected and gap-free. On the other hand, a minimal voxelized surface should not contain voxels that, if removed, make no difference in terms of separability.

In this paper we propose and analyse a methodology for the computation of voxel-based fine-grained representations of the van der Waals, Solvent-Accessible and Solvent-Excluded surfaces starting directly from their PDB entries. To the extent of our knowledge, there are no available tools which can produce voxelized surface representations of macromolecules at the desired resolutions starting from their experimentally determined structural data (PDB entries). Several surface computation and visualization tools are available to date (see Table 1), but none of them provides voxelized representations of molecular surfaces. Representing and elaborating high-resolution 3D voxel grids requires memory-demanding data structures as well as high computational resources. We deal with the memory requirements by implementing a compact representation of the voxel grid and by implementing a spatial slicing protocol previously introduced in [30]. Only one bit is used to represent each voxel in the grid, which is eight times less than the smallest elemental type (char) on most systems. The molecule is sliced with parallel planes in a user-defined number of parts and the surface is computed for each slice separately. A parallelization scheme is also proposed and discussed. The parallel computation of voxelized surfaces on top of a compact data representation is the key to reducing computation time while maintaining accuracy, as shown by experimental results.

1.1   Macromolecular Surface Definitions

A molecule can be represented as a set of possibly overlapping spheres, each one having a radius equal to the van der Waals radius of the atom it represents. The topological boundary of this set of spheres is what is known as van der Waals surface. For proteins and other macromolecules, much of the van der Waals surface is buried in the inside of the molecules and is not accessible to the solvent or possible ligands. For this reason, the van der Waals surface is rarely used in bioinformatics applications. However, it is very important because it serves as a foundation to other surface definitions, and also because it is the basis of what is known as Corey-Pauling-Koltun (CPK) model (also known as calotte model or space-filling model [41, 42]).

The Solvent-Accessible and Solvent-Excluded surfaces determine the 3D shape of the molecule in functional relationship with the external solvent. The Solvent-Accessible surface is defined as the geometric locus of the centre of a probe sphere (representing the solvent molecule) as it rolls over the van der Waals surface of the molecule. The Solvent-Excluded surface is defined as the locus of the inward-facing probe sphere as it rolls over the van der Waals surface of the molecule. This surface can be considered as a continuous sheet consisting of two parts: the contact surface and the re-entrant surface. The contact surface is part of the van der Waals surface that is accessible to a probe sphere. The re-entrant surface is the inward-facing surface of the probe when it touches two or more atoms. There is a clear relation between the SAS and the SES, as the

Table 1. Overview of some molecular surface computation tools.

Name	Surface representation			Comments
	vdW	SAS	SES	
PyMOL [31]	dot, spheres	dot, spheres	mesh	Scriptable molecular visualization system; extensible with Python
DeepView [32]	dot	dot	mesh	Tightly linked to SWISS-MODEL, an automated homology modelling server
MSMS [33]	n/a	n/a	dot, mesh	Dot surface over-sampled in some areas; can fail computing the surface of large molecules
UCSF chimera [34]	dot, spheres	dot, spheres	dot, mesh	Uses MSMS to compute the SES. Supports interactive visualization and analysis of molecular structures, density maps, assemblies, sequence alignments, docking results and trajectories
VMD [35]	dot, spheres	dot	mesh	Uses either SURF [36] or MSMS to compute the SES. Supports displaying, animating, and analyzing large biomolecular systems using 3-D graphics and built-in scripting
RasMol [37]	dot, spheres	spheres	n/a	Aimed at visualisation and generation of publication quality images
Jmol [38]	dot, spheres	dot, spheres	dot, mesh	Supports multiple molecules with independent movement, surfaces, orbitals, cavity visualization and crystal symmetry
Avogadro [39]	mesh	mesh	n/a	Advanced molecule editor; extensible via a plugin architecture
DS visualizer [40]	mesh	n/a	mesh	Commercial-grade graphics visualization tool for viewing, sharing, and analysing protein and modelling data

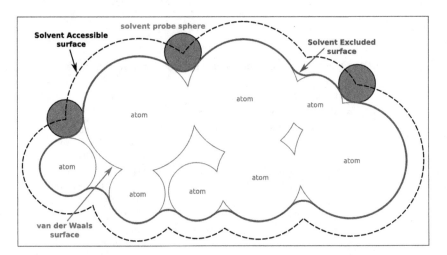

Fig. 1. The three main molecular surfaces: van der Waals (green), Solvent-Accessible (black) and Solvent-Excluded (blue). (Color figure online)

Solvent-Accessible surface is displaced outward from the Solvent-Excluded one by a distance equal to the probe radius. Figure 1 gives a graphical representation of the three molecular surfaces.

2   Materials and Methods

2.1   Surface Calculation Algorithm

The first step of the proposed methodology consists in acquiring the atomic coordinates of a macromolecule from its corresponding Protein Data Bank entry. The atomic coordinates of each atom composing the macromolecule and the relative radius are stored in a dedicated data structure. By default, the radius assignment for each atom type is based on the CHARMM27 force field [43], but users can provide their own radii information.

Pose normalisation follows the data acquisition step. The centre of gravity of the molecule is first translated to the coordinate origin. Then, a rotation is applied to the atomic coordinates in order to align the three principal axes of the molecule with the coordinate axes. The rotation matrix is determined by running Principal Component Analysis on the atomic coordinates.

The algorithm calculates the tightest axis-aligned bounding-box enclosing the whole molecule by determining the minimal and maximal atomic coordinates. Given a user-defined grid resolution parameter, the dimensions of the voxel grid which will contain the molecule are calculated. All atomic coordinates previously imported are translated, scaled and quantized to the new coordinate system defined by the voxel grid: each atom centre is mapped in its corresponding voxel in the voxel grid.

An adaptation of Bresenham's line algorithm [44] is used to efficiently determine the voxels occupied by a given atom. Each atom in the molecule is represented by a ball having a radius equal to either the atom's radius when calculating the vdW, or the atom's radius increased by the solvent-probe's radius when calculating the SAS and SES. After all the atoms composing the macromolecule have been mapped into the grid, we obtain the voxelized representation of the CPK model.

To obtain the van der Waals or the Solvent-Accessible surfaces, the boundary voxels of the voxelized representation of the CPK volumetric model of the macromolecule are extracted using an efficient 3D flood-filling algorithm [45]. The Solvent-Excluded surface is trickier to calculate because it includes the re-entrant surface portions. The proposed method is based on the Euclidean Distance Transform (EDT) algorithm for surface smoothing.

The implemented tool supports four different output formats: the Point Cloud Data file [46], OpenDX [47], Visualization Toolkit Structured Points and Visualization Toolkit PolyData [48, 49].

2.2   The Euclidean Distance Transform

A distance transform (also known as distance map or distance field), is a derived representation of a digital image (usually a binary image). Distance maps are images where the value of each voxel of the foreground is the distance to the nearest voxel of the background. Let $B \in \{0,1\}^{l \times w \times h}$ be a binary voxel grid of length l, width w and height h. There are exactly $l \times w \times h$ voxels in B, each one identified by the ordered triple $\boldsymbol{v} = (i, j, k) \in V = \{1, \dots, l\} \times \{1, \dots, w\} \times \{1, \dots, h\}$. Also, let $I_B : V \rightarrow \{0, 1\}$ be the image function of B defined as

$$I_B(i, j, k) = b_{i,j,k} \in \{0, 1\}, \tag{1}$$

where $b_{i,j,k}$ is the value of voxel (i, j, k) in B. Let V_O be the set of occupied voxels of B, i.e.

$$V_O = \{\boldsymbol{v} = (i, j, k) \in V \mid I_B(i, j, k) = 1\}. \tag{2}$$

Also, let $NBV_B : V \rightarrow V_O$, such that $\forall \boldsymbol{v} \in V$, $NBV_B(\boldsymbol{v})$ is a nearest occupied voxel of B to $\boldsymbol{v}$, that is

$$NBV_B(\boldsymbol{v}) \in \arg \min_{\boldsymbol{w} \in V_O} d(\boldsymbol{w}, \boldsymbol{v}) = \{\boldsymbol{w} \in V_O \mid \forall \boldsymbol{y} \in V_O : d(\boldsymbol{w}, \boldsymbol{v}) \leq d(\boldsymbol{y}, \boldsymbol{v})\}, \tag{3}$$

according to some distance metric d. $NBV_B(\boldsymbol{v})$ is called the nearest boundary voxel (NBV) of $\boldsymbol{v}$ in B. Clearly, if $\boldsymbol{v} \in V_O$ then $NBV_B(\boldsymbol{v}) = \boldsymbol{v}$.

Finally, the distance transform of B (also known as distance map or distance field) is defined as a real-valued voxel grid $DT_B \in \mathbb{R}^{l \times w \times h}$ such that

$$I_{DT_B}(\boldsymbol{v}) = d(\boldsymbol{v}, NBV_B(\boldsymbol{v})), \ \forall \boldsymbol{v} \in V, \tag{4}$$

where $I_{DT_B} : V \rightarrow \mathbb{R}$ is the image function of DT_B.

When the chosen distance metric is the Euclidean Distance we talk about Euclidean Distance Transform (EDT). The Euclidean distance between two points $v, w \in \mathbb{R}^3$ is given by

$$d(v, w) = \|w - v\| = \sqrt{(w_x - v_x)^2 + (w_y - v_y)^2 + (w_z - v_z)^2}. \qquad (5)$$

Squared Euclidean distance values among voxels are integers, and are often used to avoid time-consuming square root calculations.

2.3   Computing the Solvent-Excluded Surface

The Solvent-Excluded surface computation is based on the Euclidean Distance Transform. The employment of the Euclidean Distance Transform for macro-molecular surface computation was first introduced in [50]. Let us consider the voxel grid containing the SAS and its relative Euclidean Distance Transform $EDT(SAS)$. Because the SAS is displaced outward from the SES by a distance equal to the probe radius, the latter can be obtained from the $EDT(SAS)$ by removing all voxels with a distance map value smaller than the probe radius from the CPK model, and then extracting the surface voxels of the resulting voxelized volume with the above-mentioned flood-filling algorithm.

To compute the SES, we only need distance values up to one probe-sphere radius from the SAS, and only for voxels inside the volume delimited by the SAS. We implemented the region-growing Euclidean Distance Transform algorithm described in [51] which can limit the computation up to a certain distance value and only to a given subset of the voxels in the grid. Starting from the voxels in the SAS, nearest boundary voxels are propagated towards the interior of the molecule. The boundary propagation is done considering voxels by increasing distance values, until the desired *depth* inside the Solvent-Accessible volume is reached.

The processing order of the voxels is enforced by a data structure called Hierarchical Queues (HQ), made of a collection of FIFO queues labelled from 0 to $d_{\max}^2$. Ingoing voxel in the HQ are inserted in the queue with label corresponding to their squared distance value. Outgoing voxels are extracted from the first non-empty queue with the least label. This way voxels are guaranteed to be parsed by increasing distance values. A map data structure is created to contain the squared distance values of each voxel. At the end of the procedure, this map will contain the squared Euclidean Distance Transform of the input voxel grid containing the SAS.

The HQ is initialised with queue 0 containing all the voxels belonging to the Solvent-Accessible surface, and all other queues empty. The map is initialised with 0 for all voxels belonging to the SAS and with MAXINT for all other voxels. Voxels are processed in the HQ-imposed order. For each voxel extracted from the HQ, its NBV is passed to its neighbours. If this leads to a smaller distance value than the previously stored one, the map is updated with the new value and the neighbour is inserted in the HQ. This procedure allows voxels to be mislabelled with a wrong NBV at first an then be corrected in a subsequent step when a closer boundary voxel is found. The parsing order imposed by the HQ guarantees that errors are not propagated. Corrections are always processed before the initial errors propagate since they have smaller distance values and are placed in queues of smaller label.

Because the propagation is done only within a certain neighbourhood ($3 \times 3 \times 3$), certain voxels might be assigned an erroneous distance value. Erroneous distance values arise because the correct NBV are not propagated. Propagating boundary voxels to a larger neighbourhood ($5 \times 5 \times 5$) can significantly diminish the percentage of erroneous distance values, but also increases the time complexity of the algorithm. For this reason we adapted a two phase algorithm: the distance map is first computed quickly with the $3 \times 3 \times 3$ neighbourhood, and then the $5 \times 5 \times 5$ neighbourhood is used to correct errors made during the first scan. The corrections are required only for a small subset of voxels, i.e. the ones that were not propagated during the initial scan, and that have a distance value greater than a certain threshold.

2.4   The Slicing Procedure

To enable the computation of high resolution surfaces in spite of memory limitations we have developed a slicing protocol for the macromolecule. The molecule is sliced in a user-defined number of parts, and the surface is calculated separately for each part in a sequential fashion. The slicing is done with planes perpendicular to the x-axis of the Cartesian coordinate system (see Fig. 2).

Atom coordinates parsed from the PDB file are translated, scaled and quantized to the coordinate system defined by each slice. For each slice, we subtract the slice-length to the x coordinate of the translation vector $k - 1$ times, where k is the current slice index ($k = 1, 2, ..., n$). The space filling procedure is performed for each slice separately, also taking into account any portions of atoms intersecting the slice whose centres might be located outside the current slice.

The correct determination of the distance map value for a given voxel requires knowledge of all boundary voxels within one probe sphere distance from the given voxel. Voxels in the immediate proximity of the slice borders require knowledge regarding the nearby boundary voxels in the adjacent slices in order to correctly calculate their distance map values. For this reason some extra margin on the x coordinate must be considered for each slice in order for the surface computation to yield correct results and it must be greater than the scaled and quantized probe-sphere radius (see Fig. 3).

Proteins can have solvent-accessible pockets which often serve as binding sites of other molecules. When running the spatial slicing procedure, pockets could run though two or more slices. Solvent-excluded voids buried inside the molecule can also be cut by the slicing planes, and there are cases when the surface portion belonging to a cavity cut by a slicing plane is disconnected from the outer molecular surface in that given slice. Communication among slices is required in order to correctly identify disconnected cavities cut by the slicing planes as solvent accessible or solvent excluded.

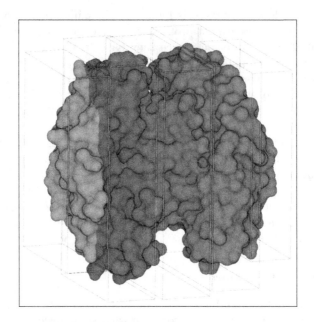

Fig. 2. Solvent-Excluded surface of 1VLA (4258 ATOM entries) [52] calculated with 5 slices, 1.4 Å probe-radius, 10^3 voxels per Å^3 resolution.

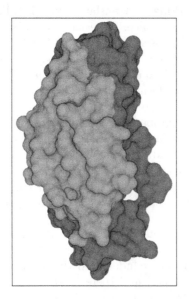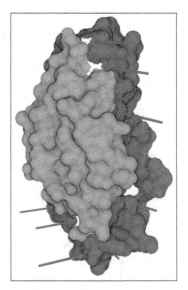

Fig. 3. Slices 1 and 2 of the SES of 1VLA, calculated with 5 slices, 1.4 Å probe-radius, 10^3 voxels per Å^3 resolution. We can see the differences between surfaces computed with (left) and without (right) the slice margin.

The procedure can be described as follows. Candidate pocket cavities are identified by checking in the margin region of each slice for free solvent-excluded voxels. The algorithm extracts all surface voxels of potential pockets from each slice using an adaptation of the same efficient 3D flood-filling procedure mentioned earlier. Pockets and solvent-excluded cavities running through two or more slices must be distinguished from each other. The algorithm extracts all surface voxels belonging to potential pockets from each slice and stores them in an apposite data structure. For each pair of adjacent slices, the border voxels of the candidate pockets are matched against their neighbours on the other slice. A candidate pocket is solvent-accessible if its border voxels have free neighbours on the adjacent slice. Otherwise, if its border voxels are matched with the border voxels of another candidate pocket on the adjacent slice, the current pocket remains undetermined as it could run through two or more slices in length. Multiple border exchange iterations are required if there are large pockets or cavities that run through more than one slice in length in the current macromolecule. The procedure ends when, at a given iteration, no new candidate pockets are recognized as solvent accessible.

2.5   Parallelization

The macromolecular surface calculation protocol with slicing introduced in the previous sections suggests an immediate parallelization scheme. The surface calculation for each slice can be executed nearly-independently from the others, as process synchronization and communication is required only for the pocket-detection and extraction procedure, in order to correctly identify pockets spanning between two or more slices.

3   Results and Discussion

We have run different tests of an MPI-based implementation of the parallel algorithm on an IBM®Power®P770 Server with 6 IBM®Power7 3.1 GHz CPUs and 640 Gb of RAM, running SUSE Linux Enterprise 11, and experimentally determined the average computation times for different input molecules at various resolutions, while calculating the three molecular surfaces. For the tests we used molecules 1GZX (4387 ATOM records) [53] and 2AEB (9568 ATOM records) [54], see Fig. 4.

3.1   Workload Distribution

To obtain an equitable workload distribution among processes, the slicing procedure should guarantee a uniform distribution of the workload. A uniform distribution of the number of atoms per slice (i.e. variable-length slices) yields better speedup values than employing a constant slice length value.

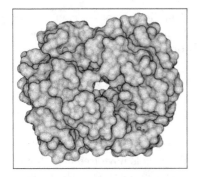 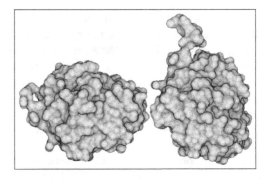

Fig. 4. SES of 1GZX (left) and 2AEB (right) computed with 1.4 Å probe-radius and 10^3 voxels per $Å^3$ resolution.

3.2 Surface Computation Time

The average surface computation times that follow were calculated over 100 runs of the corresponding configuration (PDB entry, desired molecular surface, resolution, probe radius, number of processes). We progressively increased the number of processes (from 1 to 64) and evaluated the mean computation time for each configuration (see Fig. 5). The tests were conducted for very high resolution molecular surface representations.

The overall speedup is affected by the constant margin introduced in each slice during the SES computation. At some point, while increasing the number of processors (increasing number of slices), the overhead introduced by the margins will become comparable to the slice computation time, thus vanishing the benefits of further parallelization.

3.3 Sequential Slicing

To overcome memory limitations the method implements a compact representation of the voxel grid and uses a sequential spatial slicing procedure: the molecule is sliced by a user-defined number of parallel planes perpendicular to the x-axis and the surface is computed for each slice sequentially. For instance, the calculation of the surface for 1GZX at a resolution of 9000 voxels per $Å^3$ and while dividing the molecule in 10 slices, needs nearly 5 GB of RAM, against the 2.6 GB used while dividing the molecule in 20 slices (tests were made on a desktop computer with an Intel Core i7 860 CPU and 8 GB of RAM (4 × 2GB DDR3-1333 banks) running Ubuntu 13.10 × 64), which is an easily affordable amount of memory nowadays in desktop computers. By tuning the resolution and number-of-slices parameters, various memory utilization rates can be achieved depending on the users' needs.

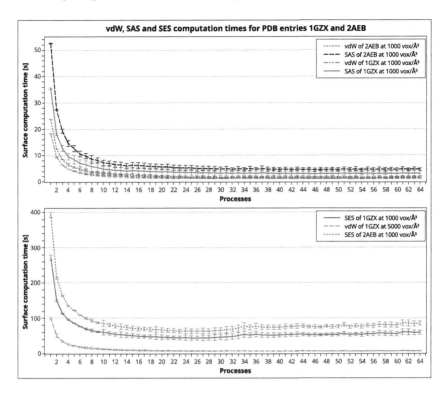

Fig. 5. Surface computation time for 1GZX and 2AEB for various configurations. Error bars represent one standard deviation of the computation time based on 100 repetitions.

3.4 Separability and Minimality

The voxelized representations of molecular surfaces produced by this approach possess both separability and minimality properties. The voxelized surface is extracted using a flood-filling algorithm: any gap would cause leakage of the flood through the discrete surface. Also, if any voxel in the resulting surface is removed, the flood-filling algorithm would propagate through that voxel towards the outside of the space-filling volume. This means that the produced surface is both separating and minimal.

4 Conclusion

We presented a methodology for the computation of discrete fine-grained voxelized macromolecular surfaces, implemented in a tool that can generate the three main molecular surfaces at high-resolutions effectively and in a timely manner. This makes it perfectly suitable for integration into other applications or pipelines of bioinformatics tools which require the computation of such surfaces at runtime. The parallel implementation introduces advantages in terms of the overall speedup, however the uniform distribution of atoms per slice may not necessarily yield a balanced workload between processes. On the other hand, the constant slice margin represents the main limitation to this parallelization scheme as it introduces constant overhead regardless of the slice size. These issues are left for future work.

Voxelized representations are well suited to represent multiple physicochemical and geometrical properties of molecular surfaces, as each voxel can describe multiple properties of a portion of the 3D space. Although only binary voxels were employed in this work, they could easily be extended to contain multiple values in order to represent several properties, such as electrostatic potentials, hydrophobicity, curvature, surface normals, etc. For instance, we are currently developing a local surface descriptor for protein-protein docking based on the voxelized representation of shape and electrostatic properties of the molecular surface. The descriptor is invariant to roto-translations and allows the efficient comparison of geometric and electrostatic complementarity between surface portions.

Linux binaries and test results are available at: http://www.dei.unipd.it/~daberdak/VoxSurf

Acknowledgements. This work has been partially supported by the University of Padova ex60 % grant "Advanced Applications in Computer Science".

The authors are grateful to Prof. Gianfranco Bilardi and Dr. Paolo Emilio Mazzon for their help in using the facilities of the Advanced Computing Paradigms Lab.

References

1. Whitley, D.C.: Van der Waals surface graphs and molecular shape. J. Math. Chem. **23**(3–4), 377–397 (1998)
2. Lee, B., Richards, F.: The interpretation of protein structures: estimation of static accessibility. J. Mol. Biol. **55**(3), 379–IN4 (1971)
3. Connolly, M.L.: Analytical molecular surface calculation. J. Appl. Crystallogr. **16**(5), 548–558 (1983)
4. Connolly, M.L.: The molecular surface package. J. Mol. Graph. **11**(2), 139–141 (1993)
5. Sanner, M.F., Olson, A.J., Spehner, J.C.: Fast and robust computation of molecular surfaces. In: Proceedings of the Eleventh Annual Symposium on Computational Geometry, SCG 1995, pp. 406–407. ACM, New York (1995)
6. Mitchell, J.C., Kerr, R., Ten Eyck, L.F.: Rapid atomic density methods for molecular shape characterization. J. Mol. Graph. Model. **19**(3–4), 325–330 (2001)
7. Kinoshita, K., Nakamura, H.: Identification of the ligand binding sites on the molecular surface of proteins. Protein Sci. **14**(3), 711–718 (2005)
8. Bock, M.E., Cortelazzo, G.M., Ferrari, C., Guerra, C.: Identifying similar surface patches on proteins using a spin-image surface representation. In: Apostolico, A., Crochemore, M., Park, K. (eds.) CPM 2005. LNCS, vol. 3537, pp. 417–428. Springer, Heidelberg (2005)
9. Albou, L.P., Schwarz, B., Poch, O., Wurtz, J.M., Moras, D.: Defining and characterizing protein surface using alpha shapes. Proteins: Struct. Funct. Bioinf. **76**(1), 1–12 (2009)

10. Venkatraman, V., Yang, Y., Sael, L., Kihara, D.: Protein-protein docking using region-based 3D Zernike descriptors. BMC Bioinform. **10**(1), 407 (2009)
11. Li, B., Kihara, D.: Protein docking prediction using predicted protein-protein interface. BMC Bioinform. **13**(1), 7 (2012)
12. Esquivel-Rodriguez, J., Filos-Gonzalez, V., Li, B., Kihara, D.: Pairwise and multimeric protein-protein docking using the LZerD program suite. Protein Struct. Predict. **1137**, 209–234 (2014)
13. Grandison, S., Roberts, C., Morris, R.J.: The application of 3D Zernike moments for the description of "Model-Free" molecular structure, functional motion, and structural reliability. J. Comput. Biol. **16**(3), 487–500 (2009)
14. Kihara, D., Sael, L., Chikhi, R., Esquivel-Rodriguez, J.: Molecular surface representation using 3D Zernike descriptors for protein shape comparison and docking. Curr. Protein Pept. Sci. **12**(6), 520–533 (2011)
15. Sael, L., La, D., Li, B., Rustamov, R., Kihara, D.: Rapid comparison of properties on protein surface. Proteins **73**(1), 1–10 (2008)
16. Huang, B., Schroeder, M.: Ligsite^{csc}: predicting ligand binding sites using the connolly surface and degree of conservation. BMC Struct. Biol. **6**(1), 1–11 (2006)
17. Kozakov, D., Brenke, R., Comeau, S.R., Vajda, S.: PIPER: an FFT-based protein docking program with pairwise potentials. Proteins: Struct., Funct., Bioinf. **65**(2), 392–406 (2006)
18. Chowdhury, R., Rasheed, M., Keidel, D., Moussalem, M., Olson, A., Sanner, M., Bajaj, C.: Protein-protein docking with F^{2}Dock 2.0 and GB-rerank. PLoS ONE **8**(3), e51307 (2013)
19. Lee, L.W., Bargiela, A.: Protein surface atoms extraction: voxels as an investigative tool. Eng. Lett. **20**(3), 217–228 (2012)
20. Lee, L.W., Bargiela, A.: An approximated voxel approach for the identification and modelling of ligand-binding sites. J. Phys. Sci. Appl. **2**(10), 399–408 (2012)
21. Levitt, D.G., Banaszak, L.J.: POCKET: a computer graphics method for identifying and displaying protein cavities and their surrounding amino acids. J. Mol. Graph. **10**(4), 229–234 (1992)
22. Hendlich, M., Rippmann, F., Barnickel, G.: LIGSITE: automatic and efficient detection of potential small molecule-binding sites in proteins. J. Mol. Graph. Model. **15**(6), 359–363 (1997)
23. Weisel, M., Proschak, E., Schneider, G.: PocketPicker: analysis of ligand binding-sites with shape descriptors. Chem. Cent. J. **1**(1), 7 (2007)
24. Li, B., Turuvekere, S., Agrawal, M., La, D., Ramani, K., Kihara, D.: Characterization of local geometry of protein surfaces with the visibility criterion. Proteins: Struct. Funct. Bioinform. **71**(2), 670–683 (2008)
25. Berman, H., Henrick, K., Nakamura, H.: Announcing the worldwide protein data bank. Nat. Struct. Mol. Biol. **10**(12), 980 (2003)
26. Sael, L., Li, B., La, D., Fang, Y., Ramani, K., Rustamov, R., Kihara, D.: Fast protein tertiary structure retrieval based on global surface shape similarity. Proteins: Struct., Funct., Bioinf. **72**(4), 1259–1273 (2008)
27. La, D., Esquivel-Rodrguez, J., Venkatraman, V., Li, B., Sael, L., Ueng, S., Ahrendt, S., Kihara, D.: 3D-SURFER: software for high-throughput protein surface comparison and analysis. Bioinformatics **25**(21), 2843–2844 (2009)
28. Axenopoulos, A., Daras, P., Papadopoulos, G., Houstis, E.: A shape descriptor for fast complementarity matching in molecular docking. IEEE/ACM Trans. Comput. Biol. Bioinf. **8**(6), 1441–1457 (2011)

29. Huang, J., Yagel, R., Filippov, V., Kurzion, Y.: An accurate method for voxelizing polygon meshes. In: 1998 IEEE Symposium on Volume Visualization, pp. 119–126. IEEE (1998)

30. Daberdaku, S., Ferrari, C.: A voxel-based tool for protein surface representation. In: Angelini, C., Bongcam-Rudloff, E., Decarli, A., Rancoita, P.M., Rovetta, S., (eds.) Computational Intelligence Methods for Bioinformatics and Biostatistics, CIBB 2015, Naples, Italy, pp. 96–101 (2015)

31. Schrödinger, L.L.C.: The PyMOL molecular graphics system, version 1.8, November 2015

32. Guex, N., Peitsch, M.C.: SWISS-MODEL and the Swiss-Pdb viewer: an environment for comparative protein modeling. Electrophoresis 18(15), 2714–2723 (1997)

33. Sanner, M.F., Olson, A.J., Spehner, J.C.: Reduced surface: an efficient way to compute molecular surfaces. Biopolymers 38(3), 305–320 (1996)

34. Pettersen, E.F., Goddard, T.D., Huang, C.C., Couch, G.S., Greenblatt, D.M., Meng, E.C., Ferrin, T.E.: UCSF chimera–a visualization system for exploratory research and analysis. J. Comput. Chem. 25(13), 1605–1612 (2004)

35. Humphrey, W., Dalke, A., Schulten, K.: VMD – visual molecular dynamics. J. Mol. Graph. 14, 33–38 (1996)

36. Varshney, A., Brooks Jr., F.P., Wright, W.V.: Computing smooth molecular surfaces. IEEE Comput. Graphics Appl. 14(5), 19–25 (1994)

37. Sayle, R.A., Milner-White, E.J.: RASMOL: biomolecular graphics for all. Trends Biochem. Sci. 20(9), 374–376 (1995)

38. Jmol: an open-source Java viewer for chemical structures in 3D. http://www.jmol. org/

39. Hanwell, M.D., Curtis, D.E., Lonie, D.C., Vandermeersch, T., Zurek, E., Hutchison, G.R.: Avogadro: an advanced-semantic chemical editor, visualization, and analysis platform. J. Cheminformatics 4(1), 17 (2012)

40. BIOVIA, Dassault Systèmes: Discovery Studio Modeling Environment, Release 4.5 (2015)

41. Corey, R.B., Pauling, L.: Molecular models of amino acids, peptides, and proteins. Rev. Sci. Instrum. 24(8), 621–627 (1953)

42. Koltun, W.L.: Precision space-filling atomic models. Biopolymers 3(6), 665–679 (1965)

43. MacKerell, A.D., Bashford, D., Bellott, M., Dunbrack, R.L., Evanseck, J.D., Field, M.J., Fischer, S., Gao, J., Guo, H., Ha, S., Joseph-McCarthy, D., Kuchnir, L., Kuczera, K., Lau, F.T.K., Mattos, C., Michnick, S., Ngo, T., Nguyen, D.T., Prodhom, B., Reiher, W.E., Roux, B., Schlenkrich, M., Smith, J.C., Stote, R., Straub, J., Watanabe, M., Wirkiewicz-Kuczera, J., Yin, D., Karplus, M.: All-atom empirical potential for molecular modeling and dynamics studies of proteins. J. Phys. Chem. B 102(18), 3586–3616 (1998)

44. Bresenham, J.: Algorithm for computer control of a digital plotter. IBM Syst. J. 4(1), 25–30 (1965)

45. Yu, W.W., He, F., Xi, P.: A rapid 3D seed-filling algorithm based on scan slice. Comput. Graph. 34(4), 449–459 (2010). Procedural Methods in Computer Graphics Illustrative Visualization

46. Rusu, R.B.: The pcd (point cloud data) file format. http://pointclouds.org/ documentation/tutorials/pcd_file_format.php. Accessed 27 Apr 2016

47. Thompson, D., Braun, J., Ford, R.: OpenDX: Paths to Visualization; Materials Used for Learning OpenDX the Open Source Derivative of IBM's Visualization Data Explorer. Visualization and Imagery Solutions, Missoula (2004)

48. Schroeder, W., Martin, K., Lorensen, B.: Visualization Toolkit: An Object-Oriented Approach to 3D Graphics, 4th edn. Kitware, New York (2006)
49. Avila, L., Kitware, I.: The VTK User's Guide. Kitware, New York (2010)
50. Xu, D., Zhang, Y.: Generating triangulated macromolecular surfaces by Euclidean distance transform. PLoS ONE **4**(12), e8140 (2009)
51. Cuisenaire, O.: Region growing Euclidean distance transforms. In: Bimbo, A. (ed.) ICIAP 1997. LNCS, vol. 1310, pp. 263–270. Springer, Heidelberg (1997)
52. Joint Center for Structural Genomics (JCSG): Crystal structure of Hydroperoxide resistance protein OsmC (TM0919) from Thermotoga maritima at 1.80 Å resolution (2004)
53. Paoli, M., Liddington, R., Tame, J., Wilkinson, A., Dodson, G.: Crystal structure of T state haemoglobin with oxygen bound at all four haems. J. Mol. Biol. **256**(4), 775–792 (1996)
54. Di Costanzo, L., Sabio, G., Mora, A., Rodriguez, P.C., Ochoa, A.C., Centeno, F., Christianson, D.W.: Crystal structure of human arginase I at 1.29 Å resolution and exploration of inhibition in the immune response. Proc. Nat. Acad. Sci. U.S.A. **102**(37), 13058–13063 (2005)

The Challenges of Interpreting Phosphoproteomics Data: A Critical View Through the Bioinformatics Lens

Panayotis Vlastaridis[1], Stephen G. Oliver[2],
Yves Van de Peer[3,4,5], and Grigoris D. Amoutzias[1(✉)]

[1] Bioinformatics Laboratory, Department of Biochemistry and Biotechnology,
University of Thessaly, Larisa, Greece
panosvlastaridis@gmail.com, amoutzias@bio.uth.gr,
amoutzias@hotmail.com
[2] Department of Biochemistry, Cambridge Systems Biology Centre,
University of Cambridge, Cambridge, UK
sgo24@cam.ac.uk
[3] Department of Plant Systems Biology, VIB,
Department of Plant Biotechnology and Bioinformatics,
Ghent University, 9052 Ghent, Belgium
yves.vandepeer@psb.vib-ugent.be
[4] Bioinformatics Institute Ghent, Technologiepark 927, 9052 Ghent, Belgium
[5] Department of Genetics, Genomics Research Institute,
University of Pretoria, Pretoria 0028, South Africa

Abstract. During the last decade, there has been great progress in high-throughput (HTP) phosphoproteomics and hundreds or even thousands of phosphorylation sites (p-sites) can now be detected in a single experiment. This success is attributable to a combination of very sensitive Mass Spectrometry instruments, better phosphopeptide enrichment techniques and bioinformatics software that are capable of detecting peptides and localizing p-sites. These new technologies have opened up a whole new level of gene regulation to be studied, with great potential for therapeutics and synthetic biology. Nevertheless, many challenges remain to be resolved; these concern the biases and noise of these proteomic technologies, the biological noise that is present, as well as the incompleteness of the current datasets. Despite these problems, the datasets published so far appear to represent a good sample of a complete phosphoproteome of some organisms and are capable of revealing their major properties.

Keywords: Phosphoproteomics · Phosphorylation · Bioinformatics · Data integration

1 The Biological Significance of Protein Phosphorylation

To understand a biological system, it is not enough to know which molecules are expressed in various conditions/states. Recent advances in high-throughput proteomics and phosphoproteomics have highlighted the importance of knowing whether the expressed proteins have their molecular functions turned on or off via post-translational modifications.

© Springer International Publishing Switzerland 2016
C. Angelini et al. (Eds.): CIBB 2015, LNBI 9874, pp. 196–204, 2016.
DOI: 10.1007/978-3-319-44332-4_15

Phosphorylation is the most abundant reversible post-translational modification (PTM) [1]. It may function as either a digital switch or as a rheostat, regulating one or more functions in a protein, including: enzyme activity, subcellular localization, complex formation or degradation. These effects are mediated via conformational changes, regulation of order/disorder transitions, and affinity change on molecular interaction surfaces [2, 3]. Phosphorylation/dephosphorylaton is also a key component of signal transduction. More than one switch of this kind may be present in a protein; they may be independent of each other, or there may be interdependencies between them or even with other types of switches [4]. Previous studies [5–7] have estimated that 1/3–2/3 of the proteins in a eukaryote are expected to be phosphorylated. Moreover, a protein may have from one to tens of p-sites. Therefore, the combinatorics behind this process as well as the potential for complexity at the molecular level is enormous.

Mutation of only one site of phosphorylation in a key protein may have dramatic effects not only for the function of that specific protein, but also for the pathways in which it is involved and even the overall phenotype of the organism [8, 9]. For example, a point mutation that results in a single amino-acid change (S42 -> A) in the yeast Cdc28 protein results in decreased cell size, whereas the mutation of another p-site may even be lethal, within a certain genetic context, or it may rescue the lethal effect of another point mutation [10]. Furthermore, point mutations of p-sites in key enzymes may alter flux through the biochemical pathways of the cell towards desired biotechnological products or properties [11, 12].

Abnormal protein phosphorylation is involved in many diseases; including cancer, diabetes, autoimmune, cardiovascular and neurodegenerative diseases [13, 14]. New generations of drugs in cancer and other diseases target this PTM, whereas there is intense interest in measuring serum or blood phosphoproteomes, for improved diagnostics [15]. Furthermore, many bacteria disrupt the host immune system by interfering with the phosphorylation networks of the host [16], whereas many viruses rely on host kinases to phosphorylate and regulate their proteins [17].

Therefore, phosphorylation appears to be an extremely attractive area of research, not only for understanding organismal complexity or how the cell is regulated, but also for therapeutics and even for synthetic biology. It holds the promise of manipulating molecular pathways and phenotypes via a few point mutations that modify a small number of critical phosphorylation sites.

The advent of high-throughput (HTP) phosphoproteomics in the last decade has revolutionized the field, enabling hundreds or even thousands of phosphorylation sites (p-sites) to be detected in a single experiment. This success may be attributed to a combination of very sensitive Mass Spectrometry instruments, better phosphopeptide enrichment techniques and bioinformatics tools that are capable of detecting phosphopeptides and localizing p-sites [18–20]. Nevertheless, many challenges remain to be addressed.

2   The Challenges of Phosphoproteomics

2.1   Biological Noise and Technical Problems

A major challenge relates to the quality of the generated phosphoproteomic datasets. As with any new HTP technology, the data which they generate are afflicted by experimental biases and noise. The various phosphopeptide enrichment techniques capture only a fraction of the complete phosphoproteome while also introducing biases [21]. Lienhard [22] has raised the possibility that, due to the high sensitivity of the new MS instruments, biologically noisy p-sites are being detected. 'Biological noise', in this case, represents phosphorylation events occurring in degenerate motifs by non-cognate kinases; i.e. frequent but of low abundance off-target phosphorylations. Also, during the process of cell-lysis, kinases and scaffold proteins may encounter target proteins from different cellular compartments, which they would not meet under normal conditions. More concern is raised by the observed low occupancy (~10 %) of the majority of p-sites for a given condition [23]. In addition, less than 20 % of p-sites identified in a single phosphoproteomic experiment are up/down-regulated when a perturbation occurs [24]. [25] exploited evolutionary information to estimate that up to 65 % of p-sites in HTP experiments could be non-functional, indicating that biological noise may indeed be a significant problem. In agreement with [22, 25], it was demonstrated that, within a compendium of 12 HTP phosphoproteomic experiments from yeast, more than half of non-redundant p-sites were identified only once, further highlighting the problem of potential false positive or non-functional p-sites in HTP datasets [5].

Another concern relates to the stringency of the criteria and algorithms used to identify phosphopeptides and to correctly localize p-sites within a phosphopeptide. Some databases, bioinformatics analyses or even prediction tools extract phosphorylation sites from supplementary material of publications without applying very stringent criteria. Such analyses often rely on the criteria set by each individual publication, and these are far from uniform. The general tendency to publish phosphoproteomic datasets with as many p-sites as possible (the more, the better) means that not very stringent filtering criteria have been applied in some of the original publications. Nevertheless, in the last few years, this problem has been ameliorated, as more software tools have appeared that try to detect phosphopeptides and also localize the p-site, by either estimating the p-site's correct localization probability or the Search engine difference scores [26]. In parallel, more and more studies have started to adopt more stringent criteria, with a cutoff of 99 % probability of correct peptide identification and 99 % probability of correct p-site localization.

The bioinformatics analysis of 12 phosphoproteomic yeast datasets revealed that the phosphoprotein and p-site overlap between two experiments from two different research groups in very similar conditions (alpha-factor treated yeast cells) was 31 % and 11 % respectively, whereas the overlap between two experiments of one research group in two different phases of the yeast cell-cycle were 54 % and 28 % respectively [5, 27–29]. Similarly, a 2010 study by the Proteome Informatics Research group from ABRF showed that for the same phosphoproteomic dataset, the average agreement of identified phosphoproteins and phosphosites by any two software packages was ~57 % and ~38 % respectively [26]. Furthermore, a replication of the same experiment may increase the detection of p-sites by 25 % [30], whereas 3–4 replicates are sufficient to reach saturation in terms of p-sites identified for a particular condition using a given technology [31].

Another issue that significantly affects p-site detection is the choice of the proteolytic enzyme employed to generate peptide fragments for analysis by MS/MS. The most commonly used proteolytic enzyme is trypsin. Nevertheless, several studies have shown that the consecutive use of more than one proteolytic enzyme may increase phospho-peptide and p-site detection by up to 40–70 % [30, 32, 33]. Clearly, the detection of phosphoproteins and phosphopeptides is still very much a protocol-dependent issue.

The availability of even more phosphoproteomic datasets for a given species, in combination with more sensitive instruments, better enrichment protocols, better local-ization software and comparative phosphoproteomics will help filter out noisy p-sites.

2.2 Incompleteness of the Datasets

Baker's yeast (*S. cerevisiae*) is the best studied unicellular eukaryote and it harbors only ~6.000 proteins [34, 35]. Thus the plethora of HTP phosphoproteomic experiments (performed on this organism under a reasonably wide range of conditions) has probably revealed the majority of proteins (3100–3800) that are regulated at some stage by phos-phorylation [5, 7]. Another analysis estimated that high-throughput phosphoproteomic studies have revealed about 80–90 % of all *S. cerevisiae* phosphoproteins [36]. Yet, we are far away from identifying the majority of p-sites in this relatively simple organism. In an updated compendium of yeast p-sites, [7] found that 45 % of p-sites identified in low-throughput (LTP) experiments were also identified by at least two independent high-throughput experiments. There are also many reports in the literature where a well-known p-site was not detectable by the high-throughput technologies. In yeast, more than 70 % of its whole proteome is detectable by MS/MS technology in a single experi-ment [37, 38]. Clearly, for a multicellular organism such as *Homo sapiens*, with a much more complex proteome and more transient or spatial (tissue- or organ-specific) expres-sion patterns, the identification of its entire phosphoproteome, estimated to involve hundreds of thousands of p-sites, is much more challenging.

Better experimental protocols and the availability of comparative phosphoproteo-mics data from several closely related species should help us to estimate how many more p-sites are missing. The basic principle is that if a p-site is found in organism X and the same amino acid or even phosphorylation motif is conserved in another species Y, based on alignment of orthologous proteins, then, the conserved amino acid of species Y could also be phosphorylated. Thus, a better estimation of the total number of p-sites in species Y may be obtained. Nevertheless, a disturbing finding that complicates this type of evolutionary analyses is that the precise positioning of p-sites is not always required for proper regulation [39, 40]. The implication is that multiple alignments of orthologous proteins may not be sufficient to fully determine the conservation of a p-site in another organism. It may be the case that the phosphorylated amino acid is not conserved in another organism, but an equivalent p-site has emerged in the vicinity. Therefore, it is not sufficient to have the phosphoproteome of one reference organism and the multiple alignments of orthologous proteins. One needs to take into consideration the "neigh-borhood" of the p-site as well.

2.3   The Challenges of Predicting Phosphorylation Sites

The increase of HTP phosphoproteomic data has stimulated research to develop bioinformatics tools to predict p-sites, either from amino-acid sequence alone, or in combination with structural and other types of information [17, 41–43]. More than 40 prediction methods have been published on this computational problem, applying artificial neural networks, support vector machines, decision trees, genetic algorithms or position-specific scoring matrices. In addition, a plethora of databases exists (see two extensive reviews on this subject, [44, 45]. There are still ongoing discussions on what is the optimal size of the sequence region around the p-site that contains enough information to build a prediction pipeline, without decreasing the signal/noise ratio and still remaining computationally tractable for the machine-learning algorithms to analyze. A crucial issue is the training datasets used for these algorithms. Abundant and high-quality p-sites, as well as very good negative datasets are needed for successful implementation. Not surprisingly, the negative datasets are very difficult to obtain since a large fraction of the phosphoproteome of an organism remains unknown. Also, gold-standard reference datasets (both positive and negative) are needed to allow the community to evaluate any new algorithm/tool and compare it to existing ones. So far, the datasets used to train such algorithms suffer from noisy p-sites and poor filtering of technical and biological noise, and do not account for the fact that kinases may mistakenly phosphorylate a serine that is very close to the cognate site [5, 22, 25]. Furthermore, most of the high-throughput p-sites have been identified with a protocol involving only trypsin. The use of two or more proteolytic enzymes sequentially (e.g. Lys-N and trypsin) resulted in datasets that were enriched in significantly different phosphorylation motifs [30]. Therefore, as more complex protocols with more than one digestion enzymes are used, more diverse phosphorylation motifs will be sampled and used for training new prediction algorithms.

3   Biological Properties of the Phosphoproteome

Despite all of the above problems, many investigations have tried to shed light on the properties of the best studied phosphoproteome, that of the model organism, *S. cerevisiae*, either from only one or from a compendium of filtered datasets. In yeast, phosphorylation occurs most frequently on serines (81 %), then on threonines (17 %), whereas tyrosines are very rarely phosphorylated (2 %) [5]. This is probably due to the lack of tyrosine specific kinases and the presence of kinases with dual (threonine/tyrosine) specificity [29, 46]. About 90 % of p-sites have been identified as being within intrinsically disordered regions, whereas between 12–17 % of p-sites are found to be either within or in the vicinity (10 amino acids) of a conserved and characterized structural or functional domain [5, 41]. Most of the identified phosphoproteins have a small number of p-sites, whereas there exists a very small number of proteins with many p-sites. Phosphoproteins are more ancient and are under tighter regulatory control, with shorter protein half-lives, more ubiquitination, more genetic interactions and more protein-protein interactions than non-phosphoproteins [5, 47, 48]. In addition, as the number of phosphorylation sites in a protein increases, the chance of this gene surviving after a gene or (especially) a whole-genome duplication increases as well [49]. Thus, protein phosphorylation may shape the evolution of genomes.

Interestingly, there is no strong correlation between the number of kinases targeting a phosphoprotein and the number of p-sites in that protein. Apparently, a kinase may phosphorylate more than one p-site in the same protein or a p-site may be phosphorylated by more than one closely related kinase [5, 43]. In addition, p-sites tend to cluster together and this is not due to false detection/localization of p-sites [5, 40, 50].

Despite the noise that is present in the current phosphoproteomic datasets and their acknowledged incompleteness, the conclusions of computational analyses done so far with limited datasets are not necessarily invalid. As demonstrated in [5], several conclusions remain robust, even when creating large and high-quality compendia of p-sites. Nevertheless, efficient filtering of noise will substantially increase confidence and resolution in the conclusions of related computational analyses and thus, it will allow new discoveries.

Acknowledgements. This work is supported and implemented under the "ARISTEIA II" Action of the "operational programme education and lifelong learning" and is co-funded by the European Social Fund (ESF) and National Resources (code 4288 to G.D.A). G.D.A acknowledges additional support by research grants from the Postgraduate Programme 'Applications of Molecular Biology-Genetics, Diagnostic Biomarkers', code 3817 of the Department of Biochemistry & Biotechnology, University of Thessaly, Greece. S.G.O. acknowledges support from the Wellcome Trust (grant no. 104967/Z/14/Z). Y.V.d.P acknowledges the Multidisciplinary Research Partnership "Bioinformatics: from nucleotides to networks" Project (no. 01MR0310 W) of Ghent University.

References

1. Krüger, R., Kübler, D., Pallissé, R., Burkovski, A., Lehmann, W.D.: Protein and proteome phosphorylation stoichiometry analysis by element mass spectrometry. Anal. Chem. **78**, 1987–1994 (2006)
2. Nishi, H., Shaytan, A., Panchenko, A.R.: Physicochemical mechanisms of protein regulation by phosphorylation. Front. Genet. **5**, 270 (2014)
3. Strumillo, M., Beltrao, P.: Towards the computational design of protein post-translational regulation. Bioorg. Med. Chem. **23**, 2877–2882 (2015)
4. Cohen, P.: The regulation of protein function by multisite phosphorylation–a 25 year update. Trends Biochem. Sci. **25**, 596–601 (2000)
5. Amoutzias, G.D., He, Y., Lilley, K.S., Van de Peer, Y., Oliver, S.G.: Evaluation and properties of the budding yeast phosphoproteome. Mol. Cell. Proteomics MCP **11**, M111.009555 (2012)
6. Cohen, P.: The origins of protein phosphorylation. Nat. Cell Biol. **4**, E127–E130 (2002)
7. Sadowski, I., Breitkreutz, B.-J., Stark, C., Su, T.-C., Dahabieh, M., Raithatha, S., Bernhard, W., Oughtred, R., Dolinski, K., Barreto, K., Tyers, M.: The PhosphoGRID Saccharomyces cerevisiae protein phosphorylation site database: version 2.0 update. Database J. Biol. Databases Curation. **2013**, bat026 (2013)
8. Amoutzias, G.D., Bornberg-Bauer, E., Oliver, S.G., Robertson, D.L.: Reduction/oxidation-phosphorylation control of DNA binding in the bZIP dimerization network. BMC Genom. **7**, 107 (2006)

9. Papadopoulou, N., Chen, J., Randeva, H.S., Levine, M.A., Hillhouse, E.W., Grammatopoulos, D.K.: Protein kinase A-induced negative regulation of the corticotropin-releasing hormone R1alpha receptor-extracellularly regulated kinase signal transduction pathway: the critical role of Ser301 for signaling switch and selectivity. Mol. Endocrinol. Baltim. Md. **18**, 624–639 (2004)

10. Zhang, K., Lin, W., Latham, J.A., Riefler, G.M., Schumacher, J.M., Chan, C., Tatchell, K., Hawke, D.H., Kobayashi, R., Dent, S.Y.R.: The Set1 methyltransferase opposes Ipl1 aurora kinase functions in chromosome segregation. Cell **122**, 723–734 (2005)

11. Oliveira, A.P., Sauer, U.: The importance of post-translational modifications in regulating Saccharomyces cerevisiae metabolism. FEMS Yeast Res. **12**, 104–117 (2012)

12. Oliveira, A.P., Ludwig, C., Picotti, P., Kogadeeva, M., Aebersold, R., Sauer, U.: Regulation of yeast central metabolism by enzyme phosphorylation. Mol. Syst. Biol. **8**, 623 (2012)

13. Deschênes-Simard, X., Kottakis, F., Meloche, S., Ferbeyre, G.: ERKs in cancer: friends or foes? Cancer Res. **74**, 412–419 (2014)

14. Reimand, J., Wagih, O., Bader, G.D.: The mutational landscape of phosphorylation signaling in cancer. Sci. Rep. **3**, 2651 (2013)

15. Khadjavi, A., Barbero, G., Destefanis, P., Mandili, G., Giribaldi, G., Mannu, F., Pantaleo, A., Ceruti, C., Bosio, A., Rolle, L., Turrini, F., Fontana, D.: Evidence of abnormal tyrosine phosphorylated proteins in the urine of patients with bladder cancer: the road toward a new diagnostic tool? J. Urol. **185**, 1922–1929 (2011)

16. Jers, C., Soufi, B., Grangeasse, C., Deutscher, J., Mijakovic, I.: Phosphoproteomics in bacteria: towards a systemic understanding of bacterial phosphorylation networks. Expert Rev. Proteomics **5**, 619–627 (2008)

17. Schwartz, D., Church, G.M.: Collection and motif-based prediction of phosphorylation sites in human viruses. Sci. Signal **3**, rs2 (2010)

18. Doll, S., Burlingame, A.L.: Mass spectrometry-based detection and assignment of protein posttranslational modifications. ACS Chem. Biol. **10**, 63–71 (2015)

19. Engholm-Keller, K., Larsen, M.R.: Technologies and challenges in large-scale phosphoproteomics. Proteomics **13**, 910–931 (2013)

20. Olsen, J.V., Mann, M.: Status of large-scale analysis of post-translational modifications by mass spectrometry. Mol. Cell. Proteomics MCP. **12**, 3444–3452 (2013)

21. Bodenmiller, B., Mueller, L.N., Mueller, M., Domon, B., Aebersold, R.: Reproducible isolation of distinct, overlapping segments of the phosphoproteome. Nat. Methods **4**, 231–237 (2007)

22. Lienhard, G.E.: Non-functional phosphorylations? Trends Biochem. Sci. **33**, 351–352 (2008)

23. Olsen, J.V., Vermeulen, M., Santamaria, A., Kumar, C., Miller, M.L., Jensen, L.J., Gnad, F., Cox, J., Jensen, T.S., Nigg, E.A., Brunak, S., Mann, M.: Quantitative phosphoproteomics reveals widespread full phosphorylation site occupancy during mitosis. Sci. Signal **3**, ra3 (2010)

24. Soufi, B., Kelstrup, C.D., Stoehr, G., Fröhlich, F., Walther, T.C., Olsen, J.V.: Global analysis of the yeast osmotic stress response by quantitative proteomics. Mol. Biosyst. **5**, 1337–1346 (2009)

25. Landry, C.R., Levy, E.D., Michnick, S.W.: Weak functional constraints on phosphoproteomes. Trends Genet. TIG **25**, 193–197 (2009)

26. Lee, D.C.H., Jones, A.R., Hubbard, S.J.: Computational phosphoproteomics: from identification to localization. Proteomics **15**, 950–963 (2015)

27. Gruhler, A., Olsen, J.V., Mohammed, S., Mortensen, P., Faergeman, N.J., Mann, M., Jensen, O.N.: Quantitative phosphoproteomics applied to the yeast pheromone signaling pathway. Mol. Cell. Proteomics MCP **4**, 310–327 (2005)

28. Holt, L.J., Tuch, B.B., Villén, J., Johnson, A.D., Gygi, S.P., Morgan, D.O.: Global analysis of Cdk1 substrate phosphorylation sites provides insights into evolution. Science **325**, 1682–1686 (2009)
29. Li, X., Gerber, S.A., Rudner, A.D., Beausoleil, S.A., Haas, W., Villén, J., Elias, J.E., Gygi, S.P.: Large-scale phosphorylation analysis of alpha-factor-arrested Saccharomyces cerevisiae. J. Proteome Res. **6**, 1190–1197 (2007)
30. Gauci, S., Helbig, A.O., Slijper, M., Krijgsveld, J., Heck, A.J.R., Mohammed, S.: Lys-N and trypsin cover complementary parts of the phosphoproteome in a refined SCX-based approach. Anal. Chem. **81**, 4493–4501 (2009)
31. Sharma, K., D'Souza, R.C.J., Tyanova, S., Schaab, C., Wiśniewski, J.R., Cox, J., Mann, M.: Ultradeep human phosphoproteome reveals a distinct regulatory nature of Tyr and Ser/Thr-based signaling. Cell Rep. **8**, 1583–1594 (2014)
32. Choudhary, G., Wu, S.-L., Shieh, P., Hancock, W.S.: Multiple enzymatic digestion for enhanced sequence coverage of proteins in complex proteomic mixtures using capillary LC with ion trap MS/MS. J. Proteome Res. **2**, 59–67 (2003)
33. Wiśniewski, J.R., Mann, M.: Consecutive proteolytic digestion in an enzyme reactor increases depth of proteomic and phosphoproteomic analysis. Anal. Chem. **84**, 2631–2637 (2012)
34. Goffeau, A., Barrell, B.G., Bussey, H., Davis, R.W., Dujon, B., Feldmann, H., Galibert, F., Hoheisel, J.D., Jacq, C., Johnston, M., Louis, E.J., Mewes, H.W., Murakami, Y., Philippsen, P., Tettelin, H., Oliver, S.G.: Life with 6000 genes. Science **274**(546), 563–567 (1996)
35. Oliver, S.G., van der Aart, Q.J., Agostoni-Carbone, M.L., Aigle, M., Alberghina, L., Alexandraki, D., Antoine, G., Anwar, R., Ballesta, J.P., Benit, P.: The complete DNA sequence of yeast chromosome III. Nature **357**, 38–46 (1992)
36. Beltrao, P., Trinidad, J.C., Fiedler, D., Roguev, A., Lim, W.A., Shokat, K.M., Burlingame, A.L., Krogan, N.J.: Evolution of phosphoregulation: comparison of phosphorylation patterns across yeast species. PLoS Biol. **7**, e1000134 (2009)
37. De Godoy, L.M.F., Olsen, J.V., Cox, J., Nielsen, M.L., Hubner, N.C., Fröhlich, F., Walther, T.C., Mann, M.: Comprehensive mass-spectrometry-based proteome quantification of haploid versus diploid yeast. Nature **455**, 1251–1254 (2008)
38. Wu, R., Dephoure, N., Haas, W., Huttlin, E.L., Zhai, B., Sowa, M.E., Gygi, S.P.: Correct interpretation of comprehensive phosphorylation dynamics requires normalization by protein expression changes. Mol. Cell. Proteomics MCP. **10**, M111.009654 (2011)
39. Landry, C.R., Freschi, L., Zarin, T., Moses, A.M.: Turnover of protein phosphorylation evolving under stabilizing selection. Front. Genet. **5**, 245 (2014)
40. Moses, A.M., Hériché, J.-K., Durbin, R.: Clustering of phosphorylation site recognition motifs can be exploited to predict the targets of cyclin-dependent kinase. Genome Biol. **8**, R23 (2007)
41. Iakoucheva, L.M., Radivojac, P., Brown, C.J., O'Connor, T.R., Sikes, J.G., Obradovic, Z., Dunker, A.K.: The importance of intrinsic disorder for protein phosphorylation. Nucleic Acids Res. **32**, 1037–1049 (2004)
42. Ingrell, C.R., Miller, M.L., Jensen, O.N., Blom, N.: NetPhosYeast: prediction of protein phosphorylation sites in yeast. Bioinformatics **23**, 895–897 (2007)
43. Mok, J., Kim, P.M., Lam, H.Y.K., Piccirillo, S., Zhou, X., Jeschke, G.R., Sheridan, D.L., Parker, S.A., Desai, V., Jwa, M., Cameroni, E., Niu, H., Good, M., Remenyi, A., Ma, J.-L.N., Sheu, Y.-J., Sassi, H.E., Sopko, R., Chan, C.S.M., De Virgilio, C., Hollingsworth, N.M., Lim, W.A., Stern, D.F., Stillman, B., Andrews, B.J., Gerstein, M.B., Snyder, M., Turk, B.E.: Deciphering protein kinase specificity through large-scale analysis of yeast phosphorylation site motifs. Sci. Signal **3**, ra12 (2010)

44. Xue, Y., Gao, X., Cao, J., Liu, Z., Jin, C., Wen, L., Yao, X., Ren, J.: A summary of computational resources for protein phosphorylation. Curr. Protein Pept. Sci. **11**, 485–496 (2010)
45. Trost, B., Kusalik, A.: Computational Prediction of Eukaryotic Phosphorylation Sites. Bioinformatics **27**, 2927–2935 (2011)
46. Manning, G., Plowman, G.D., Hunter, T., Sudarsanam, S.: Evolution of protein kinase signaling from yeast to man. Trends Biochem. Sci. **27**, 514–520 (2002)
47. Chi, A., Huttenhower, C., Geer, L.Y., Coon, J.J., Syka, J.E.P., Bai, D.L., Shabanowitz, J., Burke, D.J., Troyanskaya, O.G., Hunt, D.F.: Analysis of phosphorylation sites on proteins from Saccharomyces cerevisiae by electron transfer dissociation (ETD) mass spectrometry. Proc. Natl. Acad. Sci. U. S. A. **104**, 2193–2198 (2007)
48. Yachie, N., Saito, R., Sugiyama, N., Tomita, M., Ishihama, Y.: Integrative features of the yeast phosphoproteome and protein-protein interaction map. PLoS Comput. Biol. **7**, e1001064 (2011)
49. Amoutzias, G.D., He, Y., Gordon, J., Mossialos, D., Oliver, S.G., Van de Peer, Y.: Posttranslational regulation impacts the fate of duplicated genes. Proc. Natl. Acad. Sci. U. S. A. **107**, 2967–2971 (2010)
50. Schweiger, R., Linial, M.: Cooperativity within proximal phosphorylation sites is revealed from large-scale proteomics data. Biol. Direct. **5**, 6 (2010)

Bioinformatics Challenges and Potentialities in Studying Extreme Environments

Claudio Angione[1]([✉]), Pietro Liò[2], Sandra Pucciarelli[3], Basarbatu Can[4],
Maxwell Conway[2], Marina Lotti[5], Habib Bokhari[6], Alessio Mancini[3],
Ugur Sezerman[4], and Andrea Telatin[7]

[1] School of Computing, Teesside University, Middlesbrough, UK
c.angione@tees.ac.uk
[2] Computer Laboratory, University of Cambridge, Cambridge, UK
[3] School of Biosciences and Veterinary Medicine, University of Camerino,
Camerino, Italy
[4] Epigenetiks Genetik Biyoinformatik Yazilim A.S., Istanbul, Turkey
[5] Department of Biotechnology and Biosciences, State University of Milano Bicocca,
Milan, Italy
[6] COMSATS Institute of Information Technology, Islamabad, Pakistan
[7] BMR Genomics, Padova, Italy

Abstract. Biological systems show impressive adaptations at extreme environments. In extreme environments, directional selection pressure mechanisms acting upon mutational events often produce functional and structural innovations. Examples are the antifreeze proteins in Antarctic fish and their lack of hemoglobin, and the thermostable properties of TAQ polymerase from thermophilic organisms. During the past decade, more than 4000 organisms have been part of genome-sequencing projects. This has enabled the retrieval of information about evolutionary relationships among all living organisms, and has increased the understanding of complex phenomena, such as evolution, adaptation, and ecology. Bioinformatics tools have allowed us to perform genome annotation, cross-comparison, and to understand the metabolic potential of living organisms. In the last few years, research in bioinformatics has started to migrate from the analysis of genomic sequences and structural biology problems to the analysis of genotype-phenotype mapping. We believe that the analysis of multi-omic information, particularly metabolic and transcriptomic data of organisms living in extreme environments, could provide important and general insights into the how natural selection in an ecosystem shapes the molecular constituents. Here we present a review of methods with the aim to bridge the gap between theoretical models, bioinformatics analysis and experimental settings. The amount of data suggests that bioinformatics could be used to investigate whether the adaptation is generated by interesting molecular inventions. We therefore review and discuss the methodology and tools to approach this challenge.

Keywords: Multi-omic · Multi-layer networks · Adaptation · Metabolism · Extreme environments

C. Angione, P. Lió and S. Pucciarelli—Contributed equally to this work.

© Springer International Publishing Switzerland 2016
C. Angelini et al. (Eds.): CIBB 2015, LNBI 9874, pp. 205–219, 2016.
DOI: 10.1007/978-3-319-44332-4_16

1   Introduction

Population genetics and multi-omics systems biology have independently witnessed increasing research attention in recent years [1,2]. For instance, mathematical models for investigating genotype-phenotype relations have been developed for specific organisms, mainly bacteria [3,4]. However, predictions of cellular behavior cannot disregard bioinformatics methodologies that estimate the capability of adaptation of the cell to varying environmental conditions [5]. Methods for studying the molecular response to adaptation are still lacking, and would require a multi-scale and multi-omic combination of tools commonly employed in bioinformatics, but currently used referring only to single scales or to a single omic.

This paper reviews molecular and bioinformatics methods for studying molecular adaptation. The review is divided into distinct methods/software blocks describing existing software tools and techniques (Fig. 1), further reviewed in the following sections. Note that we focus mainly on the analysis of organisms from extreme environments as they possess distinct properties that allow deciphering the basis of environmental adaptation. In fact, the evolution of genome and phenome depends on the robustness of an organism and on its ability to adapt to varying conditions [6]. For this reason, as we discuss in the following, innovations are often found in organisms living in extreme environments.

The sequence of functional blocks in the figure leads to the identification of pathways, genes and proteins involved in adaptation. Each block contains distinct methodologies which could be implemented in one or more existing software tools, reviewed in the relevant sections. For the sake of clarity, each block is numbered and described below in the paper.

Throughout the paper, we will stress the need for multi-omic tools. Analogously, since many tools rely on networks to represent relations between biological entities, we will propose the use of multi-layer networks instead of single-layer networks. The strengths of our design are the following: (i) use and calibration of multi-omic and multi-layer information; (ii) use of pathway information; (iii) machine learning, bioinformatics and multi-objective optimization integrated in a powerful and novel inferential engine. The paper is structured according to the three main figures. First, we follow Fig. 1 to describe how experimental techniques produce data that need models, bioinformatics and machine learning for useful interpretations. Then, we describe the main problem of the present study, i.e. the relationship between molecular changes and selection (Figs. 2 and 3). We list below the main definitions of multi-omic terms employed in this paper:

– Multi-omics approach: Method that combines several omics data, such as genomics, proteomics and transcriptomics.
– Multi-layer network: A set of networks with a 1:1 correspondence between nodes, but different edges. This can be a useful methodology to represent relations between biological networks, e.g. from genes to proteins, and from pathways to metabolic fluxes. Networks are often a natural way to model many types of biological knowledge at different levels.

– Multi-scale model: Typically, a model that has been created by combining techniques that are valid at different scales of organization. For instance, combining a number of individual flux balance models (traditionally fine-grained single cell models) to model an entire population (which would normally be done using coarse-grained agents).

The main tools described in this paper are:

– METRADE: pipeline for building and optimizing genome-scale multi-omic models that accounts for metabolism, gene expression and codon usage at both transcriptional and translational levels. Freely available as a MATLAB toolbox [7].
– Colombos v3.0: database integrating publicly available transcriptomics data for several prokaryotic model organisms [8].
– MMETSP: database providing over 650 assembled, functionally annotated, and publicly available transcriptomes. These transcriptomes largely come from some of the more abundant and ecologically significant microbial eukaryotes in the ocean [9].
– Panoga: software used to identify the affected pathways in organisms living in extreme conditions [10]. It takes the list of significantly altered genes and their significance values and maps them to a protein-protein interaction network. Panoga is also a web-server for identification of SNP targeted pathways from genome-wide association study data. The web-server is freely available at: http://panoga.sabanciuniv.edu/.

2 Sampling, Next Generation Sequencing (NGS) and Ribotyping to Detect Microbial Diversity and Adaptations

One of the most important parts in studying environmental adaptation is the acquisition of data needed to perform multi-omic analysis. It is essential to carefully plan the number and location for up-taking environmental samples that will be collected, in order to avoid a number of gaps for future analysis. Furthermore, it is worth developing a protocol detecting even the least prevalent type of microorganisms from relative abundant index from different sources. Sampling biases such as selection of least turbulent sites, depth, width as well as time of the year and later on DNA extraction methods for direct isolation without culturing depending must be taken into consideration during the analysis. The analysis of metagenomes from samples collected in extreme environments, such as volcanoes, glaciers, or deep ocean waters (Fig. 1, panel 1) represents a valuable resource to study the molecular mechanisms underlying environmental adaptation. An extreme environment can also be found in the human body, e.g. gut microbiota during dietary extremes and exercise, or during metabolic diseases, for which models are available [11].

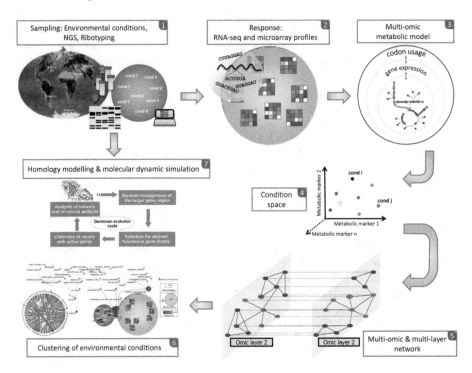

Fig. 1. Sampling in extreme ecosystems and bioinformatics methodological applications. The response to extreme environmental conditions (for instance Arctic and Antarctic regions, glaciers, deep ocean seawater, volcanoes and arid areas, and also similarly extreme but less exotic locations such as gut microbiota) is sampled for different associated species (1), and measured through expression profiling (2). To evaluate the environmental conditions and detect their community structure, a multi-omic model (3) can be applied to the species' metabolism, taking into account gene expression and codon usage. This model lets us map the input genotype and environment to the external behavior. The set of possible states for the organism as a response to the set of growth conditions is called *condition space* (4). Conditions can be measured for various biomarkers and therefore on various levels of omic information, e.g., a gene expression profile on the transcriptomic layer and a profile of flux rates on the fluxomic layer. Their interaction can be modeled using multi-layer networks (5). Statistical estimators and community detection methods defined on the multi-omic model can be used to investigate the pathway basis of the relationships between conditions and species in the association (6). After sampling, in parallel, homology modeling and molecular dynamic simulation can be applied to calculate structure flexibility and binding affinity of molecules of interest at different temperatures (7).

According to the extreme environment under consideration, different molecular, physiological and phenotypic strategies can be unraveled by applying multi-omic approaches. For example, in [12], a comprehensive survey of the distribution of bacteria from 213 samples was generated from 60 stations along the horizontal and vertical salinity gradients of the Baltic Sea. This represented the first detailed taxonomic study of an indigenous brackish water microbiome composed by a diverse combination of freshwater and marine clades that appears to have adapted to the brackish conditions. Furthermore, by applying whole-genome shotgun sequencing to microbial populations collected *en masse* from the Sargasso Sea near Bermuda, it was possible to discover 148 previously unknown bacterial phylotypes and to identify over 1.2 million previously unknown genes, suggesting substantial oceanic microbial diversity [13]. Microbial community profiling is also benefiting from advanced Bayesian techniques that have proven efficient strategies when multiple species are present in the mixture sampled [14].

Different populations within the same species may adapt differently to specific environmental conditions. These ecotypes or ecospecies are usually genetically distinct geographic subspecies of organisms that typically exhibit different phenotypes. However, microbial ecotypes cannot always be recognized by obvious phenotypic differences. In the last years, several genotypic methods usually based on the small subunit (SSU) ribosomal RNA (rRNA) analysis, or the rRNA internal transcribed spacer (ITS) regions (ribotyping) have greatly enhanced our capacity to quickly identify microbial species and sometimes populations from environmental samples. Also, they can be used to compare the distribution of various microorganisms isolated from animals, humans and food.

The analysis of SSU/ITS RNA sequences is also a powerful tool to characterize symbiote/host association and to identify whether a species is widespread in its distribution, or has dispersed through recent human-mediated events. For example, SSU/ITS RNA sequences were used to assess that a ciliate species of *Stentor* genus was introduced in the Lake Garda by anthropogenic activities [15]. Recently, SSU RNA phylogenetic analysis was used to characterize a bacterial consortium associated to *Euplotes focardii*, a strictly psychrophilic marine bacteria isolated from Terra Nova Bay, in Antarctica [16]. This study indicates that the consortium is also represented by Antarctic bacteria that were probably acquired by *E. focardii* after the colonization of the Antarctic marine habitat and may have contributed to its adaptation to the extreme conditions of this environment.

Extreme environments played a key role in shaping processes that are currently used in molecular biology. At extreme temperatures, it is in fact more difficult to keep a stable DNA replication process; this explains why innovations are often found in organisms living in extreme conditions. For instance, *Taq* polymerase, which is frequently a key step for the polymerase chain reaction, originates from *Thermus aquaticus*, a thermophilic microorganism. The high discriminatory power and reproducibility of polymerase chain reaction ribotyping (PCR RT) is also used for studying outbreaks at a local level, like in healthcare centers. Nosocomial infections are one of the leading causes of death among hospitalized patients and remain a major problem in all hospitals across the world. Many types of microorganisms cause infections in humans. Therefore, understanding the microbial diversity is a fundamental goal in healthcare. An increase in incidence of a PCR RT in hospitals could provide useful data for monitoring changes in type prevalence rates and control outbreaks [17]. In this survey, 14 *Clostridium difficile* have been isolated from nosocomial patients. PCR RT was used to prove if these microorganisms were identical. Result showed that among the isolations there was a predominant *C. difficile* lineage spreading in the hospital, with 5 out 14 identical ribotyping patterns. PCR RT has rapidly become the most widely used, straightforward and affordable typing method to detect diversities among the same microorganism species [18].

3   Omic Datasets to Measure Cellular Response to Varying Environments

Due to reduced costs and improved technology, data collection has witnessed a massive growth in speed and efficiency. For instance, multi-omic datasets can be used in association with multi-omic models to further extend, optimize and refine them [19], as well as to give insights into mechanisms of adaptation to different environmental conditions (Fig. 1).

By combining network inference algorithms and experimental data derived from 445 *Escherichia coli* microarrays, Faith et al. [20] identified 1079 regulatory interactions, 741 of which were new regulators of amino acid biosynthesis, flagella biosynthesis, osmotic stress response, antibiotic resistance, and iron regulation. This approach contributed to the understanding on how organisms can adapt to changing environments.

A more comprehensive dataset of gene expression levels measured in various environmental conditions, named Colombos v3.0, has been recently published by Meysman et al. [8]. Colombos includes *E. coli* microarray profiles for over 4000 conditions, measured using microarrays (Affymetrix *E. coli* Genome 2.0) with raw hybridization of intensities, and RNA-seq (Illumina MiSeq) with short read sequences. The expression profiles, measured on different platforms, have been then homogenized, and the conditions have been fully annotated.

RNA-seq and microarray techniques have revolutionized gene expression studies and allowed large-scale parallel measurement of whole genome expression. Both approaches represent valuable tools for the identifications of genes that are up- or down- regulated to respond to extreme conditions. High-resolution RNA-Seq transcriptome analysis of *Deinococcus gobiensis* following UV irradiation indicated the induction of genes involved in photoreactivation and recombinational repair, together with a subset of previously uncharacterized genes [21]. The investigation of the unknown genes and pathways required for the extreme resistance phenotype will highlight the exceptional ability of *D. gobiensis* to withstand environmental harsh conditions [21], providing the groundwork for the understanding of the general mechanisms of adaptation to extreme environments.

The Marine Microbial Eukaryotic Transcriptome Sequencing Project MMETSP[1] provided over 650 assembled, functionally annotated, and publicly available transcriptomes. These transcriptomes largely come from some of the more abundant and ecologically significant microbial eukaryotes in the ocean,

---

[1] http://marinemicroeukaryotes.org/.

and allowed the creation of a valuable benchmark against which environmental data can be analyzed [9]. By exploiting MMETSP datasets, researchers are allowed to study the evolutionary relationships among marine microbial eukaryotic clades, such as ciliates [22], and within the overall eukaryotic tree of life (Keeling et al., 2014). Furthermore, the interpretation of metatranscriptomic data generated from marine ecosystems allows us to explore the physiology and adaptation of diverse microbial eukaryotes from marine ecosystems [9].

Microarray transcriptional profiling of Arctic *Mesorhizobium* strain N33 allowed the identification of the most prominent up- and down- regulated genes under eight different temperature conditions, including both sustained and transient cold treatments, compared with cells grown at room temperature [23]. Up-regulated genes encode proteins involved in metabolite transport, transcription regulation, protein turnover, oxidoreductase activity, cryoprotection (mannitol, polyamines), fatty acid metabolism, and membrane fluidity [23]. Some genes were significantly down-regulated and classified in secretion, energy production and conversion, amino acid transport, cell motility, cell envelope and outer membrane biogenesis functions. This transcriptional profiling suggests that one of the strategy to survive under cold stress conditions is to adjust cellular function and save energy by reducing or ceasing cell growth rate.

4   Multi-omic Models Can Predict Cellular Activity

Several computational algorithms have been developed to analyze genes and gene sets in a multi-omic fashion [24]. The main goals are detecting dependencies among genes over different conditions and unraveling gene expression programs controlled by the dynamic interactions of hundreds of transcriptional regulators [20] (Fig. 1, panel 3).

The dataset by Faith et al. [20] was mapped to a multidimensional objective space through METRADE [7], a comprehensive tool for multi-omic flux balance analysis (Fig. 1, panel 4). The Colombos dataset has been exploited to predict growth rates and secretion of chemicals of interest (acetate, formate, succinate and ethanol) by mapping each environmental condition onto a multi-omic *E. coli* model that includes underground metabolism [25]. More specifically, using a multi-level linear program and a multi-omic extension of flux-balance analysis (FBA), gene expression was mapped onto a model of *Escherichia coli*. As a result, condition-specific models of *E. coli* were generated, and their predicted growth rate and production of byproduct were assessed.

A hybrid method combining multi-omic FBA and Bayesian inference was recently proposed [26] with the aim of investigating the cellular activities of a bacterium from the transcriptomic, fluxomic and pathway standpoints under different environmental conditions. The authors integrate an augmented FBA model of *E. coli* and a Bayesian factor model to regard pathways as latent factors between environmental conditions and reaction rates. Then, they determine the degree of metabolic pathway responsiveness and detect pathway cross-correlations. They also infer pathway activation profiles as a response to a set of environmental conditions. Finally, they use time series of gene expression profiles combined with their hybrid model in order to investigate how metabolic pathway responsiveness vary over time.

In two research works, Taffi and colleagues proposed a computational framework that integrates bioaccumulation information at the ecosystem level with genome-scale metabolic models of PCB degrading bacteria [27,28]. The authors applied their methods to the case study of the polychlorinated biphenyls (PCBs) bioremediation in the Adriatic food web. Remarkably, they were able to discover species acting as key players in transferring pollutants in contaminated food web. In particular, the role of the bacterial strain *Pseudomonas putida* KT2440, known to be able to degrade organic compound, in the reduction of PBCs in the trophic network, was assessed in different scenarios. Interestingly, one aspect of their analysis involved a scenario computed by using a synthetic strain of *Pseudomonas* performing additional aerobic degradation pathways. Combining these computational tools allows designing effective remediation strategies for contaminated environments, which can present challenges of natural selection, and provides at the same time insights into the ecological role of microbial communities within food webs.

5   Genome-Scale Modeling and Community Detection of Extreme Environmental Conditions

Omics technologies facilitate the study of organisms living at extreme conditions from different perspectives. They provide insights at genomic level, transcription level, protein level, and metabolites level. When compared to omics data from the organisms living at normal conditions, these may help understand mechanism of adaptation to extreme conditions. However, each dataset represents one portion of the whole picture and to understand the whole mechanism it is vital to integrate the data and reveal the mechanisms supported by diverse range of omics data. One way to integrate omics data is through identification of the pathways affected by each data source, and through combining the significance of affected pathways via Fisher's z-score. Panoga [10] is one of the methods that is used to identify the affected pathways in organisms living in extreme conditions. It takes the list of significantly altered genes and their significance values and maps them to a protein-protein interaction network. Then it searches for active subnetworks containing most of the affected genes. Affected KEGG pathways from the set of genes in the active subnetworks are determined and assigned significance values based on hypergeometric distribution. Combination of significance values of all the affected pathways for each type of omics data reveals affected pathways by all the data available.

In the past decade, genome-scale metabolic modeling has been successfully applied also for studying large-scale metabolic networks in microbes, with the aim of guiding rational engineering of biological systems, with applications in industrial and medical biotechnology, including antibiotic resistance [29]. Even though antibiotics remain an essential tool for treating animal and human diseases in the

21st century, antibiotic resistance among bacterial pathogens has garnered global interest in limiting their use, and to provide actionable strategies to search and support development of alternative antimicrobial substances [30]. It is interesting to note that bacterial strains such as *Arthrobacter* and *Gillisia* sp. CAL575, producing an array of molecules with potential antimicrobial activity vs human pathogenic *Burkolderia cepacia* complex strains were isolated from Antarctica [31]. These strains represent useful models to unravel metabolic pathways responsible for the production of bioactive primary and/or secondary metabolites [32].

Using methods for community detection in networks (Fig. 1, panels 5–6), environmental conditions can be grouped according to their predicted response, which is measured in the metabolic network using multi-omic models. Interestingly, this response can be measured on different omic levels. For instance they can be evaluated on the transcriptomic and fluxomic levels, each of which can constitute a layer of a multi-layer network. This approach would enable the study of the response individually on each omic layer, but also globally, e.g. by using a network fusion approach, where layers can be weighted and the multi-layer network can be fused to a single-layer network. Finally, although not covered here, another important approach to study metabolic networks is stochastic simulation based on an approach pioneered by Gillespie [33], and relying on molecular counts to simulate the evolution of populations of chemical species [34].

6   Protein Homology Modeling and Directed Evolution

The proteome forms the primary link between the genome and the metabolome. As such, understanding protein function is extremely important to taking a truly multi-omic view where we understand the causal interactions between layers, rather than just finding correlations. Computational models allow us to predict protein folding, and hence functionality, and are particularly useful when combined with directed evolution, which can allow us to explore entirely new structures and their properties.

Computational methods such as homology modeling and molecular dynamic simulation can be employed for protein engineering and design. Directed evolution of enzymes and/or bacterial strains can be exploited for industrial processes [35]. For instance, protein modeling and molecular dynamic simulation can be applied to molecules from psychrophilic organisms to unravel the molecular mechanisms responsible for cold-adaptation. In [36], a computational structural analysis based on molecular dynamics (MD) was performed for three β-tubulin isotypes from the Antarctic psychrophilic ciliate *Euplotes focardii*. Tubulin eterodimers (the building block of microtubules composed of α-tubulin and β-tubulin) from psychrophilic eukaryotes can polymerize into microtubules at $4\,^{\circ}C$, a temperature at which microtubules from mesophiles disassemble. The structural analysis based on MD indicated that all isotypes from *E. focardii*, with respect to those of mesophilic organisms, display different flexibility properties in the regions involved in the formation of longitudinal and lateral contacts during microtubule polymerization. A higher flexibility of these regions may facilitate the formation of lateral and longitudinal contacts among heterodimers for the formation of micro-tubules in an energetically unfavorable environment. Given that the protein structure could be thought of as a unit of phenotype, homology modeling analysis plays a major role in the generation of testable hypothesis on selection processes (disruptive, directional, stabilizing; see Fig. 2) acting at the level of genomic coding regions. One of the most important parameters influencing selection processes is the temperature. Molecular dynamics studies provide important insights into the mechanism of activity and stability of enzymes working in extreme conditions. If the three dimensional structure of the enzyme is known, one can conduct molecular dynamics runs at variable temperatures and compare the root mean square deviation (RMSD), root mean square fluctuation (RMSF) and the radius of gyration values of the enzyme at room temperatures, elevated temperatures and at low temperatures. This approach would allow tracing the unfolding mechanism and the flexibility of the enzyme. When compared with enzymes that are working at room temperatures, these runs would reveal crucial factors for activity and

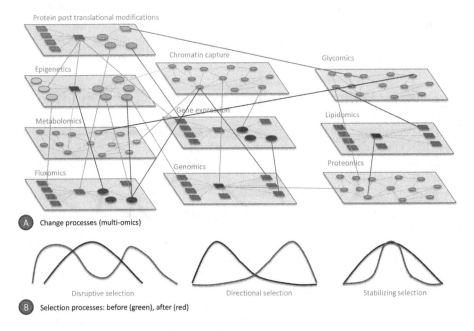

Fig. 2. Multi-omic changes - selection problem statement. (A) This is an extension of Fig. 1, panel 5; we show how multi-omics enlarge the molecular changes events affecting the phenotypic variations. For the sake of clarity, we show only few connections between the different omic levels. The changes are then filtered by selection processes. (B) We show the three main types of selection pressure: disruptive, directional and stabilizing. In green we show the allele distribution before the selection and in red the distribution after the selection process. It is noteworthy that with adaptation there is an increase of specificity, i.e. the number of many-to-many relationships decreases and the number of one-to-one relationships increases. (Color figure online)

stability at extreme conditions. One such study that used a MD-based approach to study the mechanism of temperature stability of thermophilic lipases, showed the importance of tryptophans involved in dimer formation and enhancement of aggregation tendency [37]. Another study conducted on the same family of lipases also revealed the importance of these tryptophans for coordination of zinc ions and the dependence of the thermostability to zinc concentrations [38].

Bioinformatics and mathematical methods play a crucial role in the experimental design and in understanding the pathways of the natural and artificial evolution of protein properties. Focused databases and computational tools to study and design evolution pathways have been developed. A promising collaboration between computational methods and evolutionary mutagenesis is envisaged in the field of de novo protein design, in which folds and functions not yet existing in nature are simulated computationally. Coupling computational design with directed evolution can help improve the performance of new proteins, as shown by designing and evolving proteins able to catalyze reactions not accessible to natural enzymes [39]. Interestingly, concepts originally developed in protein design studies such as protein folding funnels and fitness adaptive landscapes [40–42] could be further extended to multi-omic information.

7   Multi-omic Adaptive Landscapes

One of the biggest challenges of multi-omic bioinformatics is the estimation of mutual relationships between different omics. As shown in Fig. 3, one omic level Z may show a different structure over time. This can depend on a previous structure at the same level, but also on the structure of the interacting levels X and Y. By hypothesizing a linear relation between two omic levels, and a perturbation term P affecting X, in a discrete time domain we can define the interdependencies between omic levels as

$$\mathbf{x}(t + T) = \mathbf{\Lambda}\mathbf{x}(t) + \mathbf{p}(t), \tag{1}$$

where $\mathbf{x} = (X, Y, Z)^{\mathsf{T}}$, $\mathbf{\Lambda} = (\lambda_{ij})_{i,j=1,2,3}$, $\mathbf{p} = (P, 0, 0)^{\mathsf{T}}$. Note that, due to the large availability of genotype/phenotype data, the parameters $(\lambda_{ij})_{i,j=1,2,3}$ can be calculated with parameter estimation techniques (e.g. minimization of root mean square error).

From a reverse engineering point of view, reconstructing systems from multi-omic data will require identifying correlations between selective micro-level identifiers and macro-level properties (for example the physiological level). To achieve this aim, we need to model structure and dynamics of the funnel to figure out requisite dimensionality (danger of losing dimensions instead of merely losing detail) and covariates of the problem space, given that many possible paths may have led to the same observed outcome. The funnel reconstruction is affected by the uncertainty due to multi-omic data analytics from different sources (quality, and conditions), i.e. the identifying the structure and dimensionality of metadata.

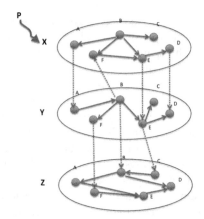

Fig. 3. Interdependency among omic layers in a multi-layer network. One of the challenges is to estimate the mutual relationships between omics, i.e. the causal relationships. In general terms, the control and target is an important bioinformatics challenge. Nodes represent genes forming a network, but they can also represent different features. A perturbation P may affect one or more layers. The overall response of each layer is produced by the combination of X, Y and Z, according to the interdependency among layers (see Eq. 1).

Although the different omics are subtly coupled, one meaningful and useful perspective from the operational standpoint, is the concept of multi-layer networks. Network theory investigates the global topology and structural patterns of the interactions among the constituents of a multi-omic adaptive system. Networks are the most natural way to model many types of biological knowledge at different levels. Recently, complex network theory has been extended towards multiple networks. Multilayered networks can be used for the representation and quantification of the interactions arising from the combined action of omics in response to mutational events and selection pressures. The architecture of complex networks is a natural embedding for fitness-changes diffusion processes as a function of natural selection constraints, e.g. those observed during environmental changes.

8   Conclusion

In this paper, we reviewed available methods to study selection and adaptation processes to extreme environments by means of multi-omic experimental and bioinformatics approaches. To date, a great deal of biological information has already been acquired through application of individual 'omics' approaches. However, "multi-omic technology", coupled with or mapped to multi-layer networks, will enable the integration of knowledge at different levels: from genes to proteins, from pathways to metabolic fluxes. We have also discussed how the investigation of the correlation between changes at the level of omics and the selection processes may provide a useful approach to study the genotype-phenotype mapping.

We argue that extreme conditions could provide better insights into genotype-phenotype mapping, because extreme phenotypes are required to adapt to extreme environments. Large adaptations such as the production of vast amounts of antifreeze protein are much easier to detect and understand than the far more subtle changes one expects in more hospitable environments, where adaptation is primarily concerned with slight changes to an already near-optimal phenotype.

Multi-omic and multi-layer models represent a novel and powerful tool to generate discrete and testable biological hypotheses. We also argue that the study of life in extreme environments will provide useful clues to general laws of biological adaptations and easier conditions to test mechanistic hypotheses.

Acknowledgements. This research work was supported by the European Commission Marie Sklodowska-Curie Actions H2020 RISE Metable - 645693.

References

1. Nielsen, R., Slatkin, M.: An Introduction to Poulation Genetics: Theory and Applications. Sinauer Associates, Sunderland (2013)
2. Yurkovich, J.T., Palsson, B.O.: Solving puzzles with missing pieces: the power of systems biology [point of view]. Proc. IEEE **104**(1), 2–7 (2016)
3. Karr, J.R., Sanghvi, J.C., Macklin, D.N., Gutschow, M.V., Jacobs, J.M., Bolival, B., Assad-Garcia, N., Glass, J.I., Covert, M.W.: A whole-cell computational model predicts phenotype from genotype. Cell **150**(2), 389–401 (2012)
4. Carrera, J., Estrela, R., Luo, J., Rai, N., Tsoukalas, A., Tagkopoulos, I.: An integrative, multi-scale, genome-wide model reveals the phenotypic landscape of escherichia coli. Mol. Syst. Biol. **10**(7), 735 (2014)
5. Romero, I.G., Ruvinsky, I., Gilad, Y.: Comparative studies of gene expression and the evolution of gene regulation. Nat. Rev. Genet. **13**(7), 505–516 (2012)
6. Koonin, E.V., Wolf, Y.I.: Constraints and plasticity in genome and molecular-phenome evolution. Nat. Rev. Genet. **11**(7), 487–498 (2010)
7. Angione, C., Lió, P.: Predictive analytics of environmental adaptability in multi-omic network models. Sci. R. **5**, 15147 (2015)
8. Meysman, P., Sonego, P., Bianco, L., Qiang, F., Ledezma-Tejeida, D., Gama-Castro, S., Liebens, V., Michiels, J., Laukens, K., Marchal, K., et al.: Colombos v2. 0: an ever expanding collection of bacterial expression compendia. Nucleic Acids Res. **42**(D1), D649–D653 (2014)
9. Keeling, P.J., Burki, F., Wilcox, H.M., Allam, B., Allen, E.E., Amaral-Zettler, L.A., Virginia Armbrust, E., Archibald, J.M., Bharti, A.K., Bell, C.J., et al.: The marine microbial eukaryote transcriptome sequencing project (MMETSP): illuminating the functional diversity of eukaryotic life in the oceans through transcriptome sequencing. PLoS Biol. **12**(6), e1001889 (2014)
10. Bakir-Gungor, B., Egemen, E., Sezerman, O.U.: PANOGA: a web server for identification of SNP-targeted pathways from genome-wide association study data. Bioinformatics **30**(9), 1287–1289 (2014)

11. Karlsson, F., Tremaroli, V., Nielsen, J., Bäckhed, F.: Assessing the human gut microbiota in metabolic diseases. Diabetes **62**(10), 3341–3349 (2013)
12. Herlemann, D.P., Labrenz, M., Jürgens, K., Bertilsson, S., Waniek, J.J., Andersson, A.F.: Transitions in bacterial communities along the 2000 km salinity gradient of the Baltic sea. ISME J. **5**(10), 1571–1579 (2011)
13. Craig Venter, J., Remington, K., Heidelberg, J.F., Halpern, A.L., Rusch, D., Eisen, J.A., Dongying, W., Paulsen, I., Nelson, K.E., Nelson, W., et al.: Environmental genome shotgun sequencing of the Sargasso sea. Science **304**(5667), 66–74 (2004)
14. Morfopoulou, S., Plagnol, V.: Bayesian mixture analysis for metagenomic community profiling. Bioinformatics **31**(18), 2930–2938 (2015)
15. Pucciarelli, S., Buonanno, F., Pellegrini, G., Pozzi, S., Ballarini, P., Miceli, C.: Biomonitoring of Lake Garda: identification of ciliate species and symbiotic algae responsible for the black-spot bloom during the summer of 2004. Environ. Res. **107**(2), 194–200 (2008)
16. Pucciarelli, S., Devaraj, R.R., Mancini, A., Ballarini, P., Castelli, M., Schrallhammer, M., Petroni, G., Miceli, C.: Microbial consortium associated with the antarctic marine ciliate euplotes focardii: an investigation from genomic sequences. Microb. Ecol. **70**(2), 484–497 (2015)
17. Mancini, A., Verdini, D., Vigna, G.L., Recanatini, C., Lombardi, F.E., Barocci, S.: Retrospective analysis of nosocomial infections in an Italian tertiary care hospital. New Microbiol. 39(3) (2016)
18. Knetsch, C.W., Lawley, T.D., Hensgens, M.P., Corver, J., Wilcox, M.W., Kuijper, E.J.: Current application and future perspectives of molecular typing methods to study Clostridium difficile infections. Euro Surveill. **18**(4), 20381 (2013)
19. Angione, C., Costanza, J., Carapezza, G., Lió, P., Nicosia, G.: Multi-target analysis and design of mitochondrial metabolism. PLoS One **10**(9), e0133825 (2015)
20. Faith, J.J., Hayete, B., Thaden, J.T., Mogno, I., Wierzbowski, J., Cottarel, G., Kasif, S., Collins, J.J., Gardner, T.S.: Large-scale mapping and validation of escherichia coli transcriptional regulation from a compendium of expression profiles. PLoS Biol. **5**(1), e8 (2007)
21. Yuan, M., Chen, M., Zhang, W., Wei, L., Wang, J., Yang, M., Zhao, P., Tang, R., Li, X., Hao, Y., et al.: Genome sequence and transcriptome analysis of the radioresistant bacterium deinococcus gobiensis: insights into the extreme environmental adaptations. PloS one **7**(3), e34458 (2012)
22. Zhao, Y., Yi, Z., Gentekaki, E., Zhan, A., Al-Farraj, S.A., Song, W.: Utility of combining morphological characters, nuclear and mitochondrial genes: an attempt to resolve the conflicts of species identification for ciliated protists. Mol. Phylogenet. Evol. **94**, 718–729 (2016)
23. Ghobakhlou, A.-F., Johnston, A., Harris, L., Antoun, H., Laberge, S.: Microarray transcriptional profiling of arctic mesorhizobium strain N33 at low temperature provides insights into cold adaption strategies. BMC Genomics **16**(1), 383 (2015)
24. Sass, S., Buettner, F., Mueller, N.S., Theis, F.J.: RAMONA: a web application for gene set analysis on multilevel omics data. Bioinformatics **31**(1), 128–130 (2014)
25. Angione, C., Conway, M., Lió, P.: Multiplex methods provide effective integration of multi-omic data in genome-scale models. BMC Bioinform. **17**(4), 257 (2016)
26. Angione, C., Pratanwanich, N., Lió, P.: A hybrid of metabolic flux analysis and bayesian factor modeling for multiomics temporal pathway activation. ACS Synth. Biol. **4**(8), 880–889 (2015). doi:10.1021/sb5003407
27. Taffi, M., Paoletti, N., Angione, C., Pucciarelli, S., Marini, M., Liò, P.: Bioremediation in marine ecosystems: a computational study combining ecological modeling and flux balance analysis. Front Genet. **5**, 319 (2014)

28. Taffi, M., Paoletti, N., Liò, P., Pucciarelli, S., Marini, M.: Bioaccumulation modelling and sensitivity analysis for discovering key players in contaminated food webs: the case study of PCBs in the adriatic sea. Ecol. Model. **306**, 205–215 (2015)
29. Milne, C.B., Kim, P.-J., Eddy, J.A., Price, N.D.: Accomplishments in genome-scale in silico modeling for industrial and medical biotechnology. Biotechnol. J. **4**(12), 1653–1670 (2009)
30. Nolte, O.: Antimicrobial resistance in the 21st century: a multifaceted challenge. Protein Pept. Lett. **21**(4), 330–335 (2014)
31. Fondi, M., Orlandini, V., Maida, I., Perrin, E., Papaleo, M.C., Emiliani, G., Pascale, D., Parrilli, E., Tutino, M.L., Michaud, L., et al.: Draft genome sequence of the volatile organic compound-producing antarctic bacterium arthrobacter sp. strain TB23, able to inhibit cystic fibrosis pathogens belonging to the burkholderia cepacia complex. J. Bacteriol. **194**(22), 6334–6335 (2012)
32. Orlandini, V., Maida, I., Fondi, M., Perrin, E., Papaleo, M.C., Bosi, E., Pascale, D., Tutino, M.L., Michaud, L., Lo Giudice, A., et al.: Genomic analysis of three sponge-associated arthrobacter antarctic strains, inhibiting the growth of burkholderia cepacia complex bacteria by synthesizing volatile organic compounds. Microbiol. Res. **169**(7), 593–601 (2014)
33. Gillespie, D.T.: Exact stochastic simulation of coupled chemical reactions. J. Phys. Chem. **81**(25), 2340–2361 (1977)
34. Cardelli, L., Kwiatkowska, M., Laurenti, L.: Stochastic analysis of chemical reaction networks using linear noise approximation. In: Roux, O., Bourdon, J. (eds.) CMSB 2015. LNCS, vol. 9308, pp. 64–76. Springer, Heidelberg (2015)
35. Adrio, J.L., Demain, A.L.: Microbial enzymes: tools for biotechnological processes. Biomolecules **4**(1), 117–139 (2014)
36. Chiappori, F., Pucciarelli, S., Merelli, I., Ballarini, P., Miceli, C., Milanesi, L.: Structural thermal adaptation of β-tubulins from the antarctic psychrophilic protozoan euplotes focardii. Proteins Struct. Funct. Bioinf. **80**(4), 1154–1166 (2012)
37. Timucin, E., Sezerman, O.U.: Zinc modulates self-assembly of bacillus thermocatenulatus lipase. Biochemistry **54**(25), 3901–3910 (2015)
38. Timucin, E., Cousido-Siah, A., Mitschler, A., Podjarny, A., Sezerman, O.U.: Probing the roles of two tryptophans surrounding the unique zinc coordination site in lipase family i. 5. Proteins Struct. Funct. Bioinf. **84**(1), 129–142 (2016)
39. Khersonsky, O., Röthlisberger, D., Wollacott, A.M., Murphy, P., Dym, O., Albeck, S., Kiss, G., Houk, K.N., Baker, D., Tawfik, D.S.: Optimization of the in-silico-designed kemp eliminase KE70 by computational design and directed evolution. J. Mol. Biol. **407**(3), 391–412 (2011)
40. Meini, M.R., Tomatis, P.E., Weinreich, D.M., Vila, A.J.: Quantitative description of a protein fitness landscape based on molecular features. Mol. Biol. Evol. **32**(7), 1774–1787 (2015)
41. Kauffman, S.A.: The Origins of Order: Self Organization and Selection in Evolution. Oxford University Press, New York (1993)
42. Svensson, E., Calsbeek, R.: The Adaptive Landscape in Evolutionary Biology. OUP, Oxford (2012)

Improving Genome Assemblies
Using Multi-platform Sequence Data

Pınar Kavak[1,2](✉), Bekir Ergüner[1], Duran Üstek[3], Bayram Yüksel[4],
Mahmut Şamil Sağıroğlu[1], Tunga Güngör[2], and Can Alkan[5](✉)

[1] Advanced Genomics and Bioinformatics Research Group (İGBAM), BİLGEM,
The Scientific and Technological Research Council of Turkey (TÜBİTAK),
41470 Kocaeli, Gebze, Turkey
pinarkavak@gmail.com
[2] Department of Computer Engineering, Boğaziçi University, 34342 İstanbul,
Bebek, Turkey
[3] Department of Medical Genetics, İstanbul Medipol University,
34810 İstanbul, Beykoz, Turkey
[4] TÜBİTAK - MAM - GMBE (The Scientific and Technological Research Council
of Turkey, Genetic Engineering and Biotechnology Institute), 41470 Kocaeli,
Gebze, Turkey
[5] Department of Computer Engineering, Bilkent University, 06800 Ankara,
Bilkent, Turkey
calkan@cs.bilkent.edu.tr

Abstract. Accurate *de novo* assembly using short reads generated
by next generation sequencing technologies is still an open problem.
Although there are several assembly algorithms developed for data generated with different sequencing technologies, and some that can make
use of hybrid data, the assemblies are still far from being perfect. There
is still a need for computational approaches to improve draft assemblies.
Here we propose a new method to correct assembly mistakes when there
are multiple types of data generated using different sequencing technologies that have different strengths and biases. We exploit the assembly
of highly accurate short reads to correct the contigs obtained from less
accurate long reads. We apply our method to Illumina, 454, and Ion Torrent data, and also compare our results with existing hybrid assemblers,
Celera and Masurca.

Keywords: *de novo* assembly · Assembly improvement · Next generation multi-platform sequencing

1   Scientific Background

Since the introduction of high throughput sequencing (HTS) technologies, traditional Sanger sequencing is being abandoned especially for large-scale sequencing
projects. Although cost effective for data production, HTS also imposes increased

© Springer International Publishing Switzerland 2016
C. Angelini et al. (Eds.): CIBB 2015, LNBI 9874, pp. 220–232, 2016.
DOI: 10.1007/978-3-319-44332-4_17

cost for data processing and computational burden. In addition, the data quality is in fact lower, with greater error rates, and short read lengths for most platforms. One of the main algorithmic problems in analyzing HTS data is the *de novo* assembly: i.e. "stitching" billions of short DNA strings into a collection of larger sequences, ideally the size of chromosomes. However, "perfect" assemblies with no gaps and no errors are still lacking due to many factors, including the short read and fragment (paired-end) lengths, sequencing errors in basepair level, and the complex and repetitive nature of most genomes. Some of these problems in *de novo* assembly can be ameliorated through using data generated by different sequencing platforms, where each technology has "strengths" that may be used to fix biases introduced by others.

There are three kinds of assemblers mainly used to do genome assembly: (i) greedy assemblers [1–3], (ii) overlap-layout-consensus (OLC) graph based assemblers [4–6] and (iii) de Bruijn graph based assemblers [7–11]. Greedy assemblers follow a greedy approach such that: given one read or contig, at each step assembler adds one more read or contig with the largest overlap. The problem of greedy assemblers is that they can get stuck at local maxima. Therefore they are generally used for small genome assemblies. Since they also use more memory and are slower, it is not feasible to assemble large genomes with greedy assemblers. OLC graph based assemblers work well when the long reads are available for assembly. They generate all-against-all pairwise alignments and build the graph by representing reads as nodes and overlaps between reads as edges. They obtain the consensus assembly by following a Hamiltonian path on the graph. Assemblers that are based on de Bruijn graphs are designed primarily for short reads. They use a k-mer graph approach instead of calculating all-against-all pairwise alignments. They build the graph by using k-mers as edges and the overlaps between k-mers as nodes. They follow an Eulerian path through the k-mer graph to find a consensus assembly. Several assemblers use multiple read libraries [12,13,15,16] for better assembly construction. CABOG [12] was initially designed for Sanger sequencing, and then it was revised to use 454 data, but it also accepts Illumina data to generate a hybrid assembly. Masurca [13] is able to assemble Illumina reads together with longer 454 and Sanger reads. MIRAest [15] can use Sanger, 454, Illumina, Ion Torrent and corrected PacBio data for hybrid assembly. It works on small genomes. Cerulean [16] uses long PacBio reads and short Illumina reads to construct a hybrid assembly. It uses ABySS [10] assembler to generate *assembly graphs* with paired end Illumina reads. Then, as input, it uses these *assembly graphs* and also long PacBio read alignments to the assembled contigs.

Additionally, strategies to merge different assemblies using different data sources into a single coherent assembly are described in the literature (e.g. [18]). Our method differs from that of [18], in data types. [18] works on Illumina, 454 and ABI SOLID data, where we work on Illumina, 454 and Ion Torrent data. Also pre- and post-processing steps of the two methods differ. [18] at first assembles 454, Illumina and SOLID data separately with different assemblers and then assembles the resulting contig collection again with another assembler.

In this work, we propose a method to improve draft assemblies (i.e. produced using a single data source, and/or single algorithm) by incorporating data generated by different HTS technologies, and by applying novel correction methods. To achieve better improvements, we exploit the advantages of both short but low-error-rate reads and long but erroneous reads. We show that correcting the contigs built by assembling long reads through mapping short and high quality read contigs produces the best results, compared to the assemblies generated by algorithms that use hybrid data all at once. With this study, we also have the opportunity to compare Ion Torrent and Roche/454 reads in terms of assembly performances.

2   Materials and Methods

We cloned a part of human chromosome 13 into a bacterial artificial chromosome (BAC), and sequenced it separately using Illumina, Roche/454, and Ion Torrent platforms (Table 1). We also obtained a "gold standard" reference assembly for this BAC using GRCh37-guided assembly generated by Mira [14] using Roche/454 data, which we then corrected using the Illumina reads [17]. Since Roche/454 and Ion Torrent platforms have similar sequencing biases (i.e. problematic homopolymers), we separated this study into two different groups: Illumina & 454 and Illumina & Ion Torrent. We applied the same method on the two groups and evaluated them separately which gave us the opportunity to compare Roche/454 and Ion Torrent data. The flowchart of the pipeline is depicted in Fig. 1.

Table 1. Properties of the data

Technology	Length range	Mean length	Mean base qual (phred s.)	Paired
Illumina	101 bp (all reads have equal length)	101 bp	38	paired
Roche/454	40 bp-1027 bp	650 bp	28	single-end
Ion Torrent	5 bp-201 bp	127 bp	24	single-end

Technology: The name of the sequencing technology used to produce the reads. **Length range**: Minimum and maximum lengths of the generated reads. **Mean length**: The mean length among all reads. **Mean base qual**: The average phred score sequence quality of all reads. Calculated by summing up all phred scores of the bases in a read and dividing it to sequence length of the read, over all reads. **Paired**: Represents whether the sequencing is performed as paired-end or single-end.

2.1   Pre-processing

Pre-processing steps consist of the following:

- First, we discard the reads that have low average quality value (phred score 17, i.e. $\geq 2\%$ error rate).
- Then, we remove the reads with high N-density (with $>10\%$ of the read consisting of Ns) from consideration. Ns would destroy the assembly contiguity.

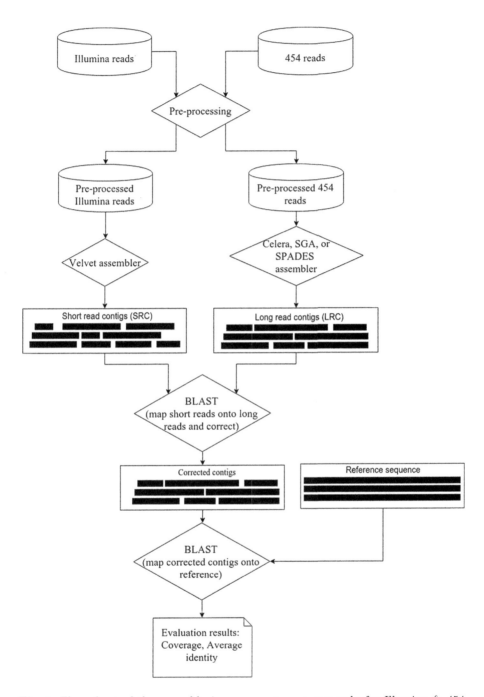

Fig. 1. Flow chart of the assembly improvement processes only for Illumina & 454. Same is valid for Illumina & Ion torrent.

– Third, we trim the groups of bases at the beginning and/or at the end of
  the read that seem to be non-uniform according to sequence base content (A,
  T, G, C) (See Fig. 2). These regions would cause erroneous structures in the
  assembly.
– Finally, we apply the pre-processing operations of each assembler we used.

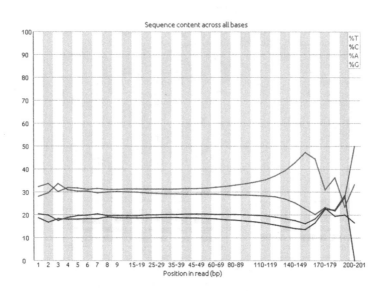

Fig. 2. Non-uniform A, T, G, C regions of Ion Torrent reads. First 8 bases and the
bases after the 130th base are trimmed in pre-processing.

2.2   Assembly

After the pre-processing step, we used several assembly tools suitable to assemble different types of data: We used Velvet [7], a de Bruijn graph based assembler
that is designed to assemble short reads for assembling the Illumina reads. Considering the trimmed beginning and/or end parts of 101bp long paired-end reads
from Illumina, and after testing kmers 21 and 31, we decided to use k = 51 for
short read assembly. We ran Velvet with shortPaired mode with insert size
400bp, expected coverage 80, coverage cutoff 2, and minimum contig length 100.
N50 value of the resulting short read contigs was 8,865 bp. We used two different OLC assemblers: Celera [5] and SGA [6] to assemble the long read data sets
(Roche/454 and Ion Torrent) separately. We ran Celera assembler in unmated
mode and with default parameters to assemble 454 and Ion Torrent reads. N50
value of the assembly obtained with 454 and Ion Torrent reads with Celera was
1,308 bp and 1,284 bp, respectively. We also used SGA assembler in unmated
mode for the same data sets. We obtained N50 values of 505 bp and 117 bp

for Roche/454 and Ion Torrent data, respectively. In addition, we also used a de Bruijn graph based assembler, SPAdes [8], to assemble the long read data. Again, we applied default parameters. N50 values of the assemblies obtained with 454 and Ion Torrent reads with SPAdes were 212 bp and 259 bp, respectively.

We mapped all draft assemblies to the E. coli reference sequence using BLAST [19]'s MegaBLAST [20] task to identify and discard E. coli contamination due to the cloning process. We discarded any contig that mapped to the E. coli reference sequence with sequence identity ≥95 %. Finally, we obtained one short read, and three long read assemblies without contamination.

2.3 Correction

In the correction phase, we wanted to exploit the accuracy of the short read contigs (SRC) and the coverage of the long read contigs (LRC) to obtain a better assembly. Hence, we mapped all SRCs onto all LRCs of each group and corrected the LRCs according to the mapping results. First, we used BLAST [19]'s MegaBLAST [20] mapping task to map the SRC onto the LRC. We then used an in-house C++ program to process the MegaBLAST mapping results. Since MegaBLAST may report multiple mapping locations due to repeats, we only accepted the "best" mapping locations. Reasoning from the fact that short reads show less sequencing errors, we preferred the sequence reported by the SRC over the LRC when there is a disagreement between the pair. By doing this, we patched the "less fragmented" long read assemblies. If there is an overlap between different SRC mappings at the same region on the LRC, the latter overwrites the first. Figure 3 shows a visual representation of the strategy on correcting the LRCs.

Briefly, we describe our strategy in the following steps:

- If there is a mapping between a SRC and a LRC, and if the mapping does not start at the beginning of the LRC, add the unmapped prefix of the LRC.
- Next, if the mapping does not start at the beginning of the SRC (very rare situation), add the unmapped prefix of the SRC with lowercase (i.e. low confidence) letters.
- Over the mapping region between SRC and LRC, pick the SRC values.
- If the mapping does not end at the end of the SRC (rare), add the unmapped suffix of the SRC, again with lowercase letters. One may argue that it might disturb the continuity of the resulting contig, however, we observe such mapping properties very rarely. The reason for using lowercase letters is to keep track of the information that there is a disagreement between the SRC and LRC on these sections, so the basepair quality will be lower than other sections of the assembly.
- Finally, add the unmapped suffix of the LRC and obtain the corrected contig.

We repeated this process to correct each of the three long read assembly contig sets. We applied our correction strategy on each data set multiple times until there is no improvement in the *Coverage* and *Average Identity* metrics.

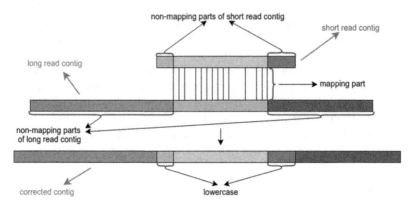

Fig. 3. Correction method: correct the long read contig according to the mapping information of the short read contig.

2.4 Evaluation

To evaluate and compare the resulting and corrected assemblies all-against-all, we mapped all of the assembly candidates, including primary assemblies and also final corrected assemblies to the "gold standard" BAC assembly. According to the alignment results, we calculated various statistics such as the number of mapped contigs, how many bases on the reference sequence were covered, how many gaps exist on the reference sequence, and the total gap length. We calculated metrics such as "Coverage" and "Average Identity" and compared the resulting assemblies with these metrics.

To calculate these statistics, we kept an array of arr_reference[0,0,0,...0], where length(arr_reference) = length(reference). We updated the contents of arr_reference according to the alignments. If there is a match at a location, we assigned the corresponding position in the array to "1", if there is a mismatch at a location, we set it as "−1", and if that location is not included in any alignment, we left it as "0" (which means a gap). We assumed deletions in the contig (query) as mismatches. We also calculated the number of insertions in the contig. Scanning the array and summing up the number of "1"s (matches), "−1"s (mismatches), "0"s (gaps) and "insertionInQuery", we obtained the number of matches, mismatches, gaps, and insertions in contig. Using these numbers, we calculated the Coverage (Eq. 1) and Average Identity values (Algorithm 1).

We also used two hybrid assemblers, Celera-CABOG [12] and Masurca [13], with Illumina & 454 and Illumina & Ion Torrent. These hybrid assemblers load all reads as input and assemble them with a hybrid method. We assembled the two data sets with these hybrid assemblers to compare our correction method with the results of them.

$$\text{Coverage} = \left(\frac{\# \text{ of covered bases}}{\text{length of the reference}} \right) \tag{1}$$

Algorithm 1. Average identity

---

while no contigs left **do**

    alignmentLength ← matches + mismatches + insertionInContig

    identity ← $\left(\frac{\text{matches}}{\text{alignmentLength}} \right)$

    avgIdentity ← avgIdentity + identity × contigLength

end while

avgIdentity ← $\left(\frac{\text{avgIdentity}}{\sum_{i=1}^{contigNum} contigLength_i} \right)$

---

3 Results

We present the results in Table 2, and interpret them in different point of views.

3.1 454 vs. Ion Torrent

Ion Torrent reads are shorter than 454 reads and they have less mean base quality (Table 1). So, we did not expect to have better assembly with Ion Torrent reads than 454 reads. The results in Table 2 agree with our expectations. In Table 2, we see that the assembly of 454 reads performs better on evaluation metrics than Ion Torrent with all kind of assemblers. The assembly of Ion Torrent reads with Celera assembler has very low coverage value: 26.94 %. The reason for the low coverage might be because Celera assembler is not designed for Ion Torrent read type (shorter reads with lower quality). Even 454 and Ion Torrent reads have similar error types at the homopolymer regions. SGA assembly with Ion Torrent reads performs better on Coverage (86.57 %) but it cannot reach to the Coverage of SGA assembly with 454 reads (99.83 %). The assembly of Ion Torrent reads has the highest coverage with SPAdes assembler (94.94 %). Correction of the Ion Torrent contigs improves the assembly quality but even after correction phase Ion Torrent corrected assembly cannot reach the results of 454 corrected assembly.

3.2 Assemblers

Table 2 shows that the assembly obtained by Velvet with only short Illumina reads showed good coverage (99.05 %) and average identity rates (97.52 %). The number of contigs obtained with Velvet assembly is 455, of which 437 map to the reference. There are 39 gaps and the total size of the gaps is 1,671 bp. Our aim was to increase the coverage, improve the average identity, decrease the number of contigs and gaps, and shrink the lengths of the gaps.

Since we observed that 454 reads resulted better assembly than Ion Torrent reads as stated in Sect. 3.1, we compared different assemblers using 454 contigs. The assembly of Celera with the 454 long reads has 97.58 % coverage and 92.59 % average identity, which are lower than Illumina-Velvet values. Number of contigs (735) is reasonable but number of gaps and total gap length are high (18 and 4,280 bp, respectively). SGA assembly using 454 reads has very high

Table 2. Results of assembly correction method on BAC data.

Name	Length	# of contigs	# of mapped contigs	# of covered bases	Coverage	Avg. identity	# of Gaps	Size of Gaps
Reference	*176.843*							
Velvet								
Ill. Velvet	197,040	455	437	175, 172	0.99055	0.97523	39	1,671
Celera								
454 Celera	908,008	735	735	172, 563	0.97580	0.92599	18	4,280
Ion Celera	39,347	27	27	47, 638	0.26938	0.96932	47	129,205
Corrected Celera[#]								
Ill-454 Celera	371,065	250	250	176, 071	0.995635	0.944558	5	772
Ill-454 Celera[2][*]	365,802	245	245	176, 343	0.9971	0.9455	4	500
Ill-Ion Celera	93,909	30	28	81, 819	0.46267	0.96327	36	95,024
Ill-Ion Celera[2]	145,262	30	28	91, 962	0.52002	0.97412	33	84,881
Ill-Ion Celera[3]	216,167	30	28	99, 645	0.56347	0.98066	34	77,198
SGA								
454 SGA	62,909,254	108, 095	101, 514	176, 546	0.99832	0.97439	1	297
Ion SGA	842,997	6, 417	6, 122	153, 092	0.86569	0.99124	197	23.751
Corrected SGA								
Ill-454 SGA	295,009	335	335	176, 757	0.99951	0.96823	5	86
Ill-Ion SGA	197,509	291	291	175, 052	0.98987	0.97501	45	1,791
Ill-Ion SGA[2]	203,064	291	291	175, 676	0.99340	0.97413	34	1,167
SPADES								
454 SPADES	12,307,761	49, 824	49, 691	176, 843	1.0	0.98053	0	0
Ion SPADES	176,561	110	107	167, 890	0.94937	0.92909	9	8,953
Corrected SPADES								
Ill-454 SPADES	290,702	298	298	176, 454	0.99780	0.96538	5	389
Ill-Ion SPADES	198,665	52	52	171, 977	0.97248	0.94215	4	4,866
Ill-Ion SPADES[2]	200,307	52	52	172, 101	0.97319	0.94230	2	4,742
Masurca								
Ill-454 Masurca	380	1	0	0	0	0	0	0
Ill-Ion Masurca	2,640	8	8	1, 952	0.01104	0.98223	9	174,891
Celera-CABOG								
Ill-454 Celera	1,101,716	891	891	174, 330	0.98579	0.92452	12	2,513
Ill-Ion Celera	0	0	0	0	0.0	0.0	0	0.0

Name: the name of the data group that constitutes the assembly; # of Contigs: the number of contigs that belong to the resulting assembly; # of Mapped Contigs: the number of contigs that successfully mapped onto the reference sequence; # of Covered bases: the number of bases on the reference sequence that are covered by the assembly; Coverage: percentage of covered reference; Avg. identity: percentage of the correctly predicted reference bases; # of Gaps: the number of gaps that cannot be covered on the reference genome; Size of Gaps: total number of bases on the gaps.
[*] "2" represents the results of the second cycle of correction, "3" represents the third cycle.
[#] A mistake is noticed on Ill-454 Celera data and the results are corrected after being published in the proceedings of CIBB2015.

coverage (99.83 %) and identity (97.43 %). It has just one gap with size 297 bp, but the number of contigs is also very high (101,514), which is an unwelcome situation. SPAdes-454 assembly also had a large number of contigs (49,824) which completely cover the reference sequence with 98.05 % average identity. SPAdes assembly resulted in lower number of contigs and had higher coverage and average identity than SGA. If we evaluate the results according to the number of contigs, Celera-454 results seem more reasonable than SGA or SPAdes results, since it returned a reasonable number of contigs even with low coverage and average identity.

3.3 Correction

We observed that the correction method improved both 454 and Ion Torrent based assemblies generated with all assemblers we tested (Table 2). In the remainder of the paper, we only mention the 454-based assemblies for simplicity.

When we applied our correction method on Celera-454 assembly using the Velvet-Illumina assembly, we achieved better coverage and average identity rates: the coverage of 454 assembly increases up to 99.56 % and the average identity rate increases up to 94.45 % on the first correction cycle. The second correction cycle increases the coverage and average identity rates to 99.71 % and 94.55 %, respectively, and the correction converges. The number of contigs decrease to 245 from 735, and the number of gaps decrease down to 4 (500 bp) from 18 (4,280 bp). Since the third correction cycle does not give better results it is not shown in Table 2.

Our correction method increased the coverage of SGA-454 assembly up to 99.99 % from 99.82 % but with less average identity and with more gaps although the total length of the gaps is decreased. Correction using the short read assembly decreased the number of contigs down to a reasonable number (335). Corrected SGA assembly has the largest coverage rate among all, and also with more identity than Velvet-Illumina assembly.

The number of contigs in SPAdes assembly also decreased to 298 from 49,691 using our correction method. With the decrease in number of contigs, the coverage also decreased (99.78 %) as well as the average identity (96.53 %). The number of gaps increased to 5 from 0 with a total size of 389.

In summary, we obtained substantial assembly correction in draft assemblies by using advantages of different technologies.

3.4 Hybrid Assemblers

We also compared the results of two hybrid assemblers on our multiple type of data. We used Masurca and Celera-CABOG with default parameters given two groups of hybrid data as input: Illumina & 454 and Illumina & Ion Torrent. Hybrid assemblers Masurca and CABOG did not show good assembly rates. We obtained zero coverage with 454 and Illumina reads using Masurca. The only contig left after the contamination removal did not map to the reference sequence. We also observed very low coverage (1.10 %) with 98.22 % average identity with Ion Torrent & Illumina reads. Therefore, we conclude that Masurca did not work very well in our case with our data types.

Similarly, we obtained zero coverage with Ion Torrent & Illumina using CABOG. All of the resulting contigs obtained from the assembly were removed as contamination. However, CABOG performed substantially better with Illumina & 454, and generated assembly with 98.58 % coverage and 92.45 % average identity. The assembly composed of 891 contigs and 12 gaps with total gap length of 2,513 bp. Still, the performance of CABOG was not better than the corrected assembly results described above.

3.5 Combination of the Data from all Platforms

We combined data from all 3 platforms to generate a new assembly in order to see if we have better coverage or accuracy on the results. The results are presented in Table 3. Our method is originally designed for two data types. It corrects one data type's contigs with the other data type's contigs, so we needed to combine three types of contigs sequentially. As mentioned in Sect. 2.3 our method accepts that the corrector data is more accurate than the corrected data. If there is a map between the two, it replaces the values of the corrected data with the values of the corrector data. For that reason, while working with 3 data combination, we decided to use Velvet-Illumina contigs which are built by the highest accurate reads as the last corrector. On Table 3, it is seen that Celera-454 contigs increase the coverage rate of Celera-Ion Torrent contigs (from 26.93 % to 84.26 %) although decreasing the average identity rate from 96.93 % to 94.51 %. Correcting the resulting contigs with Velvet-Illumina contigs increases the coverage (96.32 %) and average identity rates (95.50 %) even higher. The coverage and average identity rates are improved on the second and third cycles too. Correcting Ion SPADES, with 454 SPADES gives higher coverage (99.82 %) and average identity (97.33 %) rates than correcting them with only Velvet Illumina contigs (97.24 % and 94.21 % respectively). After using Velvet Illumina contigs for the last correction, the results are improved approximately by 0.1 % and 0.01 % respectively. Correcting Ion SGA contigs with 454 SGA contigs was not possible because of memory limitations of BLAST mapping with such huge data. Instead, we used corrected version of "454 SGA contigs with Illumina Velvet contigs" to correct the Ion SGA contigs. The coverage is higher than both Ill-Ion SGA and Ill-454 SGA, average identity is lower than Ill-454 SGA.

Table 3. Results with combination of 3 data types

Name	Length	# of contigs	# of mapped contigs	# of covered bases	Coverage	Avg. Identity	# of gaps	Size of gaps
Reference	*176.843*							
Corrected Ion Celera								
454-Ion Celera	500, 251	27	27	149, 021	0.84267	0.94515	63	27, 822
Ill-"454-Ion Celera"	570, 865	27	27	170, 348	0.96327	0.95503	16	6, 495
Ill-"454-Ion Celera"[2]*	575, 726	27	27	172, 516	0.97553	0.95541	12	4, 327
Ill-"454-Ion Celera"[3]	578, 727	27	27	174, 535	0.98694	0.95555	10	2, 308
Corrected Ion SPADES								
454-Ion SPADES	11, 224, 602	60	60	176, 540	0.99828	0.97334	6	303
Ill-"454-Ion SPADES"	9, 543, 712	45	45	176, 712	0.99925	0.97347	1	131
Corrected Ion SGA								
Ill-"454"-Ion SGA	281, 155	212	212	176, 769	0.99958	0.96562	4	74
Masurca(all)								
Ill-454-Ion Masurca	3, 398	7	5	1, 477	0.00835	0.99363	5	175366
Celera-CABOG(all)								
Ill-454-Ion Celera	575, 642	485	485	164, 621	0.93088	0.94664	39	12, 222

Name: the name of the data group that constitutes the assembly; # of Contigs: the number of contigs that belong to the resulting assembly; # of Mapped Contigs: the number of contigs that successfully mapped onto the reference sequence; # of Covered bases: the number of bases on the reference sequence that are covered by the assembly; Coverage: percentage of covered reference; Avg. identity: percentage of the correctly predicted reference bases; # of Gaps: the number of gaps that cannot be covered on the reference genome; Size of Gaps: total number of bases on the gaps.
* "2" represents the results of the second cycle of correction, "3" represents the third cycle.

We also used the hybrid assemblers Masurca and CABOG with default parameters with the combination of three data. Masurca resulted in very low coverage 0.8 % as it did before with the dual combinations. CABOG resulted in lower coverage and higher average identity compared to Ill-454 combination and higher in both compared to Ill-Ion Torrent combination. Hybrid assembler still did not result in as high coverage and average identity as obtained with the correction method.

We note that exploiting all the data gives us more accurate results especially when we are using a diverse data which has different strengths and weaknesses. However, one must be careful about the weaknesses and strengths of the data and where and in which order to use each of them.

4   Conclusion

In this paper, we presented a novel method to improve draft assemblies by correcting high contiguity assemblies using the relatively more fragmented contigs obtained using high quality short reads. Assembling short and long reads separately using both de Bruijn and OLC graph based assemblers according to data types and then using correction methods gives better results than using only hybrid assemblers. Using three data types together for correction or as the input of the hybrid assemblers rather than using only two of them gives more accurate results.

However, the need to develop new methods that exploit different data properties of different HTS technologies, such as short/long reads or high/low quality of reads, remains. In this manner, as future work, our correction algorithm can be improved by exploiting the paired end information of the short, high quality reads after the correction phase to close the gaps between corrected contigs.

Funding. The project was supported by the Republic of Turkey Ministry of Development Infrastructure Grant (no: 2011K120020), BİLGEM TÜBİTAK (The Scientific and Technological Research Council of Turkey) grant (no: T439000), and a TÜBİTAK grant to Can Alkan(112E135).

References

1. Warren, R.L., Sutton, G.G., Jones, S.J.M., Holt, R.A.: Assembling millions of short DNA sequences using SSAKE. Bioinformatics **23**(4), 500–501 (2007)
2. Dohm, J.C., Lottaz, C., Borodina, T., Himmelbauer, H.: SHARCGS, a fast and highly accurate short-read assembly algorithm for de novo genomic sequencing. Genome Res. **17**(11), 1697–1706 (2007)
3. Jeck, W.R., Reinhardt, J.A., Baltrus, D.A., Hickenbotham, M.T., Magrini, V., Mardis, E.R., Dangl, J.L., Jones, C.D.: Extending assembly of short DNA sequences to handle error. Bioinformatics **23**(21), 2942–2944 (2007)
4. Donmez, N., Brudno, M.: Hapsembler: an assembler for highly polymorphic genomes. In: Proceedings of the 15th Annual International Conference on Research in Computational Molecular Biology, pp. 38–52 (2008)

5. Myers, E.W., Sutton, G.G., Delcher, A.L., Dew, I.M., Fasulo, D.P., Flanigan, M.J., Kravitz, S.A., Mobarry, C.M., et al.: A whole-genome assembly of drosophila. Science **287**(5461), 2196–2204 (2000). doi:10.1126/science.287.5461.2196

6. Simpson, J., Durbin, R.: Efficient de novo assembly of large genomes using compressed data structures. Genome Res. **22**, 549–556 (2012). doi:10.1101/gr.126953. 111

7. Zerbino, D.R., Birney, E.: Velvet: algorithms for de novo short read assembly using de Bruijn graphs. Genome Res. **18**(5), 821–829 (2000). doi:10.1101/gr.074492.107

8. Bankevich, A., Nurk, S., Antipov, D., Gurevich, A.A., Dvorkin, M., Kulikov, A.S., Lesin, V.M., Nikolenko, S.I., et al.: SPAdes: a new genome assembly algorithm and its applications to single-cell sequencing. J. Comput. Biol. **19**(5), 455–477 (2012). doi:10.1089/cmb.2012.0021

9. Butler, J., MacCallum, I., Kleber, M., Shlyakhter, I.A., Belmonte, M.K., Lander, E.S., Nusbaum, C., Jaffe, D.B.: ALLPATHS: de novo assembly of whole-genome shotgun microreads. Genome Res. **18**(5), 810–820 (2008). doi:10.1101/gr.7337908

10. Simpson, J.T., Wong, K., Jackman, S.D., Schein, J.E., Jones, S.J.M., Birol, İ.: ABySS: a parallel assembler for short read sequence data. Genome Res. **19**(6), 1117–1123 (2009)

11. Chaisson, M.J., Brinza, D., Pevzner, P.A.: De novo fragment assembly with short mate-paired reads: does the read length matter? Genome Res. **19**(2), 336–346 (2008)

12. Miller, J.R., Delcher, A.L., Koren, S., Venter, E., Walenz, B.P., Brownley, A., Johnson, J., Li, K., Mobarry, C., Sutton, G.: Aggressive assembly of pyrosequencing reads with mates. Bioinformatics **24**(24), 2818–2824 (2008). doi:10.1093/bioinformatics/btn548

13. Zimin, A., Marçais, G., Puiu, D., Roberts, M., Salzberg, S.L., Yorke, J.A.: The MaSuRCA genome assembler. Bioinformatics **29**(21), 2669–2677 (2013). doi:10.1093/bioinformatics/btt476

14. Chevreux, B., Wetter, T., Suhai, S.: Genome sequence assembly using trace signals and additional sequence information. In: Computer Science and Biology: Proceedings of the German Conference on Bioinformatics (GCB), vol. 99, pp. 45–56 (1999)

15. Chevreux, B., Pfisterer, T., Drescher, B., Driesel, A.J., Müller, W.E., Wetter, T., Suhai, S.: Using the miraEST assembler for reliable and automated mRNA transcript assembly and SNP detection in sequenced ESTs. Genome Res. **14**(6), 1147–1159 (2004)

16. Deshpande, V., Fung, E.D., Pham, S., Bafna, V.: Cerulean: A hybrid assembly using high throughput short and long reads (2013). arXiv:1307.7933 [q-bio.QM]

17. Ergüner, B., Ustek, D., Sağroğlu, M.: Performance comparison of next generation sequencing platforms. In: Poster presented at: 37th International Conference of the IEEE Engineering in Medicine and Biology Society (2015)

18. Wang, Y., Yao, Y., Bohu, P., Pei, H., Yixue, L., Zhifeng, S., Xiaogang, X., Xuan, L.: Optimizing hybrid assembly of next-generation sequence data from enterococcus faecium: a microbe with highly divergent genome. BMC Syst. Biol. **6**(Suppl 3), S21 (2012). doi:10.1186/1752-0509-6-S3-S21

19. Altschul, S., Gish, W., Miller, W., Myers, E., Lipman, D.J.: Basic local alignment search tool. J. Mol. Biol. **215**(3), 403–410 (1990)

20. Zhang, Z., Schwartz, S., Wagner, L., Miller, W.: A greedy algorithm for aligning DNA sequences. J. Comput. Biol. **7**(12), 203–214 (2000)

Validation Pipeline for Computational Prediction of Genomics Annotations

Davide Chicco[1][✉] and Marco Masseroli[2]

[1] Princess Margaret Cancer Centre, University of Toronto, Ontario, Canada
davide.chicco@gmail.com
[2] Dipartimento di Elettronica Informazione e Bioingegneria,
Politecnico di Milano, Milan, Italy
masseroli@elet.polimi.it

Abstract. Controlled biomolecular annotations are key concepts in computational genomics and proteomics, since they can describe the functional features of genes and their products in both a simple and computational way. Despite the importance of these annotations, many of them are missing, and the available ones contain errors and inconsistencies; furthermore, the discovery and validation of new annotations are very time-consuming tasks. For these reasons, recently many computer scientists developed several machine-learning algorithms able to computationally predict new gene-function relationships. While several of these methods have been easily adapted from different domains to bioinformatics, their validation remains a challenging aspect of a computational pipeline. Here, we propose a validation procedure based upon three different sub-phases, which is able to assess the precision of any algorithm predictions with a reliable degree of accuracy. We show some validation results obtained for Gene Ontology annotations of Homo sapiens genes that demonstrate the effectiveness of our validation approach.

Keywords: Validation · Gene Ontology · Biomolecular annotations · Receiver Operating Characteristic · ROC curves · Genomic and Proteomic Data Warehouse (GPDW)

1 Introduction

In computational biology, a *controlled biomolecular annotation* is the association of a gene or gene product with a biological functional feature expressed through a controlled term, which can be part of a terminology or a controlled vocabulary structured within an ontology, such as the Gene Ontology (GO) [1]. Thus, the annotation states that the gene has the functional feature represented by the controlled term.

For instance, the pair *<SLC1A6, L-glutamate transmembrane transporter activity>* represents the annotation of the *SLC1A6* gene to the *L-glutamate transmembrane transporter activity* molecular function. Despite their biological importance, there are some issues with available annotations, such as the

© Springer International Publishing Switzerland 2016
C. Angelini et al. (Eds.): CIBB 2015, LNBI 9874, pp. 233–244, 2016.
DOI: 10.1007/978-3-319-44332-4_18

presence of erroneous or missing ones [2]. Thus, computational algorithms and software tools able to produce ranked lists of reliably predicted annotations are a very useful contribution [3].

In the past, we designed and developed several algorithms towards this goal. We started from a state-of-the-art algorithm based on truncated *Singular Value Decomposition* (tSVD) and developed some variants [4]. Then, in [5] we designed an algorithm to choose the best truncation level for the tSVD, while in [6,7] we developed weighted variants of the tSVD method. In [8,9] we designed and tested some *topic modeling* techniques, and in [10] we took advantage of a deep neural network approach. We compared many of these algorithms in [11], and also merge together different prediction techniques in [12].

Other scientists have dealt with this scientific task in the past. Khatri et al. used principal component analysis (PCA) through singular value decomposition (tSVD) as well in [13–15], while King et al. used decision trees and Bayesian networks in [16]. Tao and colleagues took advantage propounded to use a k-nearest neighbour (k-NN) classifier in [17]. Barutcuoglu and colleagues, used a support vector machine (SVM) algorithm in [18]. Many other algorithms were used in the past to predict GO annotations, we cannot report them all here for lack of paper space (a detailed and complete literature review is available in the Related Works section of [19]).

All these methods can be viewed as matrix-completion approaches, in which the method attempts to recover a matrix with some underlying structure from noisy observations. The input of these methods is $A = [a_{ij}]$, an $m \times n$ matrix, where each row corresponds to a gene and each column corresponds to a Gene Ontology feature term ($a_{ij} = 1$ if gene i is annotated to feature term j, $a_{ij} = 0$ otherwise). Moreover, let θ be a fixed threshold value. The prediction algorithm elaborates the matrix A to produce an output matrix $\tilde{A}$, with the same dimensions of A, where each likelihood value $\tilde{a}_{ij}$ is used to categorize an annotation: $\langle gene_i, feature_j, \tilde{a}_{ij} \rangle$. A high $\tilde{a}_{ij}$ value indicates that the probability for $gene_i$ to be associated with the feature $feature_j$ is high. Every approach described in the previously cited papers constructs the input matrix A, elaborates it with a machine-learning algorithm, and finally generates the output matrix $\tilde{A}$ containing the predicted functional annotations.

All these prediction pipelines, as well as of any other similar project, share a common final pivotal step: the validation of results. Since biomolecular annotations are always incomplete (because our knowledge of biology is incomplete), we do not have a *ground-truth* or *gold-standard* on which to rely; this makes us unable to take advantage of the usual computational methods widely used for validation in other applied machine-learning domains (such as computer vision or signal processing). To deal with this issue, we developed a method which assembles three different validation procedures that, together, lead to a reliable determination of the predicted annotation accuracy.

Here, we illustrate this method and the three techniques that it includes: the *analysis of the Receiver Operating Characteristic (ROC) curves*, the *comparison between available annotation versions*, and the *review of the scientific literature*.

To the best of our knowledge, no other complete paper has been published about the validation of predicted GO annotations in the past. Khatri and colleagues briefly mention in [13–15] their validation techniques, based upon the analysis of literature and on a search made on the updated GO datasets, without providing details about it. King et al. used a receiver operating characteristic (ROC) performance evaluation followed by an analysis on an updated organim database in [16]. Also Tao and colleagues used looked for the Homo sapiens GO annotations they predicted on a newer version of the Homo sapiens database, through a procedure they called *historical rollback validation* [17].

After this Introduction, Sect. 2 illustrates our method and the included validation procedures. Section 3 shows some example results of the proposed validation method and Sect. 4 concludes.

2   Methods

In this section we describe the validation procedures that we assembled and implemented to test the effectiveness of annotation prediction computational methods: (Sect. 2.1) ROC curve analysis, (Sect. 2.2) comparison between different versions of available annotations, and (Sect. 2.3) evaluation against the literature using available web tools.

2.1   Receiver Operating Characteristic (ROC) Curve Analysis

A ROC curve is a graphical plot which depicts the performance of a binary classifier system while its discrimination threshold τ varies [20]. Although usually in the biomolecular annotation prediction field a reference *gold standard* is not available, it can be used to compare output predicted annotations against input ones, instead of against the unavailable reference gold standard.

Our ROC curves depict the trade-off, for all possible values of τ, between the *TPrate* and the *FPrate*, where:

$$TPrate = \frac{TP}{TP+FN} \qquad FPrate = \frac{FP}{FP+TN} \qquad (1)$$

and *TP*, *FP*, *TN* and *FN* represent the number of true positive, false positive, true negative and false negative predictions. Notice that, in statistical terms, $TPrate = sensitivity$ and $FPrate = fallout = 1 - specificity$. Our ROC curves are built with the *TPrate* on the y axis and with the *FPrate* on the x axis.

Thus, this ROC curve analysis is an efficient tool to understand the similarity between the input and output annotations of an annotation prediction method. A ROC curve showing a high *area under the curve* (AUC) corresponds to having many TPs (annotations present in the input and confirmed as present in the output) and many TNs (annotations absent in the input and confirmed as absent in the output). This means that the input annotation matrix is very similar to the output annotation matrix, and the output annotation profiles strongly reflect the input ones. On the contrary, a low AUC means a lot of differences between the input and output annotations.

Given the comparison with the input annotations instead of with a gold standard, a good prediction should have a fairly high AUC. We consider a prediction insufficiently acceptable when its AUC is lower than $\omega = 2/3$. We chose this heuristic value to indicate that at least 66.67 % of the output annotation matrix should be equivalent to the input matrix, since usually most of available annotations are correct although some errors and several missing annotations generally exist.

Despite the effectiveness of this ROC AUC analysis, our two other validation methods (annotation version comparison and literature review) are more useful and efficient.

2.2   Annotation Version Comparison

When an updated version of the controlled annotations used as input to a prediction method is available, the tally of the new annotations predicted (FPs) that are found confirmed in the updated version of the analyzed annotations provides an important validation. Note however that it can give only a lower estimate of the predicted annotation accuracy, since correctly predicted annotations could not be present in the available updated annotations just because they have not been discovered yet, or simply because they have not yet been included in the available annotations.

We take advantage of the Genomic and Proteomic Data Warehouse (GPDW) [21–23], a knowledge base which integrates numerous, multi-organism, gene and protein controlled annotation data from many different sources, including the Entrez Gene and GO databases. Relevant features of the GPDW are its periodical updates of the contained data and the storage in the GPDW of their outdated versions [24]. We leverage them by retrieving from the GPDW different, time distant versions of the available gene GO annotations, and using them as analyzed annotations and updated annotations for validation comparison, respectively. Our validation procedure behaves as follows:

1. We save the FP annotation list into a table in the analyzed GPDW database version, with *geneOID*, *termOID* and the prediction likelihood *value* as fields.
2. Since the OID codes are unique IDs only within the specific GPDW version, we first enrich the FP list with additional fields to unequivocally identify annotation genes and terms also in the updated GPDW version. Towards this aim, we execute a SQL query on the analyzed GPDW version to add other fields, such as *gene_source_id*, *gene_source_name*, *term_source_id* and *term_source_name*, to the FP list table. The *gene_source_name* field includes the name of the source database of the gene (e.g. Entrez Gene), *gene_source_id* is the unique ID of the gene inside the source database, *term_source_name* is the name of the source database of the annotation term (e.g. Gene Ontology), and *term_source_id* is the unique ID of the term inside the source database. Thus, the *gene_source_name* and *gene_source_id* fields together unequivocally

identify a gene; and the *term_source_name* and *term_source_id* fields together unequivocally identify a feature term. We create a new FP_list_enriched table with the FP annotations and their values for these additional fields.

3. We read this FP_list_enriched table from the analyzed GPDW version and copy it into the updated GPDW version considered.

4. We execute a SQL query that retrieves all indirect, less specific, ontological annotations contained in the updated GPDW version and store them into a new UnfoldedAnnotations table. In the GPDW, the direct, more specific, annotations are in the *gene2biological_function_feature* table, while their indirect ones can be found with a JOIN operation between this table and the *gene2biological_function_feature_unfolded* table.

5. We execute a SQL query that counts how many direct annotations in the FP_list_enriched table are found in the *gene2biological_function_feature* table of the updated GPDW version, by joining the unique fields *gene_source_id*, *gene_source_name*, *term_source_id* and *term_source_name* of the two tables.

6. We execute a SQL query that counts how many parental annotations in the FP_list_enriched table are found in the UnfoldedAnnotations table, by joining the unique fields *gene_source_id*, *gene_source_name*, *term_source_id* and *term_source_name* of these two tables.

7. Finally, we also count how many of the predicted (FP) gene GO annotations that are found confirmed in the updated version of the GPDW have evidence *IEA* or *ND*.

We report a flow chart of this annotation prediction validation procedure in Fig. 1.

2.3 Literature Evaluation Through Web Tools

The third and last step of our annotation prediction validation procedure is based upon searching literature resources for information supporting the predicted annotations. It is the only step not fully automated in our pipeline.

The sources integrated in the GPDW mainly contain data from validated experiments, whose results are published in the literature. Yet, given the numerous research groups working independently all over the world and the many different journals in which results are published, some validated annotations published in the literature may have not yet been included in annotation databases. Thus, a literature review to search for confirmation of the annotations predicted by a computational method can provide effective additional validation results. For this last step of our validation procedure, we leverage the main online biomedical literature repository, PubMed [25], and the AmiGO [26] and GeneCards [27] web tools.

2.4 Evaluation

We applied all the described validation techniques to the gene GO annotations that we predicted with the methods described in [4]. Such methods are all based

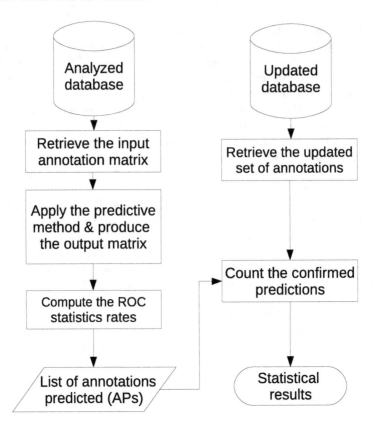

Fig. 1. Flow chart of the annotation prediction validation procedure based on the comparison of annotation versions.

on the popular tSVD, also known as *principal component analysis*. We re-use the tests made by Khatri and colleagues [13–15], based on the tSVD with a heuristic fixed truncation level (SVD-Khatri), and compared their results to those obtained with a tSVD variant that we developed (SVD-us), where the best truncation level is chosen through a ROC optimization algorithm [5]. We also compared two other variants of the tSVD, named SIM1 and SIM2, both described in [4].

For the tests, we used as input the GO annotations of *Homo sapiens* genes available in the July 2009 version of the GPDW [23] (i.e. 14,341 annotations of 7,868 genes and 684 GO Cellular Components (CC), 15,467 annotations of 8,590 genes and 2,057 GO Molecular Functions (MF), and 21,048 annotations of 7,902 genes and 2,528 GO Biological Processes (BP)).

For the result validation with the annotation version comparison techniques we used the corresponding gene GO annotations available in the March 2013 version of the GPDW (i.e. 31,135 annotations of 12,033 genes and 1,021 GO Cellular Components, 25,396 annotations of 10,460 genes and 2,603 GO

Molecular Functions, and 64,212 annotations of 11,681 genes and 7,295 GO Biological Processes).

3   Results

Using the three validation procedures defined, we compared the results and evaluated the performance of four different annotation prediction methods:

- the tSVD as used by Khatri et al. [13] (with fixed truncation level $k = 500$ for all datasets),
- the tSVD with truncation level chosen by our automatic algorithm [5],
- the SIM1 with truncation level and cluster number chosen by our automatic algorithms [5], and
- the SIM2 with truncation level and cluster number chosen by our automatic algorithms [5] and using the Resnik similarity measure [4].

We describe the evaluation results in the next sections.

3.1   ROC Curve Analysis

We generated the ROC curves for the considered prediction methods and input datasets, and report their AUCs in Table 1. Almost all ROC AUCs are greater than $\omega = 66.67\%$, which is the minimum "reliability" threshold that we consider for the predictions. Only the ROC AUC generated by the SVD-Khatri method for the GO CC gene annotations did not reach that threshold; thus, we do not explore those predicted annotations further.

Table 1. ROC AUCs for the three *Homo sapiens* gene GO annotation datasets and four prediction methods considered. The AUC percentage is always greater than the minimum reliability threshold ω, which we heuristically set at 66.67%, except for the SVD-Khatri method applied to the GO CC dataset.

Method	Cellular component	Molecular function	Biological process
SVD-Khatri	58.98 %	90.06 %	77.24 %
SVD-us	83.44 %	85.40 %	75.99 %
SIM1	80.94 %	83.58 %	70.20 %
SIM2	81.66 %	83.32 %	68.65 %

3.2   Annotation Version Comparison

In the first three cases in Table 2 we report the results obtained with a single GO sub-ontology dataset as input and output, while the results obtained with the complete whole GO dataset (CC $\cup$ MF $\cup$ BP) are in the last case in Table 2. Based upon this Table content, we can observe what follows:

Table 2. Results comparison of the methods tSVD with truncation level k fixed to 500 as in Khatri et al. [13] (*SVD-Khatri*), tSVD with our automatically determined optimal truncation level k (*SVD-us*), SIM1 and SIM2. The τ threshold minimizes the sum $FPs + FNs$. C: number of clusters for SIM1 and SIM2. SIM2 uses Resnik's similarity. *APs*: number of annotations predicted; *anDB*: number of predicted annotations found in the July 2009 GPDW version; *upDB* (*upDB%*): number (percentage) of predicted annotations found in the March 2013 updated GPDW version (percentage over the predicted annotations). The most important values are **bolded**: the percentages of APs found on the updated GPDW version. The values of the ROC AUC of these tests are in Table 1; the SVD-Khatri, SVD-us, SIM1 and SIM2 methods are described in [4].

Method	k	τ	C	APs	anDB	upDB	upDB%
Homo sapiens, GO Cellular Component - CC							
SVD-Khatri	500	0.45		0	0	0	**0.00**
SVD-us	378	0.49		8	0	4	**50.00**
SIM1	378	0.49	2	8	0	4	**50.00**
SIM2	378	0.49	2	8	0	4	**50.00**
Homo sapiens, GO Molecular Function - MF							
SVD-Khatri	500	0.48		108	0	4	**5.56**
SVD-us	607	0.48		81	2	5	**6.17**
SIM1	607	0.48	5	13	0	1	**7.69**
SIM2	607	0.48	5	30	0	3	**10.00**
Homo sapiens, GO Biological Process - BP							
SVD-Khatri	500	0.48		358	1	48	**13.51**
SVD-us	1,413	0.45		64	2	12	**18.75**
SIM1	1,413	0.45	2	35	1	10	**28.57**
SIM2	1,413	0.45	5	14	0	8	**57.14**
Homo sapiens, whole GO (CC ∪ MF ∪ BP)							
SVD-Khatri	500	0.45		794	196	234	**29.47**
SVD-us	1,905	0.43		112	3	51	**45.54**
SIM1	1,905	0.43	2	116	3	45	**38.79**
SIM2	1,905	0.43	2	111	3	49	**44.14**

(a) Our tSVD method always outperforms the Khatri tSVD method with fixed truncation: the percentage of annotations predicted (AP) found confirmed in the updated GPDW version (table last column) is greater for all datasets.

(b) Our SIM1 method always outperforms the tSVD methods (greater percentage of annotations found confirmed in the updated GPDW version), except for the CC dataset, where it has the same performance as our tSVD method, and for the (whole CC ∪ MF ∪ BP) dataset, where the SVD-us outperforms all the other methods.

(c) Our SIM2 method always outperforms the SIM1 and tSVD methods, except for the CC dataset, where they all have the same results. The complete whole GO dataset (CC ∪ MF ∪ BP) shows an increased number of validated predicted annotations, which are much more than the ones predicted in the single GO sub-ontology tests. In addition, in this complete dataset the SVD-us method outperforms all the other methods.

3.3   Literature Evaluation

Once we had the lists of the annotations predicted by our methods, we looked for confirmation of their existence in the literature, as described previously. In the scientific literature, GeneCards or AmiGO resources, out of the total 153 annotations (CC: 8, MF: 81, BP: 64) predicted with the tSVD method with our best truncation level for each single GO sub-ontology, we found the 8 (5.30 %) annotations (MF: 4, BP: 4) reported in Table 3; only one of them, i.e. <*ITGA6, Cell-matrix adhesion*> (in **bold** in Table 3) was not in the updated GPDW version. Out of the total 56 annotations predicted through the SIM1 method (CC: 8, MF: 13, BP: 35), 2 (3.57 %) annotations (1 MF and 1 BP) were found. Out of the total 52 annotations predicted through the SIM2 method (CC: 8, MF: 30, BP: 14), 4 (7.69 %) annotations were found (3 MF and 1 BP).

Through the literature analysis, out of the total 153 annotations predicted by our tSVD method, 21 (13.73 %) were validated in the updated GPDW version, and we found only 1 additional annotation in the literature. Given the time required to perform the literature evaluation, this result may seem very limited; this is why we consider the first two validation procedures (*ROC curve analysis* and *annotation version comparison*) to be more useful and reliable, particularly the latter one.

Table 3. *Homo sapiens* gene GO annotations predicted by our tSVD method (SVD-us, Table 2) and confirmed in the literature evaluation. If an annotation was available in the latest Gene Ontology annotation version, its evidence is reported (IEA: Inferred from Electronic Annotation, EXP: Inferred from Experiment, TAS: Traceable Author Statement). The single annotation not found in the annotation version comparison analysis is in **bold**.

tSVD with best truncation level chosen by our automatic algorithm				
Sub-ontology	Gene symbol	GO term ID	GO term name	Evidence
MF	SLC1A6	GO:0005313	L-glutamate transmembrane transporter activity	IEA
MF	HDAC6	GO:0004407	Histone deacetylase activity	IEA
MF	POR	GO:0004128	Cytochrome-b5 reductase activity	IEA
MF	NT5M	GO:0008253	5'-Nucleotidase activity	EXP
BP	ITGA6	GO:0007155	Cell adhesion	IEA
BP	**ITGA6**	**GO:0007160**	**Cell-matrix adhesion**	**IEA**
BP	CPA2	GO:0006508	Proteolysis	IEA
BP	AHR	GO:0006805	Xenobiotic metabolic process	TAS

4   Conclusions

Validation of functional annotation predictions in biology is always a difficult task. Available annotations continuously increase while scientists discover new aspects of biology; furthermore, some of the available annotations may contain errors, which could be corrected in their subsequent versions. A *gold-standard* to use in the validation is not available, and creating a reasonably ample and unbiased one is a daunting task. So, stating if a machine learning prediction algorithm is performing well is quite difficult.

In the past, we designed and implemented several techniques for prediction of Gene Ontology annotations, with algorithms from in linear algebra, clustering, weighting schemes, topic modeling and deep learning [4–11]. In this paper, we illustrated here three validation procedures that we assembled and used to validate the GO annotations of *Homo sapiens* genes predicted through some computational learning methods. These three techniques mutually compensate for each others' strengths and weaknesses; although they are not fully innovative, all together represent an useful tool to state the quality of biomolecular annotations predicted through any computational algorithm.

Despite our evaluation of the presented validation procedures considers only GO annotations, such procedures are not bound to the Gene Ontology, or even to the biological domain, but can be used in any scientific validation in which a full *gold-standard* does not exist, or is always changing. In the future, we plan to improve the use of our overall validation method by additionally automating the literature evaluation step, through the use of text mining techniques, as well as to integrate it in our software suite for biomolecular annotation prediction [28] and in the Bio-SeCo (Search Computing) platform [29, 30].

Acknowledgments. This work was partially supported by the "Data–Driven Genomic Computing (GenData 2020)" PRIN project (2013–2015), funded by Italy's Ministry of Education, Universities and Research (MIUR). Authors thank Coby Viner (University of Toronto) for his help in the English proof-reading of this article.

References

1. The Gene Ontology Consortium, Creating the Gene Ontology resource: Designand implementation. Genome Res. **11**(8), 1425–1433 (2001)
2. Karp, P.D.: What we do not know about sequence analysis and sequence databases. Bioinformatics **14**(9), 753–754 (1998)
3. Pandey, G., Kumar, V., Steinbach, M.: Computational Approaches for Protein Function Prediction: A Survey. Department of Computer Science and Engineering, University of Minnesota, Twin Cities (2006)
4. Chicco, D., Tagliasacchi, M., Masseroli, M.: Biomolecular annotation prediction through information integration. In: Proceedings of CIBB 2011 - 8th International Meeting on Computational Intelligence Methods for Bioinformatics and Biostatistics, Gargnagno sul Garda, Italy, pp. 1–9 (2011)

5. Chicco, D., Masseroli, M.: A discrete optimization approach for SVD best truncation choice based on ROC curves. In: Proceedings of IEEE BIBE - the 13th IEEE International Conference on Bioinformatics and Bioengineering, pp. 1–8. IEEE, Chania (2013)
6. Pinoli, P., Chicco, D., Masseroli, M.: Improved biomolecular annotation prediction through weighting scheme methods. In: Proceedings of CIBB - 10th International Meeting on Computational Intelligence Methods for Bioinformatics and Biostatistics, Nice, France, pp. 1–9 (2013)
7. Pinoli, P., Chicco, D., Masseroli, M.: Weighting scheme methods for enhanced genomic annotation prediction. In: Formenti, E., Tagliaferri, R., Wit, E. (eds.) CIBB 2013. LNCS, vol. 8452, pp. 76–89. Springer, Heidelberg (2014)
8. Pinoli, P., Chicco, D., Masseroli, M.: Enhanced probabilistic latent semantic analysis with weighting schemes to predict genomic annotations. In: Proceedings of IEEE BIBE - the 13th IEEE International Conference on Bioinformatics and Bioengineering, pp. 1–8. IEEE, Chania (2013)
9. Pinoli, P., Chicco, D., Masseroli, M.: Latent Dirichlet allocation based on Gibbs sampling for gene function prediction. In: Proceedings of CIBCB - the IEEE Conference on Computational Intelligence in Bioinformatics and Computational Biology, pp. 1–8. IEEE (2014)
10. Chicco, D., Sadowski, P., Baldi, P.: Deep autoencoder neural networks for Gene Ontology annotation predictions. In: Proceedings of ACM BCB, pp. 533–540. ACM (2014)
11. Pinoli, P., Chicco, D., Masseroli, M.: Computational algorithms to predict Gene Ontology annotations. BMC Bioinformatics 16(Suppl. 6), S4, 1–15 (2015)
12. Chicco, D., Masseroli, M.: Ontology-based prediction and prioritization of gene function annotations. IEEE/ACM Trans. Comput. Biol. Bioinform. 13(2), 248–260 (2016). IEEE
13. Khatri, P., Done, B., Rao, A., Done, A., Draghici, S.: A semantic analysis of the annotations of the human genome. Bioinformatics 21(16), 3416–3421 (2005)
14. Done, B., Khatri, P., Done, A., Draghici, S.: Semantic analysis of genome annotations using weighting schemes. In: Proceedings of CIBCB - the IEEE Symposium on Computational Intelligence in Bioinformatics and Computational Biology, pp. 212–218. IET, Honolulu (2007)
15. Done, B., Khatri, P., Done, A., Draghici, S.: Predicting novel human Gene Ontology annotations using semantic analysis. IEEE/ACM Trans. Comput. Biol. Bioinform. (TCBB) 7(1), 91–99 (2010)
16. King, O.D., Foulger, R.E., Dwight, S.S., White, J.V., Roth, F.P.: Predicting gene function from patterns of annotation. Genome Res. 13(5), 896–904 (2003)
17. Tao, Y., Sam, L., Li, J., Friedman, C., Lussier, Y.A.: Information theory applied to the sparse Gene Ontology annotation network to predict novel gene function. Bioinformatics 23(13), 529–538 (2007)
18. Barutcuoglu, Z., Schapire, R.E., Troyanskaya, O.G.: Hierarchical multi-label prediction of gene function. Bioinformatics 22(7), 830–836 (2006)
19. Chicco, D.: Computational Prediction of Gene Functions through Machine Learning methods and Multiple Validation Procedures, Doctoral Thesis, Politecnico di Milano (2014)
20. Fawcett, T.: ROC graphs: notes and practical considerations for researchers. ReCALL 31(HPL–2003–4), 1–38 (2004)

21. Canakoglu, A., Ghisalberti, G., Masseroli, M.: Integration of biomolecular interaction data in a genomic and proteomic data warehouse to support biomedical knowledge discovery. In: Biganzoli, E., Vellido, A., Ambrogi, F., Tagliaferri, R. (eds.) CIBB 2011. LNCS, vol. 7548, pp. 112–126. Springer, Heidelberg (2012)

22. Masseroli, M., Canakoglu, A., Ceri, S.: Integration and querying of genomic and proteomic semantic annotations for biomedical knowledge extraction. IEEE/ACM Trans. Comput. Biol. Bioinform. 13(2), 209–219 (2016). IEEE

23. Canakoglu, A., Ceri, S., Masseroli, M.: Biomolecular annotation integration and querying to help unveiling new biomedical knowledge. In: Ortuño, F., Rojas, I. (eds.) IWBBIO 2016. LNCS, vol. 9656, pp. 802–813. Springer, Heidelberg (2016)

24. Genomic and Proteomic Knowledge Base (GPKB). http://www.bioinformatics. deib.polimi.it/GPKB/

25. NCBI PubMed. http://www.ncbi.nlm.nih.gov/pubmed/

26. Carbon, S., Ireland, A., Mungall, C.J., Shu, S., Marshall, B., Lewis, S.: AmiGO: online access to ontology and annotation data. Bioinformatics 25(2), 288–289 (2009)

27. Rebhan, M., Chalifa-Caspi, V., Prilusky, J., Lancet, D.: GeneCards: a novel functional genomics compendium with automated data mining and query reformulation support. Bioinformatics 14(88), 656–664 (1998)

28. Chicco, D., Masseroli, M.: Software suite for gene and protein annotation prediction and similarity search. IEEE/ACM Trans. Comput. Biol. Bioinform. (TCBB) 12(4), 837–843 (2015)

29. Chicco, D.: Integration of bioinformatics web services through the Search Computing technology. Technical Report, TR 2012/02, Dipartimento di Elettronica e Informazione, Politecnico di Milano, Milan, Italy

30. Masseroli, M., Picozzi, M., Ghisalberti, G., Ceri, S.: Explorative search of distributed bio-data to answer complex biomedical questions. BMC Bioinformatics 15(Suppl. 1), S3, 1–14 (2014)

Advantages and Limits in the Adoption of Reproducible Research and R-Tools for the Analysis of Omic Data

Francesco Russo[1], Dario Righelli[1,2(✉)], and Claudia Angelini[1]

[1] Istituto per le Applicazioni del Calcolo "Mauro Picone",
Consiglio Nazionale delle Ricerche, Napoli, Italy
`d.righelli@na.iac.cnr.it`
[2] Dipartimento di Scienze Aziendali - Management and Innovation Systems
(DISA-MIS), Universitá di Salerno, Fisciano, Italy

Abstract. Reproducible (computational) Research is crucial to produce transparent and high quality scientific papers. First, we illustrate the benefits that scientific community can receive from the adoption of Reproducible Research standards in the analysis of high-throughput omic data. Then, we describe several tools useful to researchers to increase the reproducibility of their works. Moreover, we face the advantages and limits of reproducible research and how they could be addressed and solved. Overall, this paper should be considered as a proof of concept on how and what characteristic - in our opinion - should be considered to conduct a study in the spirit of Reproducible Research. Therefore, the scope of this paper is two-fold. The first goal consists in presenting and discussing some easy-to-use instruments for data analysts to promote reproducible research in their analyses. The second aim is to encourage developers to incorporate automatic reproducibility features in their tools.

Keywords: Reproducible research · Big-data · R

1 Introduction

In recent years, "irreproducibility" is becoming a crucial and widespread problem, especially in Medical and Life Sciences where most of the efforts are dedicated to the so-called *personalized medicine*. In this field, based on the analysis of omic data and the results of bio-medical studies, new drugs are synthesized/tested and new treatments are proposed. The hazard, in this case, consists in the concrete possibility to give a drug or a treatment to human patients without the achievement of the expected results due to potential weaknesses in the data analysis on which the therapy was based. As an example of irreproducible study, we mention the so-called *Duke Saga* [1,2]. Unfortunately, there have been several cases - like that one - in literature. More and more papers [3–10] are reporting the impossibility to reproduce

F. Russo—Now at Istitute of Protein and Biochemistry - CNR, Napoli, Italy.
F. Russo and D. Righelli—Equally contributing authors.

C. Angelini et al. (Eds.): CIBB 2015, LNBI 9874, pp. 245–258, 2016.
DOI: 10.1007/978-3-319-44332-4_19

results presented in several works in involving the analysis of omic data. Therefore, the attention towards reproducible findings and the need of transparency is increasing [11]. In fact, several researchers [12,13] are showing how irreproducible works often hamper the possibility to find possible errors and misconduct that might be hidden into scientific papers. Moreover, not only might irreproducible research create large increases of costs and delays, but it might also be extremely dangerous especially when clinical trials on human beings are based on irreproducible (and therefore possibly incorrect) results. Unfortunately, we must admit that the Life Science research community is still far from the achievement of what it might be considered as a minimum standard for reproducibility. In fact, in [22] the reproducibility levels of 441 biomedical journal papers published in the last fifteen years have been quantitatively evaluated. The results of this interesting study established that just one paper - among these ones - described a full protocol and none of them provided all the starting data publicly available. Moreover, only four works have been found suitable for a full replication. The impact of irreproducibility can result in an unsatisfactory level of scientific research quality, in an improper knowledge transfer and ethics problems related to misconduct in the presentation of desired results.

For all those reasons, Reproducible Research (RR) is becoming an indispensable feature to publish better and more reliable scientific research. In the last few years, a debate has been increasing about the concept of reproducibility [14–19]. The analysis of massive datasets of high-throughput omic data, such as Next Generation Sequencing (NGS), is increasing in terms of complexity, that involve preprocessing steps, statistical methods and data interpretations. Therefore, it become very difficult keeping track of all steps afront resulting in a lack of reproducibility. Consequently, it has been suggesting the use of tools in a way that can enhance transparency of the executions involved into the analyses presented in papers. Researchers are invited to make their raw data and the used software/pipelines available specifying all the versions and parameters used. Moreover, in describing a work, all instruments used and how they have been connected in a pipeline should be always available. However, as showed in [22] this is still rarely done.

In this work, we discuss several solutions to increase reliability, verifiability, quality, transparency and transfer knowledge of scientific findings. Therefore, the scope of this paper is two-fold. The first goal consists in presenting and discussing some easy-to-use instruments for data analysts to promote reproducible research in the analysis of omic data. The second aim is addressed to developers that decide to incorporate reproducibility features in their tools.

The rest of the paper is organized in the following way. In Sect. 2, we illustrate an overview on Reproducible (computational) Research and describe the advantages that researchers can benefit from the adoption of satisfactory reproducibility standards. In Sect. 3, we explore some R-tools useful to make reproducible a work or an analysis. In Sect. 4, we briefly present some repositories useful for data and code sharing. In Sect. 5, we illustrate tools outside R. In Sect. 6, we describe which are the limits of Reproducible Research. In Sect. 7, we draw out conclusions.

2   Overview

Reproducible (computational) Research is a concept born in the early nineties [25] obtaining a great impact within the statistical community thanks to the work of Donoho et al. [27]. The main idea was to give to readers the possibility to reproduce the entire analysis published in a paper (see Fig. 1 for a schematic representation). To this purpose, they released the *Wavelab Matlab package*, providing not only the code, but also the data and scripts used to generate analyses and simulations. The package was adopted to publish several papers, to teach case studies as well as to promote the RR spirit.

Since then the RR has been evolved in several ways with the help of modern technologies like the Hypertext Markup Language (HTML), the Portable Document Format (PDF), Python, R, etc., by preserving the same philosophy of Literate Statistical Programming (LP) [26].

The LP promote the incorporation of natural language sentences along with the computational language in order to enhance comprehensibility and to make the code human readable. The lines of code are supplemented with explanatory sentences that guide both readers and analysts to understand and verify the strategy adopted to solve a specific problem. The code is divided in small pieces called *code chunks*, preceded by a summary of the idea underlying the imple-

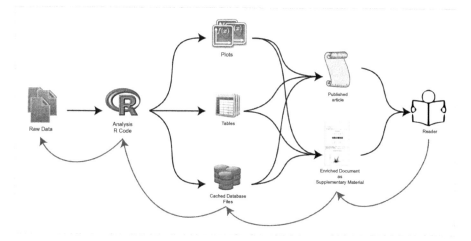

Fig. 1. This figure explains the basic idea of Reproducible Research. The raw data needs to be analysed by using several tools written in some programming language (R code in this example, but any language can be used) and by producing plots, tables and caching data. Usually, this is a very complex task involving several intermediate steps. All results (hopefully) converge in a publication (along with supplementary materials). In particular, both papers and supplementary materials can be enriched by documents that keep track of all code, data and results produced during the analysis. The final reader - thanks to this "enriched document" - can back-trace and reproduce the entire analysis made or just part of it. Figure is inspired from [28].

mentation. The summary is more than simple lines of comments and give a more complete view of the solutions used. In this manner, third-party users are able to better understand the meaning and the aim of a code chunk.

After a period of general disinterest about the RR, in the last years this concept has been recovered by the statistical community headed by Peng and Stodden [24]. Community that has produced several papers and tools, in order to establish guidelines to create a standard and providing the instruments to facilitate the adoption of RR.

Their contribution has led to an evolution of RR, expanding the initial idea of LP, highlighting that to obtain the reproducibility of a work it is necessary to release not only the source code and a description summary of a developed tool, but also the specifications about the middle steps of the analysis made with that specific tool. Hence, during a complex analysis, it is necessary to understand which tools have been used and also how the parameters has been set and which middle results they have produced.

A good example of what the RR community has produced is the *Bioconductor* project, in R language, which has reproducibility as its first goal [14]. The project aims to aggregate as much as possible packages useful to analyse high-throughput omic data, offering to the final user high quality standard software. In fact, in a complete reproducible spirit, each package released under the Bioconductor project has a *vignette*; an excellent example of RR, constituted by a mix of code and natural language describing an example of use of the package, helping the user to reproduce that specific analysis and also to adapt the code for its own analysis.

2.1   Advantages of Reproducible Research

Researchers which adopt the RR spirit in their work can benefit of many advantages [21]. In this section, we just underline three of them, such as: transparency, verification, knowledge transfer.

Transparency [22] can be easily reached - for instance - by the automatic generation of a report file that keeps track of all action performed, control version of used tools and details regarding the set of parameters chosen during the execution of a complex high-throughput omic data analysis. Even though a fully reproducibility of a research does not guarantee the correctness of its findings, a substantial improvement of its results - thanks to reproducibility - can be achieved via the possibility to check the appropriateness of statistical methodologies adopted and tools used. Presently, the peer review system cannot guarantee to find out all possible mistakes hidden in a work. Usually, most of the subtle issues and errors pass unnoticed during the revision process [13].

Therefore, transparency can assure a more technical and theoretical verification of correct uses of both instruments and methods from the scientific community. Verification helps a reader to better understand and to have a deeper insight in all the details of sets of procedures carried out during the pre-analysis and analysis process. Therefore, a reader has the possibility to learn and to re-execute the entire study described in a paper. In order to asses and verify the results of scientific publications, it is fundamental to inspect and reproduce the entire analysis carried out by authors.

Knowledge transfer is an essential reproducibility characteristic. To achieve a good level, researchers have to correctly report all steps performed by allowing other researchers to start from their data and following all the steps described making them able to obtain the same results presented in a paper. Hence, readers can have deep insight into an analysis described in a paper, to verify the authenticity of results (eventually to find out bugs and mistakes), to improve the published analysis, or reuse the pipeline in a similar content. Overall, it contributes to publish high quality scientific works and to help researchers to improve their skills and comprehension.

3   R-Tools for Reproducible Research

The R community has always pursued the RR philosophy giving instruments to incorporate RR features inside R in a simple way. A good starting point offering main guidelines to attend during software development and data analysis, is represented by [23,24], illustrating fundamental aspects on how to implement RR features in R.

Fortunately, many different tools (Refer to CRAN specific task view to have an overview https://cran.r-project.org/web/views/ReproducibleResearch.html) have been built to help developers to incorporate these characteristics inside their software for different programming languages. The *Bioconductor* project contains several packages useful for integrating RR main features, allowing both non expert and expert R user to implement RR in data analysis and software development, automating the production of enriched documents.

Moreover, a useful reference, as tool repository, is represented by the *rOpenSci* https://ropensci.org/ project, having as main aim the development and collection of R packages in order to facilitate the data and code sharing in a complete RR spirit.

3.1   R-Markdown

Since the beginning the R developers had produced tools, like *sweave*, to create enriched documents, like the R *vignette*, an example document, associated to a package, within a reproducible analysis to illustrate how to use the package.

In recent years, the *sweave* package has been mostly replaced by the *knitr* [30] package before and by *rmarkdown* package after, facing the implementation limits of previous packages and simplifying the construction of enriched documents. Thanks to these packages it is possible to create scripts written in R-Markdown language. These file types are a mix of sentences in natural language as explanations, figures, tables and *code chunks*, complete and independent code units, which can be run independently in an R console.

Using *knitr*, it is possible to compile the R-Markdown script in order to produce documents in different common format, such as: HTML, PDF and Word.

The compiled documents can be used to produce manuals, reports, tutorials and also data analysis summaries.

Anyone which is familiar with R can easily create an R-Markdown file. During the compilation all the R code is re-executed, in order to test all the steps included in the file. One of the most common suggestion, to keep in mind, is to clear the R workspace before the compilation. The empty working environment will ensure the reproducibility of the script, highlighting any error or missing variable in the code.

There are two major ways to use the *knitr* package, the standard one is by including the *knitr* package and to compile the R-Markdown script with the *knit2html* function. In such a way the package compiles the script and generates the corresponding HTML file. An easier way to create an R-Markdown script is by using the R-Studio, an R development environment tool. As described in Fig. 2, by choosing in the *new file* button the "*R Markdown*", an example script is created. It is possible to compile the R-Markdown script simply pressing the *Knitr* button. It is also possible to choose the output file format by the down arrow near to the "*knitr*" button.

The R-Markdown script can be used also to perform an entire analysis, because of its native inclusion of R code mixed to natural language. While the natural language is free to be written everywhere in the document without the need to be enclosed between special characters, the R code needs to be enclosed in special apostrophe like that:

```
```{r}

## R code here

```
```

Therefore, not only does the final document provide an open source code, but also all those lines of code, that have been actually executed during the analysis process, are clearly reported in a self-contained way, the code chunks. A reader just needs to install the needed packages and by copying & pasting the code chunks of interest can reproduce the same analysis carried out by the analyst. Such enriched document can be considered as a full detailed log file, written in human readable and friendly format usable as a supplementary material, containing executable code along with all initializations and printed results (plots, tables, arrays etc.). Moreover, each code chunk inside the file can be run independently in an R console to obtain the results shown in that report.

3.2   Caching

Caching is a mechanism to store and retrieve data. This feature can be useful in the analysis of high-throughput omic data because it helps to speed-up time consuming code re-execution, like the reports generation.

Integrating caching inside software is convenient in order to store the input and output of an entire analysis along with the parameters used. In this way,

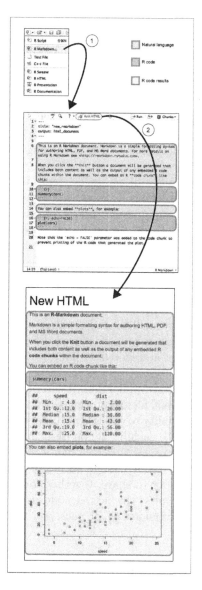

Fig. 2. This figure explains how to generate an R-Markdown script with R-Studio. By choosing the *new* button, it is possible to select the R Markdown voice. The software automatically generates an R-Markdown example script. Finally, with *Knit HTML* button the script is compiled and the HTML document is generated. The natural language comments have been highlighted in orange and the code chunks in blue. The output of the executed code chunks has been highlighted in green. (Color figure online)

caching not only does speed-up time consuming code, but it permits also to share cached data through the Internet. Caching prevents re-computation of time consuming lines of code by saving intermediate results into several objects invoked when the same data are necessary for computations.

Caching is more complex than LP, it needs advanced programming skills to be implemented in software or analysis scripts. The basic idea when using caching is to think to store data on disk when not needed, in order to free the memory, and to retrieve the specific variables when necessary.

In RR spirit, caching has to make available all the intermediate results that can be checked separately and can be used as starting points for different analyses. As a consequence, the implementation of caching allows the user to run in a faster way different types of analyses on the same stored dataset, allowing to easily modify an analysis while still preserving reproducibility.

By combining RR and caching features, a developer can achieve a better management of the entire data analysis and an automatic way of keeping track of the computational protocol used for analysing a specific dataset.

Thanks to the effort of several statisticians and developers there are many packages available in R which can be used for caching, like cacher [29], rctrack [33], filehash [34] and others [35] supporting and encouraging the caching use in reproducible research.

The common concept underlying these packages is to have an object (*saver*) devoted to the data storage on disk and another object (*loader*) devoted to retrieve the re-stored data when needed. Stored data needs a key identifier, in order to be uniquely retrieved and stored without overriding, when not expressly desired.

Moreover, depending on the application, it can be necessary to split data in order to store them in different files or storing them all together in one big file. Best packages offer the possibility to choose the preferred way of implementation. We suggest to modularize the caching file on disk, in order to facilitate the file sharing.

Anyway, caching, alone, is not enough to ensure the reproducibility of an analysis. In fact, it is necessary to share, together with the cached objects containing the intermediate results, the actual code used to generate them, the starting raw data and the versions of all packages and tools used during the analysis.

3.3   Incorporation of Reproducible Research Features Inside GUIs that Manage Big-Data

Graphical user interfaces (GUIs) are interactive tools, easy to use and very helpful for those users which do not have specific computational skill. However, the incorporation of the RR features inside GUIs is usually extremely difficult. Fortunately, there are different strategies to achieve this goal. We suggest the following one. During the usage of a particular GUI, the system writes in background all the executed lines of code in an automatically saved R-Markdown file. The file can be compiled, re-executed, and possibly showed to the user in the form of an HTML report whenever he wants.

Thus, a developer can help a user to conduct a study in a reproducible way even though the user has no knowledge of programming languages. For this scope, a developer can build a GUI that works at two levels. One is the user level, providing access to all high level functionalities. The second level operates in background and it is executed automatically by the GUI to generate and store cached object and to print a report file that contains all the code lines executed when the user performed an action. In this way, a researcher can perform an analysis in a reproducible way and benefit of a report file that describes the code lines involved in his study.

As a particular working example of a software which automatically produces fully reproducible analysis, we refer to RNASeqGUI [42,43]. This R package is a completely open source graphical user interface implemented in R. It allows to identify differentially expressed genes from RNA-Seq experiments and to support the interpretation of the results, through the pathway and gene ontology analysis. It includes several well known RNA-Seq tools, available as command line in Bioconductor.

This software, thanks to R-Markdown language, works at two levels as we have just described. Therefore, it is capable to automatically generate a dynamic report describing all the analysis carried out on a given project in a fully detailed way. The report includes all R code chunks used during the analysis, the figures and the summary of the results. These code chunks can be executed and their results are updated automatically whether some changes occur. It keeps track of all versions of the R packages used (session info), all steps, input/output parameters, file names, etc. Moreover, the report can be exported as HTML file.

4   Repositories for Public Access to Data and Code

In order to be able to reproduce an analysis, a third-party user needs to have access to code and raw data used for the analysis. To cope this need in recent decades a vast amount of web resources are born in order to give the possibility to share code and data for public accessibility.

In the bioinformatics field, examples of repositories are represented by the Gene Expression Omnibus (GEO) [36] and the Sequence Read Archive (SRA) [37]. Both give the possibility to upload and download datasets with a unique identifier, useful for sharing. The first one is for functional genomic data, while the second one is aimed to store biological sequence data.

Moreover, lots of works are based on data produced by international consortia like *The Cancer Genome Atlas* (TCGA) https://tcga-data.nci.nih.gov/tcga/, ENCODE https://genome.ucsc.edu/ENCODE/ [38] and ROADMAP epigenomics http://www.roadmapepigenomics.org/ which allow public access for download, also with interfaces of R packages.

Similarly, there are several code repositories on the Web useful to facilitate this goal. Repositories like Rpubs https://rpubs.com/ and GitHub https://github.com/, give the possibility to share the code used during an analysis and also to take trace of code versions. However, GitHub is more suited as repository for a specific tool, while Rpubs is more inspired to share pieces of codes of data analysing.

In this way, with a good combination of code and data, and a full share of them through a public repository, a third-party user is more stimulated to reproduce and to adapt a published analysis. Unfortunately, code sharing (in particular for data analysing) repositories are still too few and their development should be encouraged.

5   Reproducible Research Tools in Other Languages

We have presented so far the reproducibility feature inside the R environment. However, we can find several instruments that helps developers in most of the programming languages, such as: Python, Matlab and Java.

Python developers have the possibility to use *IPython Notebook*, freely available at http://ipython.org/notebook.html. This represents an interesting and useful computational interactive environment, very intuitive and easy to use. Thanks to this tool you can easily integrate executable code, rich text, mathematics equations and formulas, several plots and graphs and rich media as well.

For Matlab developers, there is the *Matlab Report Generator* giving the user the possibility to automatically generate enriched reports from Matlab scripts.

For Java developers, there is *ResearchAssistant* (RA): a Java library which offers the possibility to build reproducible experiments.

A completely different approach to reproducibility has been developed in Galaxy [39–41], useful for life science data analysis. Galaxy is a web platform offering a vast amount of software for genomic data analysis, creating a work-flow of made analysis, in order to trace all user actions. Moreover, Galaxy trace as much metadata as it can, that, in combination with stored work-flow, facilitates the reproducibility of analysis made.

6   Reproducible Research Issues and Limits

In the previous sections of this paper, we described the advantages gained by researchers in adopting reproducible research as main aspect in the conduction of studies in Life Science. However, several issues arise when we decide to introduce reproducibility in our work in practise.

There is no doubt about that in the short term the effort of performing an analysis that keeps track of all instruments adopted in a fully detailed way is time expensive and demanding. Nevertheless, we think that in the middle and long term reproducible analyses pay the best interests in order to gain an acceptable level of publishability, transparency, reliability and correctness of research. Analogously, the implementation of GUIs that automatically keep track of all user actions requires significant efforts.

Another issue consists in the difficulties that often many researchers face when they want to re-execute the entire analysis presented in a paper. In fact, sometimes they do not possess the required facilities (memory storage, RAM, fast processors) to perform the entire analysis of high-throughput omic data. In this particular case, caching can be helpful, but might be limited to short code chunks only.

In terms of limits, we have to admit that reproducibility is difficult to survive though the time since the tool versions change rapidly. Virtual machines can partially solve this problem. Unfortunately, after a period of time the fully reproducibility is impossible to be maintained. However, the transparency of the research still remains.

Finally, we stress that even though fully reproducible research helps a lot in the discovery of errors and misconduct, it does not assure the correctness of the results of a study. Nevertheless, it helps a lot in detecting potential errors or mis-use of statistical methodologies.

7   Conclusions

In this work, we discussed the importance of reproducible research in Life Science. We presented advantages and limits and we illustrated several R-tools which can help both data analysts and developers to make their research reproducible.

In particular, we stressed the importance to have tools that automatically generate a report file that keeps track of all actions executed by users and that provides a set of cached objects saved in a database that stores intermediate results of the executions. In this manner, scientific results can be explored by other researchers and used as starting points for alternative analyses. Moreover, the report can be attached as supporting information data and database of cached objects can be shared via the Internet. Thanks to the report and to the available caching database files, researchers promote transparency and reliability of their work.

We want to stress the importance to promote and teach how to reach acceptable levels of reproducibility. In fact, for this scope we think that training on reproducibility through seminaries and courses should be worldwide organised and encouraged. Reproducibility of data analyses should be taught as it is done for the Scientific Method. Students must learn the instruments to carry out individual and group projects in a fully reproducible way. Fortunately, we can find online lessons about reproducible research as the one promoted by the Johns Hopkins University at www.coursera.org/learn/reproducible-research taught by R. Peng, J. Leek and B. Caffo. There are also affordable courses like the one organised by the *Instituto Gulbenkian de Cincia* in Portugal (ended in 2015).

We also want to stress the importance and the urgency to adopt standard terms of reproducibility levels and conditions that should be broadly agreed and accepted by the scientific research community, universities and fund agencies.

We aim that research centers will more and more encourage and reward reproducible publications and evaluate their research staff based on the level of reproducibility of their studies, journal editors will encourage the submission of reproducible papers, funding agencies will strongly support and finance projects only if the beneficiaries of the funds accept to guarantee an high level of reproducibility of their work and finally reviewers will increasingly demand reproducibility as a mandatory characteristic of high quality scientific research essentially for a paper to be accepted. In this way, we can enhance transfer knowledge, verification of the results presented and transparency. All these features are also extremely important to face ethic problems related to misconduct in research and to improve the discovery of possibly errors or misuses of methodologies in Life Science studies, as well as in other scientific research areas.

Acknowledgments. This work was supported by the **Epigen** Project.

References

1. Editorial: An Array of Errors. The Economist (2011)
2. Baggerly, K.A., Coombes, K.C.: Deriving chemosensitivity from cell lines: forensic bioinformatics and reproducible research in high-throughput biology. Ann. Appl. Stat. **3**(4), 1309–1334 (2009)
3. Hofner, B., Schmid, M., Edler, L.: A review and guidelines for the biometrical journal. Biometrical J. **58**(2), 416–427 (2016)
4. Begley, C.G., Ellis, L.M.: Drug development: raise standards for preclinical cancer research. Nature **483**, 531–533 (2012)
5. Hothorn, T., Leisch, F.: Case studies in reproducibility. Briefings Bioinform. **12**(3), 288–300 (2011)
6. DeVeale, B., et al.: Critical evaluation of imprinted gene expression by RNAseq: a new perspective. PLoS Genet. **8**, e1002600 (2012)
7. Ioannidis, J.P.A., et al.: Repeatability of published microarray gene expression analyses. Nat. Genet. **41**, 149–155 (2009)
8. Li, M., et al.: Widespread RNA and DNA sequence differences in the human transcriptome. Science **333**, 53–58 (2011b)
9. Lin, W., et al.: Comment on widespread RNA and DNA sequence differences in the human transcriptome. Science **335**, 1302 (2012)
10. Prinz, F., et al.: Believe it or not: how much can we rely on published data on potential drug targets? Nat. Rev. Drug Discov. **10**, 712 (2011)
11. Editorial journals unite for reproducibility. Nature (2014)
12. Ioannidis, J.: Why most published research findings are false. PLoS Med. **2**, e124 (2005)
13. Witten, D.M., Tibshirani, R.: Scientific research in the age of Omics: the good, the bad, and the sloppy. JAMIA **20**(1), 125–127 (2013)
14. Gentleman, R.: Reproducible research: a bioinformatics case study. Stat. Appl. Genet. Mol. Biol. **4**(1), 1034 (2005)
15. Peng, R.D.: Reproducible research in computational science. Science **334**(6060), 1226–1227 (2011)
16. Peng, R.D.: Reproducible research and biostatistics. Biostatistics **10**(3), 405–408 (2009)

17. Ince, D.C., Hatton, L., Graham-Cumming, J.: The case for open computer programs. Nat. Perspect. **482**, 485–488 (2012)
18. Editorial: Enhancing reproducibility. Nat. Methods **10**, 367 (2013)
19. Stegmayer, G., Pividori, M., Milone, D.H.: A very simple and fast way to access and validate algorithms in reproducible research. Briefings Bioinform. **17**(1), 180–183 (2015)
20. Nekrutenko, A., Taylor, J.: Next-generation sequencing data interpretation: enhancing reproducibility and accessibility. Nat. Rev. Genet. **13**(9), 667–672 (2012)
21. Atmanspacher, H., Lambert, L.B., Folkers, G., Schubiger, P.A.: Relevance relations for the concept of reproducibility. J. Roy. Soc. Interface **11**(94), 20131030 (2014)
22. Iqbal, S.A., Wallach, J.D., Khoury, M.J., Schully, S.D., Ioannidis, J.P.A.: Reproducible research practices and transparency across the biomedical literature. PLoS Biol. **14**(1), e1002333 (2016)
23. Duvendack, M., Palmer-Jones, R.: Replication of quantitative work in development studies: experiences and suggestions. Prog. Dev. Stud. **13**(4), 307–322 (2013)
24. Stodden, V., Leisch, F., Peng, R.D. (eds.): Implementing Reproducible Research. CRC Press, Boca Raton (2014)
25. Claerbout, J., Karrenbach, M.: Electronic documents give reproducible research a new meaning. In: Proceedings 62nd Annual International Meeting of the Society of Exploration Geophysics, pp. 601–604, January 1992
26. Knuth, D.E.: Literate programming. Comput. J. **27**(2), 97–111 (1984)
27. Buckheit, J.B., Donoho, D.L.: Wavelab and reproducible research. In: Antoniadis, A., Oppenheim, G. (eds.) Wavelets and Statistics. Lecture Notes in Statistics, vol. 103, pp. 55–81. Springer, New York (1995)
28. Peng, R.D., Eckel, S.P.: Distributed reproducible research using cached computations. Comput. Sci. Eng. **11**(1), 28–34 (2009)
29. Peng, R.D.: Caching and distributing statistical analyses in R. J. Stat. Softw. **26**, 7 (2008)
30. Xie, Y.: Dynamic Documents with R and knitr, vol. 29. CRC Press, New York (2013)
31. Daring Fireball: Markdown. http://daringfireball.net/projects/markdown/
32. Markdown. http://www.aaronsw.com/weblog/00118
33. Liu, Z., Pounds, S.: An R package that automatically collects and archives details for reproducible computing. BMC Bioinform. **15**, 138 (2014)
34. Peng, R.D.: Interacting with data using the filehash package. R News **6**(4), 19–24 (2006)
35. Falcon, S.: Caching code chunks in dynamic documents. Comput. Stat. **24**(2), 255–261 (2008)
36. Edgar, R., Domrachev, M., Lash, A.E.: Gene expression omnibus: NCBI gene expression and hybridization array data repository. Nucleic Acids Res. **30**(1), 207–210 (2002)
37. Leinonen, R., Sugawara, H., Shumway, M.: The sequence read archive. Nucleic Acids Res., gkq1019 (2010)
38. ENCODE Project Consortium: An integrated encyclopedia of DNA elements in the human genome. Nature **489**(7414), 57–74 (2012)
39. Goecks, J., Nekrutenko, A., Taylor, J.: Galaxy: a comprehensive approach for supporting accessible, reproducible, and transparent computational research in the life sciences. Genome Biol. **11**(8), R86 (2010)
40. Blankenberg, D., Kuster, G.V., Coraor, N., Ananda, G., Lazarus, R., Mangan, M., Taylor, J.: Galaxy: a web based genome analysis tool for experimentalists. Curr. Protoc. Mol. Biol. **19**(10), 11–21 (2010)

41. Giardine, B., Riemer, C., Hardison, R.C., Burhans, R., Elnitski, L., Shah, P., Miller, W.: Galaxy: a platform for interactive large-scale genome analysis. Genome Res. **15**(10), 1451–1455 (2005)

42. Russo, F., Angelini, C.: RNASeqGUI: a GUI for analysing RNA-seq data. Bioinformatics **30**(17), 2514–2516 (2014)

43. Russo, F., Righelli, D., Angelini, C.: Advancements in RNASeqGUI towards a reproducible analysis of RNA-Seq experiment. BioMed Res. Int. **2016**, 11 (2016). Article ID 7972351

44. Huntley, M.A., Larson, J.L., Chaivorapol, C., Becker, G., Lawrence, M., Hackney, J.A., Kaminker, J.S.: ReportingTools: an automated result processing and presentation toolkit for high throughput genomic analyses. Bioinformatics **29**(24), 3220 (2013)

45. Hillman-Jackson, J., Clements, D., Blankenberg, D., Taylor, J., Nekrutenko, A., Galaxy, Team: Using galaxy to perform large-scale interactive data analyses. Curr. Protoc. Bioinform. **10**, 5 (2012)

NuchaRt: Embedding High-Level Parallel Computing in R for Augmented Hi-C Data Analysis

Fabio Tordini[1]([✉]), Ivan Merelli[3], Pietro Liò[2], Luciano Milanesi[3], and Marco Aldinucci[1]

[1] Computer Science Department, University of Torino, Torino, Italy
{tordini,aldinuc}@di.unito.it
[2] Computer Laboratory, University of Cambridge, Cambridge, UK
pietro.lio@cl.cam.ac.uk
[3] IBT - Italian National Research Council, Segrate, MI, Italy
{ivan.merelli,luciano.milanesi}@itb.cnr.it

Abstract. Recent advances in molecular biology and Bioinformatics techniques brought to an explosion of the information about the spatial organisation of the DNA in the nucleus. High-throughput chromosome conformation capture techniques provide a genome-wide capture of chromatin contacts at unprecedented scales, which permit to identify physical interactions between genetic elements located throughout the human genome. These important studies are hampered by the lack of biologists-friendly software. In this work we present NuchaRt, an R package that wraps NuChart-II, an efficient and highly optimized C++ tool for the exploration of Hi-C data. By rising the level of abstraction, NuchaRt proposes a high-performance pipeline that allows users to orchestrate analysis and visualisation of multi-omics data, making optimal use of the computing capabilities offered by modern multi-core architectures, combined with the versatile and well known R environment for statistical analysis and data visualisation.

Keywords: Next-generation sequencing · Neighbourhood graph · High-performance computing · Multi-Omic data · Systems biology

1   Scientific Background

Over the last decade, a number of approaches have been developed to study the organisation of the chromosome at high resolution. These approaches are all based on the Chromosome Conformation Capture (3C) technique, and allow the identification of neighbouring pairs of chromosome loci that are in close enough physical proximity (probably in the range of 10-100 nm) that they become cross-linked [1]. This information highlights the three-dimensional organisation of the chromosome, and reveals that widely separated functional elements actually

C. Angelini et al. (Eds.): CIBB 2015, LNBI 9874, pp. 259–272, 2016.
DOI: 10.1007/978-3-319-44332-4_20

result to be close to each other, and their interaction can be the key for detecting critical epigenetics patterns and chromosome translocations involved in the process of genes regulation and expression.

Among 3C-based techniques, the *Hi*-C method exploits Next-Generation Sequencing (NGS) techniques to provide a genome-wide library of coupled DNA fragments that are found to be close to each other in a spatial context. The contact frequency between the two fragments relies on their spatial proximity, and thus it is expected to reflect their distance. The output of a Hi-C process is a list of pairs of locations along all chromosomes *reads*, which can be represented as a square matrix M, where each element $M_{i,j}$ of the matrix indicates the intensity of the interactions between positions i and j.

In a previous work we proposed NuChart-II as a highly optimised, C++ application designed to integrate information about genes positions with paired-ends reads resulting from Hi-C experiments, aimed at describing the chromosome spatial organisation using a gene-centric, graph-based approach [6]. A graph-based representation of the DNA offers a more comprehensive characterization of the chromatin conformation, which can be very useful to create a representation on which other *omics* data can be mapped and characterize different spatially-associated domains.

NuChart-II has been designed using high-level parallel programming patterns, that facilitate the implementation of the algorithms employed over the graph: this choice permits to boost performances while conducting genome-wide analysis of the DNA. Furthermore, the coupled usage of C++ with advanced techniques of parallel computing (such as lock-free algorithms and memory-affinity) strengthens genomic research, because it makes possible to process much faster, much more data: informative results can be achieved to an unprecedented degree [3].

However, C++ is not widely used in Bioinformatics, because it requires highly specialised skills and does not fully support the rapid development of new interactive pipelines. Conversely, the modularity of R and the huge amount of already existing statistical packages facilitates the integration of exploratory data analysis and permits to easily move through the steps of model development, from data analysis to implementation and visualisation. In this article we discuss the integration of our C++ application into the R environment, an important step toward our objective of augmenting the usability of bioinformatics tools: we aim at obtaining a high-performance pipeline that allows users to orchestrate analysis and visualisation of multi-omics data, making optimal use of the computing capabilities offered by modern multi-core architectures, combined with the versatile and well known R environment for statistical analysis and data visualisation. The novel package has been renamed NuchaRt.

1.1   Parallelism Facilities in R

By default, it is not possible to take advantage of multiple processing elements from within the R environment. Instead, a sort of "back-end" must be registered, that effectively permits to run a portion of code in parallel. For what it concerns high-performance computing, some libraries exist that foster parallel programming in R, most of which focus on distributed architectures and clusters of computers. Worth to mention are *Rmpi* and *Snow*.

Rmpi is a wrapper to MPI and exposes an R interface to low-level MPI functions. The package provides several R-specific functions, beside wrapping the MPI API: parallel versions of the `apply()`-like functions, scripts to launch R instances at the slaves from the master and some error-handling functions to report errors form the workers to the manager. Snow (Simple Network Of Workstations) provides support for simple parallel computing on a network of workstations and supports several different low-level communication mechanisms, including private virtual machine (PVM), MPI (via Rmpi) and raw sockets. The package also provides high-level parallel functions like `apply()` and simple error-handling mechanism.

The *multicore* package builds a back-end for parallel execution of R code on machines with multiple CPUs: all jobs share the full state of R when parallel instances are spawned, so no data or code needs to be copied or initialized. Spawning uses the `fork` system call (or OS-specific equivalent) and establishes a `pipe` between master and child process, to enable inter-process communication. However, the variety of operations that can be parallelized with *multicore* is limited to simple independent math computations on a collection of indexed data items (e.g., an array).

The *doMC* package acts as an interface between *multicore* functionalities and the *foreach* package, which permits to execute looping operations on multiple processors.

R/parallel enables automatic parallelization of loops without data dependencies by exposing a single function: `runParallel()`. The implementation is based on C++, and combines low-level system calls to manage processes, threads and inter-process communications. The user defines which variable within the enclosed loop will store the calculation results after each iteration, and how these variables have to be operated and reduced.

It is worth to mention that an interface to Intel TBB for R also exists, that pretty much resembles our approach and permits to use TBB's `parallel_for` pattern to convert the work of a standard serial for loop into a parallel one, and the `parallel_reduce` construct can be used for accumulating aggregates or other values. This solution enforces a master/slave behaviour between R and C++, so that data-parallel computations can be offloaded to C++. We will shortly see that our approach pretty much resembles this latter one.

Memory Management

The notoriously "poor" memory management mechanism in R is actually a combination of multiple factors, that also include the way operating systems allocate memory. Since our development relies on Linux OS, a discussion about these factors will shed some light over this problem.

R uses a *lazy* memory reclaim policy, in a sense that it will not reclaim memory until it is actually needed. Hence, R might be holding on to memory because the OS hasn't yet asked for it back, or it does not need more space yet. In order to decide when to release memory, R uses a garbage collector (GC) that automatically releases memory to the OS when an object is no longer used. It does so by tracking how many references point to each object, and when there are no references pointing to an object it deletes that object and returns memory to the OS. This means that when we have one or more copies of a big object, explicitly removing the original object does not correspond to free memory space: until references to that object exists, the memory wont be released. Even a direct call to the GC does not force R to release memory, rather it acts as a "request", but R is free to ignore [7].

Furthermore, R has limited control over memory management mechanism: it simply uses `malloc/free` functions plus a garbage collector. One attempt to force memory to be released to the OS is the use of the `malloc_trim` function, that explicitly forces memory release, provided that a sufficiently large chunk is ready to be released. We managed to limit the drawbacks related to these weaknesses by avoiding unnecessary copies of objects and promptly freeing their memory, as soon as they are no longer needed. In this way we controlled memory leaks that cause memory fragmentation to explode.

1.2   Hi-C Data Analysis Step-by-step

The Hi-C data analysis conducted with NuChart-II walks through five main steps:

(1) data retrieval and parsing;
(2) neighbourhood graph construction;
(3) weighing of the edges as a result of data normalisation;
(4) statistical analysis;
(5) output and visualisation.

NuChart-II parses a number of options from Command Line Interface (CLI) to set up and characterise each execution. Once started, the application walks through all the steps outlined above in a "monolithic" fashion, and yields its results as a summary of the whole process: the final output is available in terms of a neighbourhood graph drawn using some plotting engine, together with formatted text files (such as `csv` files) that contain whole information necessary to examine the represented data. This include the actual sequences "contained" in edges, edges probability resulting from data normalisation, network analysis metrics and various statistical annotations.

Genomic data analysis, just like many other scientific fields, does not work as one monolithic process: different stages of data analysis are just fundamentally different, and have different parallelism patterns, memory access and data access requirements. Also, it often makes sense to run the same stage of an analysis in a number of different ways to demo nstrate the robustness of novel results or to tackle different sorts of data, for example one in which a reference genome is available, compared to one where it is not.

If we consider the possibility to map additional features on a graph — such as genes expression, CTCF binding sites or methylation profiles — we would choose a dataset from which to gather the required information and re-execute the application from the beginning, until we get our output with mapped omics data. This means that no intermediate inspection is allowed, nor we could choose some quick statistics to satisfy whatever curiosity or to banish some doubts. Despite its undeniable efficiency, this lack of modularity highlights a clear limitation in usability of the C++ implementation.

These factors led us to re-consider R as a "hosting" environment for a scalable and usable tool for Hi-C data analysis. From the early R prototype — developed within the R environment — we learned that high-performance and good memory hierarchy exploitation is hard to achieve within the R environment, due to specificities of the environment itself, and requires a substantial programming effort. Nonetheless, research during the last decade has widely explored the use of parallel computing techniques with R.

2   NuchaRt

We aim at building a tool for Hi-C data analysis that is both efficient — in terms of speed and memory resources exploitation — and usable. We decided not to use off-the-shelf libraries for parallel computing, because of the well known R's limits in memory management: our search for long-range chromatin contacts over genome-wide paired-ends reads results in a memory-bound algorithm, thus parallel memory-intensive tasks should be kept on C++ side where we can obtain a finer memory control, while we rely on R for setting up a usable working environment. Also, we already had a fully tested C++ solution to our problem, that led us to consider Rcpp [4]: it facilitates data interchange from C++ to R and vice-versa. C++ objects holding the output of a computation are made available within the R environment, ready to be used as source for advanced statistical analysis, by mean of a *wrapping* mechanism based on the templated functions `Rcpp::as<>()` and `Rcpp::wrap()`. These functions convert C++ object classes into a *S expression pointers* (called SEXP), that can be handled on the R side to construct `Lists` or `DataFrames`[1], which are essential object types in R and are used by almost all modelling functions.

In this respect, our application clearly exhibits a *master/slave* behaviour: on the R side we set up the "background" for the computation, and then we offload computationally intensive tasks to C++ (see Fig. 1). Once it terminates, the needed information is moved back to the R side and is ready to be processed, drawing from the huge R's library basket.

---

[1] We actually use `data.tables` as basic data structures for our datasets: `data.table` is an enhanced version of `data.frame` that allows to easily optimise operations for speed and memory usage.

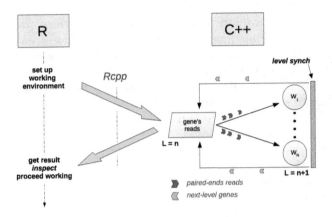

Fig. 1. Master/Slave behaviour between R and C++, on the graph construction phase

2.1   NuchaRt and Rcpp

In our context, we have dealt with four C++ objects that abstract the leading
actors of our software: SamData, Gene, Fragment and Edge. These objects contain
much of the information that is needed to build a topographical map of the DNA
from Hi-C data. In order to exchange a SamData object between C++ and R
we have specialised the templated functions above: a std::vector<SamData>
is thus treated by R as a list of Lists, while a list of Lists in R (or a
DataFrame) is managed in C++ by casting the SEXP object to a Rcpp::List
(or a Rcpp::DataFrame) object, and by subsequently filling each field of the
SamData class with the value contained in the respective field of the List.

Algorithm 1.1 Example of as and wrap usage

```
1  template<> SEXP wrap(const SamData &s) {
2     List ret = List::create( Named("Id")     = s.getId(),
3                              Named("Chr1")    = s.getChr1(),
4                              Named("Start1")  = s.getStart1(),
5                              Named("Chr2")    = s.getChr2(),
6                              Named("Start2")  = s.getStart2(),
7                              Named("Seq")     = s.getSeq()
8                            );
9     return wrap(ret);
10 }
11
12 template<> SamData as( SEXP s ) {
13    List samL = as<List>(s);
14    SamData sam;
15
16    sam.setId     ( as<long>( samL["Id"])     );
17    sam.setChr1   ( as<string>(samL["Chr1"])  );
18    sam.setStart1 ( as<long>( samL["Start1"]) );
19    sam.setChr2   ( as<string>(samL["Chr2"])  );
20    sam.setStart2 ( as<long>( samL["Start2"]) );
21    sam.setSeq    ( as<string>(samL["Seq"])   );
22
23    return sam;
24 }
```

Recalling Sect. 1.2, NuChart-II can be described as a 5 stages pipeline: from data retrieval to output and visualisation, these phases can now be broke up and used as loose modules. Phase *(1)* is responsible for data collection and early data processing: datasets are provided as static `csv`-like files, but can also be downloaded from on-line repositories. The information contained therein is parsed and processed, in order to build the data collections needed to perform the computations: unneeded fields are dropped and elements are ordered in a consistent way, while a unique identifier for each element of a collection is generated, when needed. Problems may arise if these operations are performed on the R side, as it may lead to memory overflows with big size files ($> 2GB$, as it is the case with SAM files) due to the way R objects are constructed and stored in memory. For this reason dataset whose size exceeds $2GB$ is parsed on C++ side (as it is the case for SAM files). No matter where these operations are executed, objects can be moved from C++ side to R side and vice-versa, as explained above.

Phases *(2)* and *(3)* constitute by far the most onerous parts of the application, in terms of execution time and resource consume. Both of them are suitable for being revisited in the context of loop parallelism, since their kernels can be run concurrently on multiple processors with no data dependencies involved. These phases have been thoroughly explained in our previous works [3,5,6], and we refer to those writings for a thorough explanation. Not much changes when we offload the a computation from R to C++: the very same logic is used and the `ParallelFor` skeleton permits to speed up both phases in a seamless way. Data transfer overhead is negligible, compared to the computationally intensive task that takes place.

Phase *(4)* encompasses essential features that the package ought to provide, in order to fulfil the requirements of a useful tool for genomic data interpretation. With a graph-based representation we can apply network analysis over the resulting graph: topological measures capture graph's structure for nodes and edges and highlight the "importance" of the actors. For instance, centrality metrics describe the interactions that (may) occur among local entities. Ranking of nodes by topological features (such as degree distribution) can help to prioritize targets of further studies or lead to a more local, in-deep analysis of specific chromosome locations. Here studies of functional similarity can suggest new testable hypotheses [8].

Finally, visualisation is crucial for a tool that aims at facilitating a better interpretation of genomic data. NuChart-II supplies both tabular output and graphical visualization. Concerning the latter, common plotting engines perform nicely with small-to-medium sized graphs, but cannot provide useful representations of huge graphs.

A possible approach could be to decouple visualisation from NuchaRt, and make use of external applications purposely designed for interactive visualisation of networks. One such application is *Gephi*[2], that permits to interact with the graph by manipulating structures, shapes and colors to reveal hidden properties. NuchaRt can output a resulting graph in `GraphML` format, which permit to get

---

[2] https://gephi.org/.

the most out of Gephi. In this way the user can easily browse the results of Hi-C data analysis through Gephi interface.

3   Discussion

The novel package benefits of the combined use of parallel programming techniques provided by the C++ engine, and the flexibility of the R environment, maintaining the same performance and scalability achieved in NuChart-II. Moreover, within the R environment the five steps listed in Sect. 1.2 become loose but totally compatible modules, and could be either executed in order or as services that permit to accomplish a specific task.

Results of each module are made globally available in form of `DataFrames`, and can be easily queried and inspected, exported, or saved and re-used with other, different data analysis tools. The graph can be plotted and the results can be visualised and browsed. Eventually, one can draw from the huge R's libraries basket the one that suits her need, and conduct advanced analysis over the resulting data. For instance, we also tested the *ERGM* package that permit to understand the processes of network structure emergence and tie formation: the Exponential-family Random Graph Models package provides an integrated set of tools for creating an estimator of the network through a stochastic modelling approach.

3.1   Experiments

The study of the interactions of the actor genes with the environment is of critical importance for understanding the entire system. By using the modelling functions of the package we can statistically characterize the distribution of the edges in relation to the characteristics of the nodes that represent mapped multi-omics features. We performed the analysis of the clusters of genes Human Leukocyte Antigen (HLA, Fig. 3) and Kruppel-Associated Box (KRAB, Fig. 2) in the context of four Dixon experiments (SRA:SRR400260, SRA:SRR400261, SRA:SRR400266, SRA:SRR400267) [2], to verify the correlation of the edges distribution in relation to some genomic features (hypersensitive sites, CTCF binding sites, isochores, RSSs).

The first analysed locus is located in cytoband chr19.q13.12 and concerns the clusters of Kruppel-type zinc finger genes, related to the KRAB, which are peculiar for their tandem organization. Zinc finger proteins are a family of transcription factors that regulate the gene expression, and most of these proteins are members of the KZNF family. There are 7 human-specific novel KZNFs and 10 KZNFs that have undergone pseudo-gene transformation specifically in the human lineage. 30 additional KZNFs have experienced human-specific sequence changes that are presumed to be of functional significance. Members of the KZNF family are often in regions of segmental duplications, and multiple KZNFs have undergone human-specific duplications and inversions. In Fig. 2, top panel drawings concern sequencing runs from hESC (SRR400260, SRR400261);

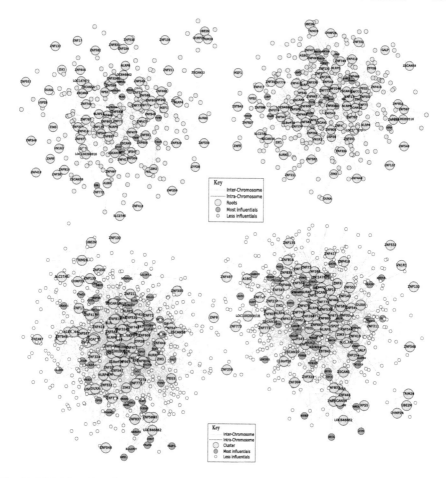

Fig. 2. Neighbourhood graphs of the KRAB cluster of genes in four different runs from the Hi-C experiments of Dixon et al.

bottom panel drawings in the same Figure concern sequencing runs from IMR90 (SRR400266, SRR400267). Seed genes are the genes given as input to the algorithm, while output genes are differentially represented according to their importance (in terms of node degree).

The second analysed gene cluster concerns the human leukocyte antigen (HLA) system, which is the name of the locus containing the genes that encode for major histocompatibility complex (MHC) in humans. It belongs to a superlocus that contains a large number of genes related to the immune system function in humans. The HLA group of genes resides on cytoband chr6.p31.21 and encodes for cell-surface antigen-presenting proteins, which have many different functions. The HLA genes are the human version of the MHC genes that are found in most vertebrates (and thus are the most studied of the MHC genes). The major HLA antigens are essential elements for the immune function. In

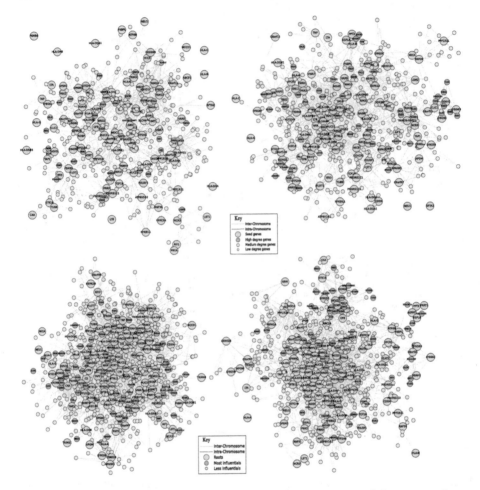

Fig. 3. Neighbourhood graphs of the HLA cluster of genes in four different runs from the Hi-C experiments of Dixon et al.

Fig. 3, top panel drawings concern sequencing runs from hESC (SRR400260, SRR400261); bottom panel drawings in the same Figure concern sequencing runs from IMR90 (SRR400266, SRR400267). Seed genes are the genes given as input to the algorithm, while output genes are differentially represented according to their importance (in terms of node degree).

The correlation between cryptic RSS sites and edges is more pronounced in the HLA cluster, in comparison to the KRAB cluster, probably due to a more consistent presence of this kind of sequences in genes related to the immune system. The correlation between hypersensitive sites (super sensitivity to cleavage by DNase) and edges, although positive, is poor, probably because the accessibility of these regions are impaired by a large number of long-range interactions. The correlation between the presence of CTCF binding sites and edges was

Table 1. Mapping CTCF binding sites, isochores, cryptic RSSs, and DNase sites on the graphs affects the edge distribution of the KRAB cluster of genes and of the HLA cluster of genes, using the ERGM package

	KRAB		HLA	
	Estimate	Std. Error	Estimate	Std. Error
SRA:SRR400260				
edges + nodecov("dnase")	0.2867	0.08451	0.1711	0.07961
edges + nodecov("ctcf")	0.6531	0.01157	0.5545	0.01253
edges + nodecov("rss")	0.5804	0.06176	0.6304	0.08196
edges + nodecov("iso")	-1.047	0.09269	-0.9406	0.09156
SRA:SRR400261				
edges + nodecov("dnase")	0.2042	0.07932	0.1706	0.07822
edges + nodecov("ctcf")	0.6629	0.04158	0.5687	0.02005
edges + nodecov("rss")	0.5378	0.03566	0.6319	0.03776
edges + nodecov("iso")	-1.015	0.09566	-0.93035	0.08969
SRA:SRR400266				
edges + nodecov("dnase")	0.2042	0.07932	0.1706	0.07822
edges + nodecov("ctcf")	0.6629	0.04158	0.5687	0.02005
edges + nodecov("rss")	0.5378	0.03566	0.6319	0.03776
edges + nodecov("iso")	-1.015	0.09566	-0.93035	0.08969
SRA:SRR400267				
edges + nodecov("dnase")	0.2042	0.07932	0.1706	0.07822
edges + nodecov("ctcf")	0.6629	0.04158	0.5687	0.02005
edges + nodecov("rss")	0.5378	0.03566	0.6319	0.03776
edges + nodecov("iso")	-1.015	0.09566	-0.93035	0.08969

clearly predictable, because linking Gene-Regulatory elements maintain different regions of the genome close to each other. On the other hand, regions with isochores seem less involved in long-range interactions, which can be quite surprising considering that these portions of the genome are considered gene-rich.

Statistical results are reported in Table 1. The network estimators are all computed using 100 iterations of stochastic modelling. A high correlation between the presence of specific genomic features and the probability of existence of an edge persists. DNase sensitivity sites are weakly correlated with the presence of an edge, while isochores are strongly anti-correlated with the presence of an edge.

Performance comparisons between the original R prototype and the actual implementation have not been conducted, because there would be no room for such debate, since the original tool often halted its execution, due to its strong limitations in memory management. For what it concerns the comparison between the C++ application and the combined R with C++ package, they report substantially similar behaviours: the graph construction execution is strongly affected by datasets size and resolution, that determine the "search space" for the BFS-like graph construction and the overall memory load. Reducing the working set ameliorates execution times and overall scalability with NuChart-II, and clearly helps in obtaining good performance when offloading the graph construction from R to C++ [3].

Figure 4 compares execution time (left) and speedup (right) in the two approaches: Figs. 4a and 4b show the performance for constructing a graph at level 1 starting from the KRAB cluster of genes using Dixon's SRR400266 experiment as Hi-C dataset. Despite similar timings and scalability, NuchaRt has slightly worse performance and shows a higher execution time. Figures 4c and 4d show a comparison of the performance during normalisation phase with NuChart-II and NuchaRt: again both implementations yield similar results, both approaching a quasi-linear scalability, even though NuchaRt's execution time is slightly higher with respect to NuChart-II's. This is likely due to the worsening of memory access time when offloading computation to C++: while the multi-threaded C++ application is running, the R environment is kept

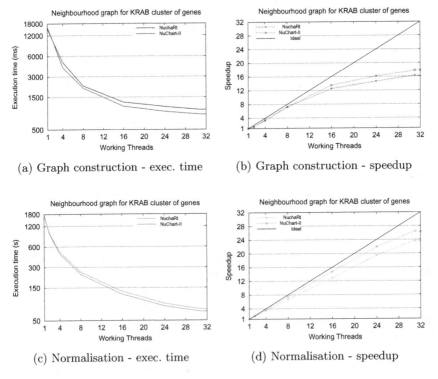

(a) Graph construction - exec. time

(b) Graph construction - speedup

(c) Normalisation - exec. time

(d) Normalisation - speedup

Fig. 4. Comparison between NuChart-II and NuchaRt during the graph construction and normalisation

alive. R stores additional information, beside the data itself, for each object created: when this small overhead is combined to the lazy memory reclaim policy adopted by R's garbage collector, and to the massive size of the dataset used for neighbourhood graph construction, resident memory consumption remains high at run-time, thus affecting memory access times and overall performance.

Our experiments where conducted on a NUMA system, equipped with 4 eight-cores E7-4820 Nehalem running at 2.0 GHz, with 18 MB L3 cache and 64 GB of main memory. The Nehalem processor has Hyper-Threading capability with 2 contexts per core, but we decided not to use it and stick to the number of physical cores: the heavy memory usage would dramatically damage performance, and likely increase chances of false-sharing among threads in the same context that share L2 cache. With this configuration the cache-coherence mechanism plays an important role in this performance degradation, where cache misses are likely frequent and cache lines updates occur frequently.

Performance differences seem to flatten when the same applications are executed on a different machine: we also conducted experiments on a workstation equipped with a single eight-cores Intel Xeon CPU E5-2650 running at 2.60 GHz. This machine features 20 MB of L3 cache with 64 GB of main memory. The SandyBridge processor also has Hyper-Threading capability allowing 2 contexts per core. Here as well we decided to not run more than 8 threads, so that the second context is not used an only physical cores are employed during computation. In this case the gap between the two solutions is reduced, though the total execution time is higher due to the limited degree of parallelism that can be achieved because of the reduced number of available cores.

4   Concluding Remarks

Embedding NuChart-II in R creates an application that can be used either to conduct a step-by-step analysis of genomic data, or as a high-performance workflow that takes heterogeneous datasets in input, processes data and produces a graph-based representation of the chromosomal information provided, supported by a rich set of default descriptive statistics derived from the topology of the graph. The graph-based approach fosters a tight coupling of topological observations to biological knowledge, which is likely to bring remarkable biological insights to the whole research community.

From a computational point of view, the ever-increasing amount of information generated by novel Bioinformatics techniques require proper solutions that permit the full exploitation of the computing power offered by modern computing systems, together with advanced tools for an efficient analysis and interpretation of genomic data. These tasks require high skills, but we believe that NuchaRt can be a valuable mean to support researchers in pursuing these objectives.

Despite the results achieved in terms of performance and usability, some problems remain partially unsolved, and are open to further investigations. Among them, our main concern is the visualisation of multi-omic graphs, which we believe is an essential feature for a usable tool aimed at facilitating genomic data analysis and interpretation, and that remains an open problem.

We are currently working on making the package compliant to Bioconductor and CRAN requirements, so that it can be easily downloaded and used by the research community. At the time of writing it is available through our research group's repository, at http://alpha.di.unito.it:8080/tordini/nuchaRt.

Acknowledgements. This work has been partially supported by the EC-FP7 STREP project "REPARA" (no. 609666), the Italian Ministry of Education and Research Flagship (PB05) "InterOmics", and the EC-FP7 innovation project "MIMOMICS".

References

1. Dekker, J., Rippe, K., Dekker, M., Kleckner, N.: Capturing chromosome conformation. Science **295**(5558), 1306–1311 (2002)
2. Dixon, J., Selvaraj, S., Yue, F., Kim, A., Li, Y., Shen, Y., Hu, M., Liu, J., Ren, B.: Topological domains in mammalian genomes identified by analysis of chromatin interactions. Nature **485**(5), 376–80 (2012)
3. Drocco, M., Misale, C., Pezzi, G.P., Tordini, F., Aldinucci, M.: Memory-optimised parallel processing of Hi-C data. In: Proceedings of International Euromicro PDP 2015: Parallel Distributed and Network-Based Processing, pp. 1–8. IEEE, March 2015. http://calvados.di.unipi.it/storage/paper_files/2015_pdp_memopt.pdf
4. Eddelbuettel, D.: Seamless R and C++ Integration with Rcpp. Springer, New York (2013). ISBN 978-1-4614-6867-7
5. Merelli, I., Liò, P., Milanesi, L.: Nuchart: An R package to study gene spatial neighbourhoods with multi-omics annotations. PLoS ONE **8**(9), e75146 (2013)
6. Tordini, F., Drocco, M., Misale, C., Milanesi, L., Liò, P., Merelli, I., Aldinucci, M.: Parallel exploration of the nuclear chromosome conformation with NuChart-II. In: Proceedings of International Euromicro PDP 2015: Parallel Distributed and Network-Based Processing. IEEE, March 2015. http://calvados.di.unipi.it/storage/paper_files/2015_pdp_nuchartff.pdf
7. Wickham, H.: Advanced R, 1st edn. Chapman and Hall/CRC, Boca Raton (2014)
8. Winterbach, W., Mieghem, P.V., Reinders, M.J.T., Wang, H., Ridder, D.: Topology of molecular interaction networks. BMC Syst. Biol. **7**, 90 (2013)

A Web Resource on Skeletal Muscle Transcriptome of Primates

Daniela Evangelista$^{(\boxtimes)}$, Mariano Avino, Kumar Parijat Tripathi, and Mario Rosario Guarracino

Laboratory for Genomics, Transcriptomics and Proteomics (LAB-GTP),
High Performance Computing and Networking Institute (ICAR),
National Research Council (CNR), Via Pietro Castellino 111, 80131 Naples, Italy
daniela.evangelista@icar.cnr.it

Abstract. Skeletal muscle represents a very well organized anatomical tissue in animals and its appearance might have predated the divergence of vertebrate and arthropods lineages about 700MYA. This diversified structure is very well visible in Primates since it differentiates according to their life styles and environmental conditions. This study focuses on *Pan troglodytes* - known as common chimpanzee - which belongs to a genus that is the most closely related to human species by which also shares a high similarity in the DNA composition. Our aim is to test the level of similarity between chimpanzee and human DNA - diversified to a functional phenotypic level to better adapt in different environmental conditions - by collecting skeletal muscle transcriptomic data from ENA (*European Nucleotide Archive*) database and performing its functional annotation analysis. We developed *PrimatesDB*, a freely available web-oriented application which contains 30,944 sequences belonging to *Pan troglodytes* skeletal muscle transcriptomic data and from which it is possible to retrieve all the information related to 12,222 transcripts. *PrimatesDB* is available at: www-labgtp.na.icar.cnr.it/PrimatesDB.

Keywords: *Pan troglodytes* · Annotations · Skeletal muscle · Transcriptome · Database

1 Introduction

We all know how similar humans are genetically with all the other primates, However, there is also a high morphological differentiation, among the whole order, due to the different environmental living conditions we have undergone to. Where and how this functional differentiation has arisen is still unclear. For this reason it would be really important to find out a link of all the information we have about genomic, transcriptomic and proteomic data of primates in order to compared them and look for the causes that might have generated all this differentiation. Moreover, shading light on how other non humans primates organisms work, might also

---

D. Evangelista and M. Avino—Contributed equally to this work.

C. Angelini et al. (Eds.): CIBB 2015, LNBI 9874, pp. 273–284, 2016.
DOI: 10.1007/978-3-319-44332-4_21

come in handy to further understand the knowledge of our species. This might also allow us to employ much more closely related organisms like apes, than using more distant species, like mice or fruit flies, to better understand the complex mechanism of certain diseases. This has been already successfully employed in studies of human pathologies tested in some primates (see below). However, a big problem to overcome, when choosing primates to study humans, would be of ethical origin, given their consequent usage in lab experiments. However, a silico approach, like we propose, could thoroughly bypass this disadvantage. We decided to start our comparative analysis picking the muscular system. Muscles might have evolved, independently, at least twice in animals from common ancestor contractile cells in sponge-grade organisms [1], once in cnidarians and cnetophores and another time in bilaterian (where we belong). In this last group, specialized forms of skeletal and cardiac muscles predated the divergence of vertebrate/arthropode lineage circa 700 millions of years ago (MYA) while smooth muscle seemed to be evolved independently to other muscles [1]. We are interested in striated muscles, which make up the skeletal musculature. In primates, muscles are anatomically adapted of their particular life style. Postcranial skeletal is, for example, adapted to a great variety of locomotor, postural and feeding activities and it is not very well specialized, like in other non-primates vertebrate, such as horses and other ungulates or even whales. However, major changes among primates compared to other vertebrate are seen in locomotor morphology [2] being, the first ones, more adapted to vertical leaping and clinging. Most nonhuman primates spend at least some time during the day in trees, therefore, grasping, when climbing, in arboreal environment is an essential component of their life. This is seen in feet and hands muscles morphology. Humans locomotors muscles are, on the other hand, adapted to a vertical postural position with evident changes in the vertebral column, pelvis, legs and feet. There have been described, at least, six different locomotion systems in primates [3]: arboreal and terrestrial quadrupedalism, knuckle-walking, leaping, suspensory climbing and bipedalism. In order to better elucidate the functional differentiation of these categories, we thought that having directly a look to expression data analysis would have been more suitable, and this is why we decided to pick transcriptomic data. Nowadays, these data are highly proliferating out there, due to the recent advent of high throughput studies, like NGS (*Next Generation Sequencing*), which are able to get the whole gene expression spectre of an organism, in a few lab experiments, at relatively low cost. However, these data are very difficult to handle due to their heaviness and complexity. To solve this problem many online tools, software, repositories and databases have been created, making difficult their immediate finding and communication to each other. Most remarkable already existing primates databases contains useful information about their ecological role in nature (All The Worlds Primates, http://www.alltheworldsprimates.org) or store genetic population data (Primate Life Histories, https://plhdb.org), or news about morphology and evolution (PRIMO, http://primo.nycep.org). Among the ones that keep primates bionformatic data we can mention IPD (Immuno Polymorphism Database, https://www.ebi.ac.uk/ipd/mhc/nhp/species.html) which aims to collect information about polymorphisms related to diseases or Primate Orthologous

Exon Database (http://giladlab.uchicago.edu/orthoExon), which includes a cat-alogue of unique orthologous exon regions in the genomes of human, chimpanzee, and rhesus macaque. In Ensembl (http://www.ensembl.org) is also possible to find out comparative data analysis regarding primates in order to construct gene trees. However, to our knowledge, none of these databases contains comparative tran-scriptomic data aimed to find out conservation and differences for functional anno-tations purposes. Therefore, we are presenting *PrimatesDB*, which is a comprehen-sive web resource - driven on a relational database - for the retrieval of primates functional annotation and other related integrated information. The first data collected and analyzed in *PrimatesDB* are skeletal muscle myoblasts transcripts [4] coming from RNA-seq experiments on common chimpanzee (*Pan troglodytes* Blumenbach). The transcripts are already genome reference-assembled and struc-tural but not functional annotated. Common chimpanzee belongs to the family Hominidae [5]. With *Pan paniscus* Schwarz (bonobo), it represents the only living species of genus *Pan* Oken and it is the most closely related species to humans, shar-ing the last common ancestor circa 6 millions years ago (MYA) [6–11], and at least 98 % DNA similarity. A recent research study of hepatitis C have shown that they sometimes become the only available source to test vaccines in humans [7]. More-over, research into how the evolution of chimpanzee is influenced by viruses, like HIV-1, can have important implications in human health advances [8]. Chimpanzee whole-genome has been already published [9]. We collected its muscular transcrip-tome, and ran it, in our lab, into an existing computational pipeline [12] called *Tran-scriptator* to obtain its functional annotation. The transcriptomic dataset was, in this way, annotated and stored in *PrimatesDB*. Currently, *PrimatesDB*, includes around 12,222 transcripts of the gene ontology and functional annotation of chim-panzee skeletal muscles transcriptome. Moreover, hyperlink services are available for Ensembl and UniProt, so that users can gain diverse insights about the tran-scripts of interest from these publicly available resources.

2   The Pipeline

PrimatesDB pipeline (Fig. 1) is divided in two main areas: (i) data retrieval (blue panel) and (ii) data visualization (orange panel). It consists of three major com-ponents: (i) BLAST analysis (ii) functional annotation (iii) retrieval and statisti-cal analysis of the data. The operations developed in the first area are carried out by using the core engine of *Transcriptator* background pipeline (lower right cor-ner). Here, the results provided are in the form of tabular reports and transferred to the LAB-GTP web server where the database, managed by phpMyAdmin soft-ware, is ready to store the entire sets of records and handle the administration of MySQL queries. Here, as second main area of the project, the web applica-tion of *PrimatesDB* is developed. It makes easier the contents visualization and through a guided path helps the users to submit their queries for data recovery. The *Graphical User Interface* (GUI) is written in PHP, JavaScript and HTML and is structured in tables based on a relational framework in order to prop-erly handle the data type for a better performance of the database with respect to speed and deployment. The *User Interface* (UI) is presented as shown in the

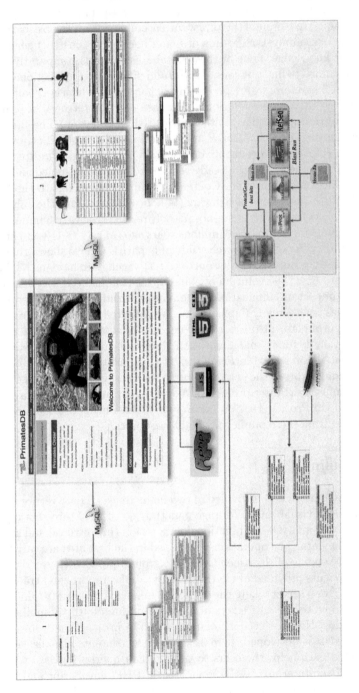

Fig. 1. *PrimatesDB* pipeline

center of the Fig. 1, where the *Home Page* is specifically reported with its hierarchical structure which provides several information related to the Order, Genus and Species of the Primates. Currently, it is possible to access to skeletal muscle data of *Pan troglodytes* which are organized in extremely easy-to-read tables for helping external users to quickly visualize and download the information. For what concern the three major components, they are based on the *Transcriptator* computational pipeline [12] which is the background pipeline of *PrimatesDB*. Here, the annotation process comprises four main parts: (i) finding the best hit in locally installed SwissProt and UniProt-Trembl database; (ii) assignment of functional annotation and gene ontology terms and their enrichment from DAVID; (iii)assignment of GOSlim terms and their analysis from QuickGO; (iv) integration and summarization of retrieved results from DAVID and QuickGO web services. BLASTX program from locally installed ncbi-blast.2.2.23 stand alone package is used (with threshold e-value 0.001, [12]) to identify the best hits for query sequences on locally installed SwissProt and UniProt-trEMBL databases: http://www.uniprot.org. The main goal of the first step is to find similar sequences within SwissProt and UniProt-trEMBL databases for the unannotated query from the user. The output of BLASTX run is an alignment file in a tsv format. This latter, using a bash script, is transformed into a protein list, which is the required input file for DAVID [13] and QuickGO [14] web services. Python client source code for DAVID web services employed in our pipeline, retrieves the functional and gene ontology annotation for every single transcript in a query dataset. In particular, the Python client for DAVID web-services uses light-weight soap client suds-0.4 module [15] while for QuickGO web-services, BioServices Python package is implemented to the pipeline. It provides access to QuickGO and a framework to easily implement web service wrappers (based on WSDL/SOAP or REST protocols). These python scripts take the input protein list file from previous step and utilize DAVID database to obtain information in the form of TableReport and SummaryReport. For a given query dataset, Python source code is implemented with default parameter for DAVID database search to obtain the TableReport obtained through DAVID web services, it is a gene centric report which lists the genes or the transcripts, their associated functions and gene ontological terms. The list of UniProt-trEMBL accession proteins was run in BAR+ [16] for validation purposes.

3   Transcriptomic Data Retrieval

Starting from a recent *Pan troglodytes* transcriptomic analysis study [4], we have downloaded the assembled dataset corresponding to the skeletal muscle myoblasts transcripts data in NCBI repository (TSA, Transcripts Shotgun Assembly) with accessions numbers range GABE01000001 - GABE01030945 (http://www.ncbi.nlm.nih.gov/nuccore/GABE00000000.1). This database contains all the transcripts already assembled by the authors to their genome reference. Unfortunately, this study does not include any functional annotation. In order to obtain biological information of this muscle specific transcriptome and expand its knowledge, we ran it into an existing computational pipeline and carried out downstream analysis for this purpose.

4   Annotation of Gene-Level Data

Presently, *PrimatesDB* not only accommodates the gene ontological information, but also other available functional annotations about domains, metabolic pathways, as well as, relevant biological information from *SwissProt* and *PIR protein* databases with respect to each and every muscle specific *Pan troglodytes* transcript. With the help of *PrimatesDB* end-users can obtain comprehensive biological information for the differentially expressed transcripts IDs. Some biological information incorporated for each transcript within *PrimatesDB* regards: (i) *Gene Ontology*: controlled vocabularies (ontologies) that describe gene products in terms of their associated biological processes (GOTERM_BP_FAT), cellular components (GOTERM_CC_FAT) and molecular functions (GOTERM_MF_FAT) in a species-independent manner (http://www.geneontology.org/GO.indices. shtml); (ii) *Functional categories*: to classify proteins from completely sequenced genomes on the basis of the orthology concept by using Clusters of Orthologous Groups of proteins, Phylogenetic classification of proteins encoded in complete genomes [17] (COGs_ONTOLOGY - http://www.ncbi.nlm.nih.gov/COG); (iii) *Domain annotation*: modular structure of the gene product, evolutionary and molecular functional aspects of the transcripts, annotations for Inter-Pro [18] (Integrated Resource of Protein Families, Domains and Sites - http://www.helmholtz-muenchen.de/en/ibis), PFAM [19] (Database of protein families - http://pfam.xfam.org) and SMART [20] (Simple Modular Architecture Research Tool - http://smart.embl-heidelberg.de) domains; (iv) *Metabolic pathway annotation*: biological pathway information from KEGG [21] (Kyoto Encyclopedia of Genes and Genomes - http://www.genome.jp/kegg/pathway.html), BBID [22] (Biological Biochemical Image Database - http://bbid.irp.nia.nih.gov) and Panther [23] (A comprehensive function information system about genes - http://www.pantherdb.org) resources.

Lastly, we also provided ENA (European Nucleotide Archive - http://www.ebi.ac.uk/ena), Ensembl (www.ensembl.org) and UniProt (Universal Protein Resource - http://www.uniprot.org) IDs entries as additional information of the collected transcripts.

5   *PrimatesDB* use case

In order to demonstrate the ease and flexibility usage of *PrimatesDB* we have selected a case study for showing the web-resource usage. By way of example, we refer to the transcript with ENA ID: GABE01006024 which is related to the Ensembl ID: ENSPTRG00000000023 and UniProt protein ID: Q5TA50 (Fig. 1/orange box). *PrimatesDB*, by connecting different databases to each other, offers the opportunity to retrieve several information related to this single transcript, such as: (i) skeletal muscle transcriptomic data from ENA database;

(ii) functional annotation from a background pipeline; (iii) information related to the transcripts ID from ENA and Ensembl repositories; (iv) protein ID knowledge from UniProt project. Depending on the user's knowledges there are three different ways to access to these data (Fig. 1/orange box): (i) by *Databases*; (ii) by *Transcriptome*; (iii) by *Search* sections. The *Databases* section hosts four Databases set which we have suitably merged in: *Domain, Gene Ontology, Pathways* and *Miscellaneous*. By typing the *transcript* or the *protein* of interest, the user can choose to select the set of database from which to get the information. Once it has been selected, the subset of repositories will be clickable and the contents accessible. By selecting the *Transcriptome* section, from the navigation menu, user is able to visualize the complete list of *Pan troglodytes* collected transcripts. Here, the header is divided into seven columns (Fig. 2): the first corresponds to the skeletal muscle ENA ID transcripts; the second one is composed

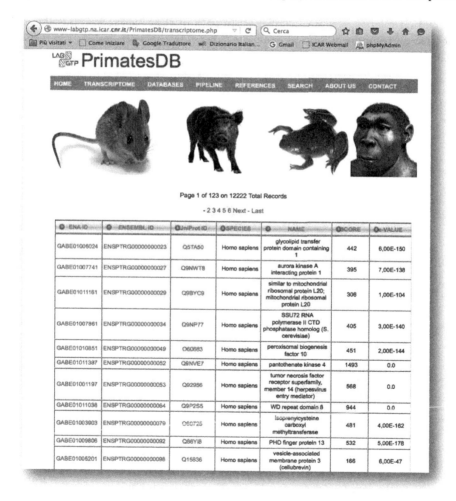

Fig. 2. Page of transcripts list. The columns ENA ID/ Ensembl ID/ UniProt ID/ are clickable buttons with redirection at the specific information of the related repositories.

of Ensembl ID, retrieved from UniProt ID. These first two columns give to the user important information about the species of interest showing which tissue specific transcripts have been already automatically annotated into the Ensembl specific organism database. The last five columns are particularly interesting for comparative studies, because they provide information about the best-hit UniProt protein (the third column) with their relative Species, Name, Score and e-value (respectively fourth, fifth, sixth and seventh column) attributes. In particular, the *Species* column shows the number of items present for in the top 15 species list - out of about 58 - for which BLASTX program obtained the proteins hits (Fig. 3). It is evident that most of the protein hits belonging to our non reference organism is present in *Homo sapiens* as well, confirming the two species are very well related with a percentage of 87 %.

Lastly, if the user has the ENA ID, or the Ensembl ID or the UniProt ID of the transcript, with the *Search* section a quick recovery way for seeking all the information related to it is also offered. In each of these cases, it is possible to access to all the information globally stored in *PrimatesDB* and be redirected to the belonging repositories.

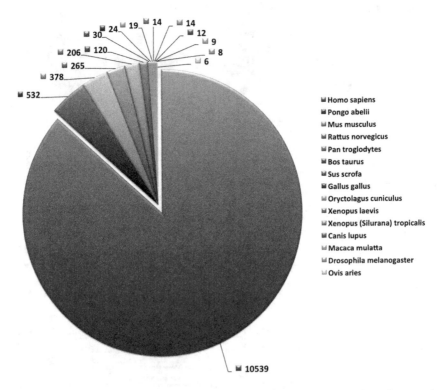

Fig. 3. Top species distribution returned by BLASTX program regarding the proteins belonging to the *Pan troglodytes* and 15 out 58 organisms taken into account

6   Database Development and Description

The *PrimatesDB* web resource is a database driven on a Relational Database
Management System. It retrieves information from the database and display
them on the related web page each time it is requested. Database updatings
are reflected by the web page, which is dinamically queried on user request.
PrimatesDB has been developed using web server Apache/2.2.26 [24]; MySQL
client version 5.3.28 - 10.04.1 (Ubuntu) [25] and the free tool phpMyAdmin
version 3.3.2 deb1 Ubuntu 0.2 [26] to handle the administration of MySQL,
over the World Wide Web, with InnoDB storage engine. The front-end side
was implemented using the scripting language PHP/5.2.6-3 [27]; the JavaScript
technology for dynamic contents [28]; the markup language HTML5 [29] and
style sheet CSS 3.0 [30]. *PrimatesDB*, although optimized for Safari, is easily
accessible and clearly visible by all browsers and smartphones.

7   Results

PrimatesDB is an open access and searchable database of complete annotation
of the predicted tissue specific transcriptome of the non-reference organism *Pan
troglodytes*. Its versatile and easily expandable structure accepts data from different sources, which are automatically processed and integrated into the platform.
The web interface allows the end-user to access several sections. *PrimatesDB*,
indeed, consists of seven sections and the core of the web portal is represented
by the *Tissue Specific Transcriptome* page. This section hosts the whole transcript list of the *Pan troglodytes* transcriptome identified by the analysis of our
Python scripts. Currently, we are in the process of implementing and increasing
the flexibility of dynamic content in the database through five database sets,
which we have suitably merged in: *Domain*, *Ontology* and *Pathways*. All other
sections were designed for all those users who want to deepen the understanding of this web application. The *PrimatesDB* web resource allows structuring
the data and displaying it in sorted and filtered tables accompanied by thorough explanations. Data were collected from literature and external databases,
then appropriately handled with ad hoc scripts. Overall, information currently
contained in *PrimatesDB* are related to 12 different functional terms of 12,222
transcripts (Fig. 2). The dataset contains 30,944 sequences, with relative lengths
ranging from 280 bp to 3100 bp, deposited into the ENA/NCBI repositories
starting from GABE01000001 - GABE01030945. A further and crucial step to
be considered at the end of an annotation analysis is validation of results, given
the possibility of getting matching errors. According to BAR+, our list of protein
resulted in a fully validated set.

8   Conclusion

The creation of a dedicated database for non-reference model organisms is an
important issue and always desirable for several reasons, especially considering

that the organism we picked, closely related to our species, can be useful in human research studies, and considering a thoroughly in silico research, which would not imply his ethical usage in the lab. In our laboratory, we developed *PrimatesDB*, a web resource for retrieving functional annotations on skeletal muscle specific transcriptome of *Pan troglodytes*, the common chimpanzee. The choice of this organism reflects the idea to start shading light of his, and his belonging orders, life style, comparing it to other species phylogenetically related - but with different morphological characteristics - in order to understand how these might have allowed them to better adapt in their specific environments. Our analysis begun with the retrieval of the specific transcriptomic data obtained from NCBI and, the usage of an home-made computational pipeline, which was used to process these data to place the functional annotations. To date, *PrimatesDB* represents a pilot study and it is useful to provide a comprehensive knowledge about the tissue specific transcriptome of *Pan troglodytes* non-reference model organism. *PrimatesDB* is a very easy-to-use web resource, freely available and without login requirements. As a modular platform, *PrimatesDB* can be easily extended and customized to future demands and developments. Indeed, we are in the process of updating *PrimatesDB* resource to make it more informative and we aim to provide functional annotation for all other transcripts. In order to deepen knowledges and increase interest about functional information of primates species, we need to start a comparative research approach by including in our analysis a further primate species to compare with common chimpanzee. Currently, we have selected the common marmoset (*Callithrix jacchus* L.) and, of this latter, we retrieved the muscle transcriptomic tissue dataset (NCBI TSA GAMQ01000001-GAMQ01033528) on which we carried out the same analysis we developed for the chimpanzee. Beyond the ease of retrieving exactly the same kind of data we had for chimpanzee, marmoset is a strategic choice given his basal position in primates phylogenetic tree compared to the more derivate position of chimpanzee. Marmoset belongs to the family Callitrichidae. It is a New World monkey (Platyrrhini parvorder) species. These ones split form Old World apes (Catarrhini parvorder, where chimpanzee belongs to) around 40 MYA colonizing the American continent [31]. This evolutionary and biogeographic distance is very much evident morphologically as well. Indeed, marmoset present a very small size body with a long tail, with no so much differences among sexes. They do posses an arboreal quadrupedalism locomotion being able to hang on trees and leap between them as well. Its whole-genome sequence has been published and available [32]. Having the opportunity to compare two species, which present quite different morphological properties and life styles, it might focus our attention on those genes, and their transcripts and functional annotation, who might exhibit rapid changes in expression, linked to many developmental differences underlying possible phenomena of directional selection in some lineages [33]. Computer science community could take example of our web resource as a model that has easily connected different context data: from phylogenetic to molecular biology and from high throughput to morphological data. We believe that by creating simple but well-structured relational databases, the intent to provide an extended panning shot for improving the knowledges of primates differentiation and evolution is possible.

Acknowledgments. This work was funded by INTEROMICS flagship Italian project, PON02-00612-3461281 and PON02-00619-3470457. Mario R. Guarracino work has been conducted at National Research University Higher School of Economics and supported by RSF grant 14-41-00039.

The research group would like to thank Giuseppe Trerotola (ICAR/CNR) for technical support.

References

1. Steinmetz, P.R.H., Kraus, J.E.M., Larroux, C., Hammel, J.U., Amon-Hassenzahl, A., Houliston, E., Wrheide, G., Nickel, M., Degnan, B.M., Technau, U.: Independent evolution of striated muscles in cnidarians and bilaterians. Nature **487**(7406), 231–234 (2012)
2. Ankel-Simons, F.: Primate Anatomy: An Introduction. Academic Press, New York (2000)
3. Fleagle, J.G.: Primate Adaptation and Evolution. Academic Press, New York (1988)
4. Maudhoo, M.D., Madison, J.D., Norgren, R.B.: De novo assembly of the chimpanzee transcriptome from NextGen mRNA sequences. GigaScience **4**(1), 1–4 (2015)
5. Wilson, D.E., Reeder, D.M.: Pan troglodytes. In: Mammal Species of the World: A Taxonomic and Geographic Reference. Johns Hopkins University Press, 3rd edn. (2005)
6. Martin, R.D.: Primate Origins, Evolution: A Phylogenetic Reconstruction. Princeton University Press, Princeton (1990)
7. Bukh, J.: A critical role for the chimpanzee model in the study of hepatitis C. Hepatology **39**(6), 1469–75 (2004)
8. de Groot, N.G., Bontrop, R.E.: The HIV-1 pandemic: does the selective sweep in chimpanzees mirror humankind's future. Retrovirology **10**(1), 53 (2013)
9. The Chimpanzee Sequencing and Analysis Consortium: Initial sequence of the chimpanzee genome and comparison with the human genome. Nature **437**(7055), 69–87 (2005)
10. Perelman, P., et al.: A molecular phylogeny of living primates. PLoS Genet. **7**, 3 (2011)
11. MacPhee, R.D. (ed.): Primates and Their Relatives in Phylogenetic Perspective. Springer, Heidelberg (2013)
12. Tripathi, K.P., Evangelista, D., Zuccaro, A., Guarracino, M.R.: Transcriptator: An automated computational pipeline to annotate assembled reads and identify non coding RNA. PLoS ONE **10**(11), e0140268 (2015)
13. Huang, D.W., et al.: Systematic and integrative analysis of large gene lists using DAVID bioinformatics resources. Nat. Protoc. **4**, 44–57 (2009)
14. Binns, D., et al.: QuickGO: A web-based tool for gene ontology searching. Bioinformatics **25**(22), 3045–3046 (2009)
15. Huang, D.W., Sherman, B.T., Lempicki, R.A.: Bioinformatics enrichment tools: toward the comprehensive functional analysis of large gene lists. Nucleic Acids Res. **37**(1), 1–13 (2009)

16. Piovesan, D., et al.: How to inherit statistically validated annotation within BAR+ protein clusters. BMC Bioinformatics **14**(Suppl 3), S4 (2013)
17. Galperin, M.Y., Makarova, K.S., Wolf, Y.I., Koonin, E.V.: Expanded microbial genome coverage and improved protein family annotation in the COG database. Nucleic Acids Res. **43**(D1), D261–D269 (2015)
18. Mitchell, L., et al.: The interpro protein families database: the classification resource after 15 years. Nucleic Acids Res. Database Issue **43**, D213–D221 (2015)
19. Finn, R.D., et al.: The Pfam protein families database. Nucleic Acids Res. Database Issue **42**, D222–D230 (2014)
20. Letunic, I., Doerks, T., Bork, P.: SMART: recent updates, new developments and status in 2015. Nucleic Acids Res. **43**, D257–D260 (2014). doi:10.1093/nar/gku949
21. Kanehisa, M., Goto, S.: KEGG: Kyoto encyclopedia of genes and genomes. Nucleic Acids Res. **28**(1), 27–30 (2000)
22. Becker, K.G., White, S.L., Muller, J., Engel, J.: BBID: The biological biochemical image database. Bioinformatics **16**(8), 745–746 (2000)
23. Thomas, P.D., et al.: PANTHER: A library of protein families and subfamilies indexed by function. Genome Res. **13**, 2129–2141 (2003)
24. The Apache HTTP Server Project. https://httpd.apache.org
25. The World's Most Popular Open Source Database. https://www.mysql.com
26. phpMyAdmin to Handle the Administration of MySQL over the Web. http://www.phpmyadmin.net
27. A Popular General-purpose Scripting Language that Is Especially Suited to Web Development. http://php.net
28. The Lightweight, Interpreted, Object-oriented Scripting Language for Web Pages. http://www.ecma-international.org
29. HTML: The Markup Language for Describing Web Documents. http://www.w3.org/TR/html5
30. Cascading Style Sheets: a Mechanism for Adding Style to Web Documents. http://www.w3.org/Style/CSS
31. Shumaker, R.W., Beck, B.B.: Primates in Question. Smithsonian Institution Press, Washington, DC (2003)
32. Worley, K.C., et al.: The common marmoset genome provides insight into primate biology and evolution. Nat. Genet. **46**(8), 850–857 (2014)
33. Romero, I.G., Ruvinsky, I., Gilad, Y.: Comparative studies of gene expression and the evolution of gene regulation. Nat. Rev. Genet. **13**(7), 505–516 (2012)

Author Index

Printed in the United States
By Bookmasters